Integration of Functional Oxides with Semiconductors

Alexander A. Demkov • Agham B. Posadas

Integration of Functional Oxides with Semiconductors

 Springer

Alexander A. Demkov
Department of Physics
The University of Texas at Austin
Austin, TX, USA

Agham B. Posadas
Department of Physics
The University of Texas at Austin
Austin, TX, USA

ISBN 978-1-4614-9319-8   ISBN 978-1-4614-9320-4 (eBook)
DOI 10.1007/978-1-4614-9320-4
Springer New York Heidelberg Dordrecht London

Library of Congress Control Number: 2013958215

Printed on acid-free paper

Springer is part of Springer Science+Business Media (www.springer.com)

Preface

Integration of Functional Oxides with Semiconductors describes the basic physical principles of oxide/semiconductor heteroepitaxy and offers a view of the current state of the field. It shows how this technology enables large-scale integration of oxide electronic and photonic devices and describes possible hybrid semiconductor/ oxide systems. The book incorporates both theoretical and experimental advances to explore the epitaxial integration of tuned functional oxides and semiconductors; to identify materials, device, and characterization challenges; and to present the incredible potential in the realization of multifunctional devices and monolithic integration of materials and devices. Intended for a multi-disciplined audience, *Integration of Functional Oxides with Semiconductors* describes processing techniques that enable atomic level control of stoichiometry and structure, and reviews characterization techniques for films, interfaces, and material performance parameters. Fundamental challenges involved in combining covalent and ionic systems, chemical interactions at interfaces, and multi-element materials that are sensitive to atomic level compositional and structural changes are discussed in the context of the latest literature. Magnetic, ferroelectric, and piezoelectric materials and the coupling between them will also be discussed. GaN, SiC, Si, GaAs, and Ge semiconductors are covered within the context of optimizing next-generation device performance for monolithic device processing.

This book would not have been possible without the support and assistance of many people. We would like to thank Kate Ziemer for her help and inspiration in the early stages of the book; Richard Hatch, Hosung Seo, and Chungwei Lin for their help with editing; Hosung Seo, Andy O'Hara, Kurt Fredrickson, and Kristy Kormondy for the artwork; and Jamal Ramdani for a critical reading of Chap. 2.

Austin, TX, USA
Austin, TX, USA
Alexander A. Demkov
Agham B. Posadas

Contents

Chapter 1
Introduction

Over a decade ago, McKee and co-workers achieved a breakthrough in the direct epitaxial growth of single crystal perovskite $SrTiO_3$ (STO) on Si(001) using 1/2 monolayer (ML) of Sr deposited on a clean Si(001) 2 × 1 surface as a template [1]. At 1/2 ML coverage, Sr atoms assume positions between Si dimer rows and inhibit the formation of an amorphous SiO_2 layer during the subsequent STO deposition in a relatively wide range of temperatures and oxygen partial pressures [1–5]. The ensuing development of crystalline epitaxial oxides on semiconductors (COS) has opened a new avenue for complementary metal oxide semiconductor (CMOS) technology for materials other than Si, e.g. Ge and GaAs. It has also ushered in the even more tantalizing possibility of growing functional oxide nanostructures utilizing ferroelectricity, superconductivity, and magnetism, in monolithic integration with Si [6–15]. This is a relatively new area with equal measure of exciting possibilities and difficult challenges. Among the fundamental aspects of monolithic integration are the crystal growth of functional oxides on semiconductors and semiconductors on oxide surfaces, and the tunability of their electronic and transport properties.

In addition to applications in logic technology, functional oxides offer new ways to store information and thus are well suited for applications in memory. Many oxide properties are sensitive to temperature, strain, electric and magnetic fields, making them attractive materials for sensors. Monolithic integration with semiconductors will enable both the sensing and logic functionalities to be incorporated on a single chip. Other potential applications include optical interconnects, automotive radar, photonics, solid state lighting, microelectromechanical systems, and photocatalysis [16–20]. For example, a schematic of a possible integrated photonic chip is depicted in Fig. 1.1. In 2013, researchers at the IBM Zürich Laboratory reported the electro-optical properties of thin barium titanate ($BaTiO_3$) films epitaxially grown on silicon substrates [21]. Remarkably, they extracted an

A.A. Demkov and A.B. Posadas, *Integration of Functional Oxides with Semiconductors*, DOI 10.1007/978-1-4614-9320-4_1, © The Author(s) 2014

Fig. 1.1 Integrated Si photonics (courtesy of IBM Research—Zürich)

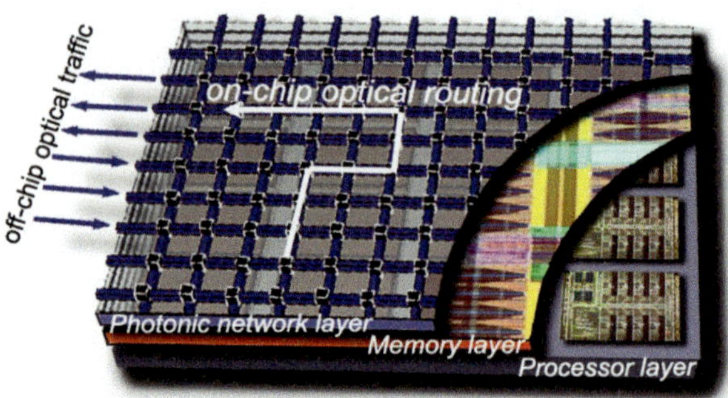

effective Pockels[1] coefficient five times larger than that of the current standard material for electro-optical devices, lithium niobate ($LiNbO_3$). The integration of electro-optically active $BaTiO_3$ (BTO) on silicon can pave the way to the realization of a new variety of photonic devices with disruptive performance. However, growing such heterostructures, as well as elucidating their atomic structure, presents a significant challenge.

The key to successful oxide-semiconductor heteroepitaxy is to achieve two-dimensional (layer-by-layer) or Frank-Van der Merwe growth. In these systems, in addition to the lattice and thermal mismatch, one has to accommodate the transition between fundamentally different types of *chemical bonding* across the interface. This bonding mismatch can be accommodated by using intermetallic Zintl compounds, as transition layers, between ionic oxides and covalent semiconductors [22]. The principal idea is to use the intrinsic charge transfer in a Zintl layer to trick the more electronegative metal to assume semi-covalent bonding, which continues into the semiconductor. There are also considerations of crystal lattice symmetry and many other additional factors. As many of the issues specific to oxide-semiconductor epitaxy are rooted in the microscopic nature of these materials, theoretical modeling, and particularly, density functional theory (DFT) have been instrumental to the rapid progress made in this field over the last decade.

In this book, we will discuss the recent progress in monolithic integration of functional oxides (mostly perovskites such as STO and BTO) on Si, Ge, and other semiconductors using primarily molecular beam epitaxy (MBE), but also other

---

[1] Linear electro-optic effect, also known as the Pockels effect, produces birefringence in an optical medium induced by a constant or varying electric field. Unlike the quadratic Kerr effect, the Pockels effect is linear in the electric field and occurs only in crystals that lack inversion symmetry. The refractive index of an isotropic (to avoid cumbersome tensor notations) electro-optic medium can be expressed as:

$$n(E) = n - \frac{1}{2}rn^3 E + O(E^2)$$

Where $n = n(0)$ is the index in the absence of the field, and rn^3 represents the field derivative of the refractive index. The coefficient r is called the linear electro-optic or Pockels coefficient.

deposition methods including pulsed laser deposition (PLD) and atomic layer deposition (ALD). The book is intended to be a self-contained introduction to the field of oxide-on-semiconductor heteroepitaxy. In this chapter we will briefly introduce the various classes of materials one has to deal with and their general properties, in particular, semiconductors, transitional metal oxides, and Zintl intermetallics. Considering the vastness of these fields of materials, and the number of books that have already been written about them, the review in this chapter is not meant to be exhaustive and the interested reader is directed to the many available references. In Chap. 2 we will focus on the specific features and challenges that set oxide-semiconductor heteroepitaxial systems apart from the more commonly discussed cases of semiconductor-semiconductor and oxide-oxide epitaxy. A good atomic-scale understanding of the materials system will be shown to be necessary, which highlights the importance of the use of microscopic theory in this field. For this reason, we attempt in Chap. 3 to provide the reader with the basic concepts of density functional theory and first principles calculations, at the minimum level needed to provide the reader with the necessary vocabulary. Epitaxial thin film deposition methods are the principal means by which these functional oxide-on-semiconductor heterostructures are achieved. In Chap. 4 we will briefly describe the basic growth methods that have been successfully used to achieve this type of monolithic integration with particular emphasis on oxide growth. Various techniques of materials characterization are also crucial in order to "see" what is happening during and after the growth, as well as to be able to analyze the properties of the resulting structure. The relevant methods of materials characterization are introduced and their basic principles are briefly described in Chap. 5. Chapter 6 is focused on the details of the epitaxial integration of STO on Si, the first and, to date, the only widely utilized direct epitaxy of a perovskite oxide on Si. In Chap. 7 we illustrate how this materials platform can be used for subsequent integration of other oxides on Si. The success of integrating perovskites on Si(001) has also stimulated work on developing growth processes for other oxide materials, other crystallographic orientations, and even other semiconductor substrates. A review of the current work is summarized in Chap. 8, including efforts at growing the opposite stack of semiconductors on oxide surfaces. In the final chapter, Chap. 9, we summarize the current status of the field and try to anticipate where this fascinating technology will go next.

1.1 Semiconductors

In their pure form semiconductors are typically insulators but can be made conductive by adding minute quantities of impurities known as dopants. Electrical conductivity may be realized by electrons, which is known as n-type doping or by holes, which is known as p-type doping. In their now classic book, Yu and Cardona identify six semiconductor classes: elemental (Si, Ge, etc.), binary compounds (GaAs, ZnS, etc.), oxides (CuO_2, ZnO, etc.), layered semiconductors (PbI_2, MoS_2, etc.), organic (polyacetylene [$(CH_2)_n$], etc.), and magnetic semiconductors (EuS, $Cd_{1-x}Mn_xTe$, etc.) [23].

Fig. 1.2 Diamond crystal
structure

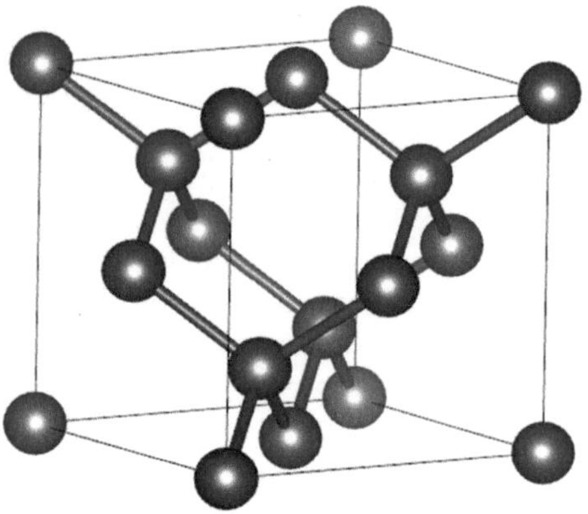

Silicon (Si) is undoubtedly the most studied elemental semiconductor owing to its role in transistor technology. Interestingly, the first transistor, for which Bardeen, Brattain and Shockley won the Nobel Prize in Physics in 1956, was fabricated in 1947 using germanium (Ge), not Si. Both elements belong to group IVA of the periodic table that starts with carbon (C $Z = 6$, Si $Z = 14$, Ge $Z = 32$ and Sn $Z = 50$). Phosphorus from group VA and sulfur, selenium and tellurium from group VIA are also semiconductors. However, as Si and Ge are the most important for our discussion we will only briefly describe these two. Both materials crystallize in the so-called diamond structure shown in Fig. 1.2. The tetrahedral bonding, characterized by the fourfold coordination and 109.5° bond angle, stems from the sp^3 hybridization of the valence electrons. This can be easily seen using the method of linear combination of atomic orbitals [24]. For example, the ground state configuration of the Si atom is $3s^2 3p^2$ with the ε_s and ε_p levels separated by approximately 7 eV in energy. In a crystal we assume that the electronic wave functions can be approximated by appropriate linear combinations of the atomic orbitals. However, it is more convenient to use the so-called hybrid orbital basis. Starting with one s and three p (x, y, and z) atomic orbitals, we can form four tetrahedral hybrids of the following form:

$$|h\rangle = \frac{1}{2}\left[|s\rangle \pm |p_x\rangle \pm |p_y\rangle \pm |p_z\rangle\right]$$

These hybrid wave functions are asymmetric with extended lobes oriented along the [111]-type axes. The energy of an isolated hybrid state is $\varepsilon_h = (\varepsilon_s + 3\varepsilon_p)/4$ meaning that each electron spends a quarter of time in the s state and three quarters in the p state. Distributing four valence electrons over four hybrids corresponds, in the case of Si, to a $3s^1 3p^3$ or, more generally, to an $s^1 p^3$ (sp^3) configuration and does cost energy. The energy is however, gained back when one considers that each Si atom in a diamond structure has four nearest neighbors along [111] directions. The extended lobes of the hybrid orbitals from two nearest neighbors are pointed towards each other and have a significant overlap and a large matrix element

Fig. 1.3 Formation of the electronic structure: from the atomic s and p orbitals to sp^3 hybrids, then to their bonding and anti-bonding combination, and finally to bands

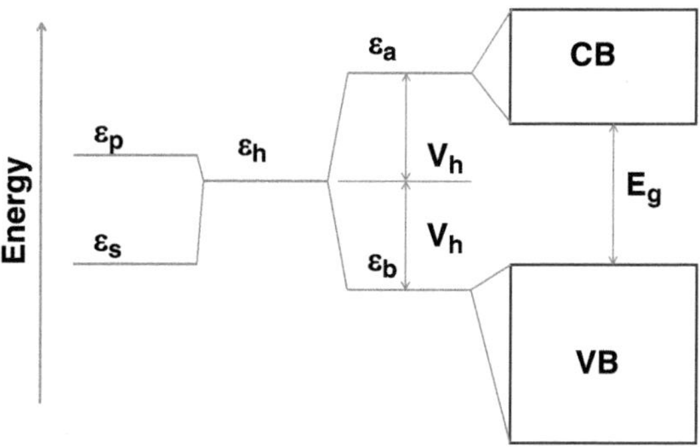

$V_h = \langle h^1 | H | h^2 \rangle$, where H is the difference between the atomic and lattice potentials [25]. The bonding combination of two such hybrid orbitals has an energy that is lower than that of the atomic state (see schematic in Fig. 1.3). This is a molecular picture of the covalent bond formation. In a crystal, bonding and anti-bonding combinations of hybrid orbitals broaden, giving origin to the occupied valence and empty conduction bands. The energy separation between the highest occupied and lowest empty states is known as the band gap (E_g) and is one of the most important properties of a semiconductor. Among group IVA semiconductors, C (diamond) has the largest band gap of 5.5 eV, followed by $E_g = 1.1$ eV in Si and $E_g = 0.7$ eV in Ge. The gaps of several common semiconductors along with their crystal structure are summarized in Table 1.1.

GaAs is probably the best known compound semiconductor formed from elements of groups IIIA and VA. It crystallizes in the cubic zincblende (sphalerite) structure shown in Fig. 1.4. It is similar to the diamond structure in terms of bond angles and coordination, and the properties of III–V compounds are similar to those of group IVA semiconductors. However, as there are two types of atoms in the lattice, the nature of the chemical bonding changes slightly. As group VA elements are more electronegative than the elements of group IIIA (Pauling electronegativity of arsenic is 2.18 vs. 1.81 for gallium), there is some charge transfer from the latter to the former, and the bonding becomes partly ionic. The ionicity increases the band gap in III–V compounds compared to that in group IVA semiconductors (with the exception of diamond). In terms of the band gap (and consequently their technological applications) there are three groups of III–V compounds; conventional such as GaAs, wide-band gap III-nitrides such as GaN, and narrow-gap materials such as InSb. GaN is important due to its applications in high power electronics and in lighting technology. It crystallizes in the hexagonal wurtzite structure shown in Fig. 1.5. The ionicity of bonding is even more pronounced in II–VI materials such as ZnS; most of these materials have band gaps larger than 1 eV, with the notable exception of Hg-containing compounds that, owing to their very small band gaps, are used in infrared technology. The group of IV–IV semiconductors is comprised of various alloys such as $Si_{1-x}Ge_x$ or $Ge_{1-x}Sn_x$, as well as the compound silicon carbide (SiC). SiC crystallizes either in the zincblende (3C) structure or in a large

Table 1.1 Semiconductors and other materials

Material	Type	Name	Crystal structure	Lattice constants at 300 K (Å)	Band gap (eV)	Thermal expansion coefficient $\alpha_T = \partial \ln a / \partial T$ (10^{-6} K^{-1})	Elastic constants (GPa)
C	Element	Carbon (diamond)	D	3.56683	5.48	$0.87 + 0.0092(T-273)$	c_{11} 1076; c_{12} 125; c_{44} 576.8
Ge	Element	Germanium	D	5.64613	0.66	$6.05 + 0.0036(T-273)$	c_{11} 128.9; c_{12} 48.3; c_{44} 67.1
Si	Element	Silicon	D	5.43095	1.12	$3.08 + 0.0019(T-273)$	c_{11} 165.7; c_{12} 63.9; c_{44} 79.6
Sn	Element	Grey tin	D	6.48920	0.08	4.7	c_{11} 69; c_{12} 29.3; c_{44} 36.2
SiC	IV–IV	Silicon carbide	W (6H)	$a = 3.086; c = 15.117$	3.05		c_{11} 570; c_{12} 108; c_{33} 547; c_{44} 159
SiC	IV–IV	Silicon carbide	ZB	$a = 4.3596$	2.36	2.9	c_{11} 289; c_{12} 234; c_{44} 55.4
AlAs	III–V	Aluminum arsenide	ZB	5.6605	2.16	$3.40 + 0.0064(T-273)$	c_{11} 190; c_{12} 53.8; c_{44} 59.5
AlP	III–V	Aluminum phosphide	ZB	5.4510	2.45	~6.1	c_{11} 140.5; c_{12} 62.03; c_{44} 70.33
AlSb	III–V	Aluminum antimonide	ZB	6.1355	1.58	4.2	c_{11} 89.4; c_{12} 44.3; c_{44} 41.6
AlN	III–V	Aluminum nitride	W	$a = 3.11; c = 4.98$	6.28	$\alpha_\perp = 5.27$ $\alpha_{//} = 4.15$	c_{11} 296; c_{12} 130; c_{13} 158; c_{33} 267; c_{44} 241
BN	III–V	Boron nitride	ZB	3.6150	6.36	1.15	c_{11} 820; c_{12} 190; c_{44} 480
BP	III–V	Boron phosphide	ZB	4.5380		3.65	c_{11} 515; c_{12} 100; c_{44} 160
GaAs	III–V	Gallium arsenide	ZB	5.6533	1.42	$5.35 + 0.0080(T-273)$	c_{11} 118.1; c_{12} 53.2; c_{44} 59.2
GaN	III–V	Gallium nitride	W	$a = 3.189; c = 5.185$	3.44	$\alpha_\perp = 3.17$ $\alpha_{//} = 5.59$	c_{11} 377; c_{12} 160; c_{13} 114; c_{33} 209; c_{44} 81.4
GaP	III–V	Gallium phosphide	ZB	5.4512	2.26	5.81	c_{11} 141.2; c_{12} 62.5; c_{44} 70.5
GaSb	III–V	Gallium antimonide	ZB	6.0959	0.73	6.7	c_{11} 88.4; c_{12} 40.3; c_{44} 43.2
InAs	III–V	Indium arsenide	ZB	6.0584	0.36	$4.33 + 0.0038(T-273)$	c_{11} 83.3; c_{12} 45.3; c_{44} 39.6
InP	III–V	Indium phosphide	ZB	5.8686	1.35	4.75	c_{11} 102.2; c_{12} 57.6; c_{44} 46
InSb	III–V	Indium antimonide	ZB	6.4794	0.18	5.37	c_{11} 66.7; c_{12} 36.5; c_{44} 30.2
InN	III–V	Indium nitride	W	$a = 3.545; c = 5.703$	0.7	$\alpha_c = 2.6$ $\alpha_a = 3.6$	c_{11} 190; c_{12} 104; c_{13} 121; c_{33} 182; c_{44} 9.9
CdS	II–VI	Cadmium sulfide	ZB	5.8320	2.42	4.7	c_{11} 67.6; c_{12} 46.3; c_{44} 29.5
CdS	II–VI	Cadmium sulfide	W	$a = 4.160; c = 6.756$	2.48	$\alpha_\perp = 5.0$ $\alpha_{//} = 2.5$	c_{11} 83.1; c_{12} 50.4; c_{13} 46.2; c_{33} 94.8; c_{55} 15.33; c_{66} 16.3

CdSe	II–VI	Cadmium selenide	ZB	6.050	1.74	3.8	c_{11} 55.4; c_{12} 37.7; c_{44} 18.9
CdTe	II–VI	Cadmium telluride	ZB	6.482	1.49	4.9	c_{11} 53.5; c_{12} 36.8; c_{44} 19.9
ZnO	II–VI	Zinc oxide	W	$a = 3.429; c = 5.2042$	3.37	$\alpha_\perp = 4.75$ $\alpha_{//} = 2.92$	c_{11} 206; c_{13} 118; c_{33} 211; c_{44} 44.3; c_{66} 44.0
ZnS	II–VI	Zinc sulfide	ZB	5.420	3.54	$6.70 + 0.0128(T\text{-}313)$	c_{11} 104; c_{12} 65; c_{44} 46.2
ZnS	II–VI	Zinc sulfide	W	$a = 3.82; c = 6.26$	3.91	$\alpha_\perp = 5.9 - 6.5$ $\alpha_{//} = 4.4 - 4.6$	c_{11} 123.4; c_{12} 58.5; c_{13} 45.5; c_{33} 28.8; c_{44} 32.45; c_{66} 139.6
ZnSe	II–VI	Zinc selenide	ZB	5.668	2.7	7.4	c_{11} 81; c_{12} 48.8; c_{44} 44.1
PbS	IV–VI	Lead sulfide	RS	5.9362	0.37	$18.81 + 0.0074(T\text{-}273)$	c_{11} 126.1; c_{12} 16.24; c_{44} 17.09
PbTe	IV–VI	Lead telluride	RS	6.4620	0.32	19.80	c_{11} 105.3; c_{12} 7.0; c_{44} 13.22
PbSe	IV–VI	Lead selenide	RS	6.117	0.27	19.40	c_{11} 123.7; c_{12} 19.3; c_{44} 15.91
Cu$_2$O	Oxide	Cuprous oxide	Cubic	4.27	2.17	-2.4 below 240 K; 0 at RT; 1.59 above RT	c_{11} 121; c_{12} 105; c_{44} 10.9
CuO	Oxide	Cupric oxide	Monoclinic		1.2	7.4	

The table lists the most commonly used semiconductor materials arranged by type. The table includes the crystal structure, room temperature lattice constants, band gap at room temperature, thermal expansion coefficients, and elastic constants. For the crystal structure, the following abbreviations are used: *D* diamond, *ZB* zincblende, *W* wurtzite, *RS* rocksalt

Fig. 1.4 Zincblende
crystal structure

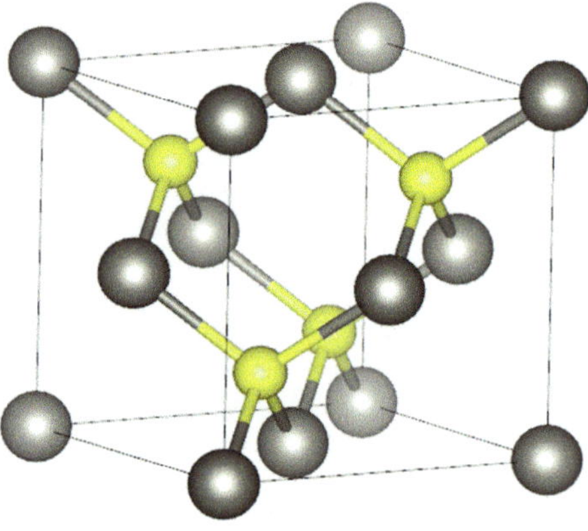

Fig. 1.5 Wurtzite unit cell

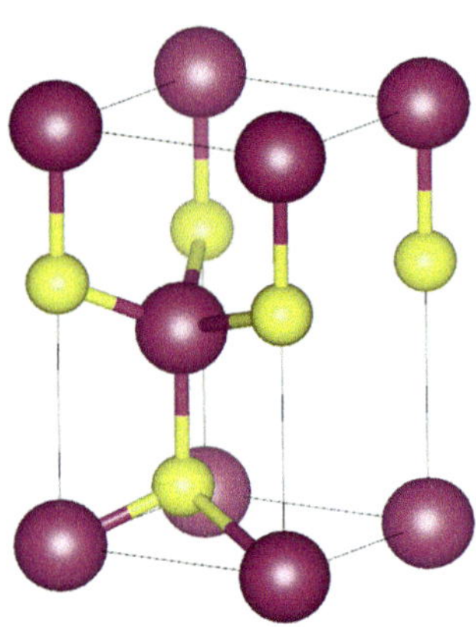

number of hexagonal polymorphs, the most common of which are the 4H and 6H
polymorphs [26]. SiC is a large band gap material thanks to the large difference in
electronegativity between carbon (2.55) and silicon (1.90).

Last but not least, we should mention the oxide semiconductors. Though most
oxides are insulating, some, such as Cu_2O, CuO or ZnO, are semiconducting.
Cubic cuprous oxide Cu_2O historically was the original material for rectifier
semiconductor diodes. Its applications date back to the mid-1920s. Recently,
there has been a great interest in wurtzite ZnO, in part, because of possible lasing
applications in the UV regime. This renewed interest is fueled by the availability
of high-quality substrates and reports of p-type conduction and ferromagnetic
behavior when doped with transitions metals, both of which remain controversial.
ZnO is not new to the semiconductor field, with studies of its crystal structure
dating back to 1935 by Bunn [27]. It has many industrial applications owing to its
piezoelectric properties and band gap in the near ultraviolet. Currently, it is being
considered for applications in optoelectronic devices. The main difficulty in the
widespread technological development of ZnO has been the lack of reproducible

and low-resistivity *p*-type ZnO. In the last few years, reports of *p*-type conductivity resulted in new hopes of using ZnO for optoelectronic applications. An excellent review of this fast-moving field has recently been published by Özgür et al. [28]. These authors discuss the mechanical, chemical, electrical, and optical properties of ZnO, in addition to the technological issues such as growth, defects, *p*-type doping, band-gap engineering, devices, and nanostructures.

1.2   Transition Metal Oxides

We now briefly discuss the basic properties of oxides containing transition metals with *d* and *f* electrons. As this field is truly immense, we will limit our discussion to oxides with the perovskite crystal structure. The principles outlined can be easily adapted to rocksalt, rutile, corundum, wurtzite, spinel and other oxide crystal structures. For more detailed accounts, we refer the reader to several excellent monographs on oxides [29, 30], their surfaces [31], and their relevant physical properties such as magnetism [32, 33] and ferroelectricity [34, 35].

The salient feature of the transition metal oxides is the presence of atoms possessing an electronic structure consisting of an incomplete inner shell and a complete outer shell. The inner shells are $3d$ (*the group of iron*), $4f$ (*rare earths*), $4d$ (*the group of palladium*) and $5d$ (*the group of platinum*), and the outer shells are $4s$, $5s$ and $6s$. The number of electrons in the incomplete d- or f-shell is indicated by a superscript d^1, d^2, f^2, *etc*. It has to be remembered that though the $4s$ state is higher in energy than the $3d$ state in hydrogen, in the case of heavier atoms, these two states are close in energy as the $4s$ level is shifted down. The effect can be understood as follows: since for the s-electron $l = 0$, it doesn't experience the centrifugal potential

$$\frac{l(l+1)\hbar^2}{2mr^2},$$

and as a result, can penetrate deeper inside the core, where the nuclear charge is screened less. This additional attraction lowers the energy of the s-state. This effect results in a competition between the s and d shells for electrons. For example, in the group of iron, the configuration of Cr is $3d^5 4s^1$ and not $3d^4 4s^2$ as one might expect. Including spin, the d shell may contain up to ten electrons (two for each value of the magnetic quantum number m_l). In a free atom the d-electrons are shielded by the filled outer s-shell electrons. The central field picture requires some refinement when the fine structure of the atomic spectra is considered. One needs to include the spin-orbit interaction in a more accurate treatment. In $3d$ transition metals the spin-orbit interaction is rather small having a magnitude of 10–100 meV, although this can be much larger in the rare earths. The effect of the combination of spin-orbit and exchange interactions is captured by Hund's rules. In accordance with Hund's rules for a given electronic configuration, the lowest energy term has the maximum

Fig. 1.6 ABO_3 perovskite
structure

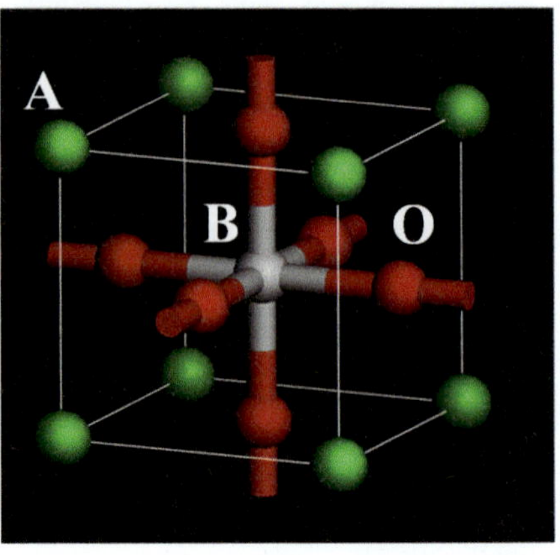

possible spin and angular momentum. If the shell is more than half-filled, the lowest
energy term corresponds to the highest total angular momentum, and if the shell is
less than half-filled, the lowest energy term corresponds to the smallest total angular
momentum. The first two rules minimize the Coulomb energy and the last rule
minimizes the energy of spin-orbit coupling.

The perovskite structure (shown in Fig. 1.6), including distortions derived from
it, is very common among the transition metal oxides for compounds having
chemical formula ABO_3. The extensive number of combinations of elements that
form ABX_3 compounds having the perovskite or distorted perovskite structure can
be seen in Fig. 1.7 [36]. In the Appendix, we provide a table of the lattice
parameters (as well as the electrical and magnetic nature) of most ABO_3 compounds that adopt the perovskite structure. The two most common perovskite
distortions (rhombohedral and orthorhombic) are also shown in the Appendix,
including a crystal structure phase diagram (at room temperature) as a function of
ionic radii of the A and B ions.

The transition metal ion is at the B-site in the center of a cube formed by the
A-site cations. Oxygen atoms are located in the center of each face forming an
octahedral cage around the transition metal. The transition metal forms covalent
bonds with oxygen and loses its outer electrons. If the transition metal ion belongs
to group IV-B (e.g. Ti), the formal charge of the BO_6 cage is 2^- thus a group II-A
A-site cation is needed to fulfill stoichiometry. STO is an example of such an
arrangement. The bonding-antibonding splitting of oxygen p-states and transition
metal d-states is known as the charge transfer gap and is another of the characteristic energy scales of the oxide. In STO, for example, it defines the fundamental
band gap.

Once the transition metal ion is placed in the oxide, the outer s-shell electrons are
stripped away, and the ion starts feeling the electrostatic field of the host crystal.
This field is usually known as the crystal field. The crystal field sets another
important energy scale. In the ideal perovskite structure, the electrostatic field has
cubic symmetry. The magnetic quantum number m_l describes the orbital moment

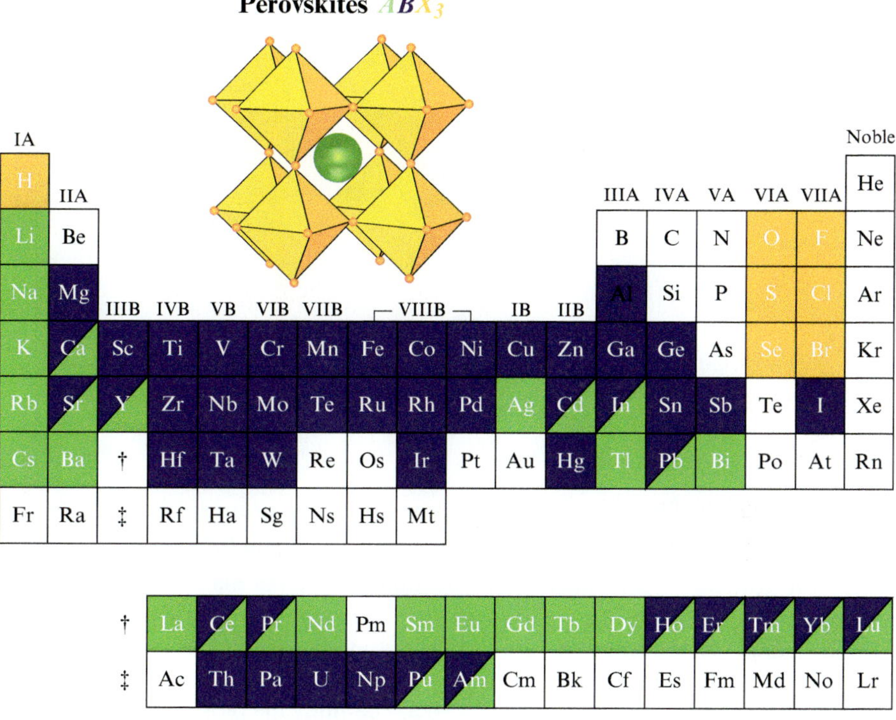

Fig. 1.7 Elements of the periodic table that can combine to form a perovskite structure. Reproduced with permission from [36]

projection on the z-axis, and in the case of spherical symmetry, all directions are equivalent such that in a free atom, the energy is degenerate with respect to m_l. However, the lowering of the full rotational symmetry to cubic results in the splitting of the fivefold degenerate d-shell into a doublet e_g and a triplet t_{2g}. In the simplest picture of the ligand field theory this is understood as follows. The orbitals forming the doublet (d_{z^2} and $d_{x^2-y^2}$) point directly at the negatively charged oxygen ions and experience stronger Coulomb repulsion than the triplet states (d_{xy}, d_{xz} and d_{yz}) that point between the negatively charged oxygen ions. This shifts the e_g states higher in energy with respect to the t_{2g} states. This is the crystal field splitting customarily called *10Dq*. Combining the covalency with the crystal field effect (molecular orbital theory), we arrive at the situation depicted in Fig. 1.8 where the charge transfer gap separates bonding and anti-bonding orbitals. Bonding states are always filled, while the anti-bonding states designated with the star symbol accommodate the electrons of the transition metal in accordance with its oxidation state. For example, in STO and BTO all Ti d-electrons are used to form the Ti-O bonds and the anti-bonding (star) orbitals are empty, making the system "closed shell" or $3d^0$. For transition metals with more than four valence electrons, the situation is more complicated. For example, Fe^{3+} corresponds to $3d^5$ in our notation and can

Fig. 1.8 Molecular orbital
theory of transition metal in
a cubic crystal field

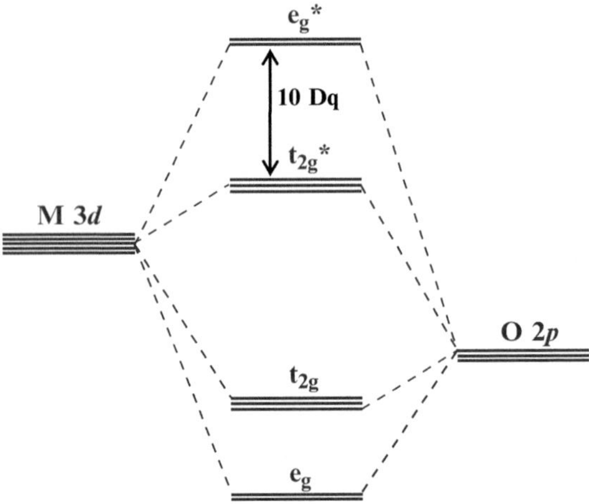

almost completely fill the six t_{2g}^* orbitals, resulting in a low spin state $S = 1/2$. However, if the crystal field splitting $10Dq$ separating t_{2g}^* and e_g^* orbitals is comparable in magnitude to the exchange interaction J, iron can maximize its spin (the Hund's rule) by promoting two electrons into the e_g state, thus reaching a high spin state $S = 5/2$. Stable local moments may lead to various forms of magnetic order, such as ferri-, ferro- or antiferromagnetism [37].

So far we have identified three energy scales defining the properties of the oxide: charge transfer gap, crystal field splitting, and exchange interaction. Another important physical consideration is provided by the Jahn-Teller theorem that relates the electron and lattice degrees of freedom. It states that when the lattice symmetry is high such that there exists degenerate electronic states, the adiabatic potential surface (the total energy for any given set of ion positions) has no minimum with respect to non-totally symmetric displacements. This means that any structural distortion that would lift the degeneracy would most likely occur as that would reduce the total energy. In perovskites, this usually means a distortion of the BO_6 octahedron by an elongation or contraction along one axis and the opposite distortion in the perpendicular plane. Mn^{3+} is a good example of a Jahn-Teller ion. In an octahedral crystal field Mn^{3+} ($3d^4$) has one electron in the degenerate e_g state. A tetragonal distortion results in the splitting of the e_g level. The filled t_{2g} state splits into a singlet and a doublet but without any net change in energy. However, the singly occupied $d_{3z^2-r^2}$ shifts in energy below the empty $d_{x^2-y^2}$ thus lowering the overall energy. On the other hand, Mn^{4+} has an empty e_g state and does not show the Jahn-Teller instability. The magnitude of the Jahn-Teller splitting is governed by the strength of the electron-phonon (vibronic) coupling.

This interplay of the local electronic structure and the crystal lattice is the origin of the rich physics of the transition metal oxides. In particular, the group of iron is especially interesting as the local picture that we have developed (advocated by Van Vleck) [38] does not take into account the itinerant electron (band structure)

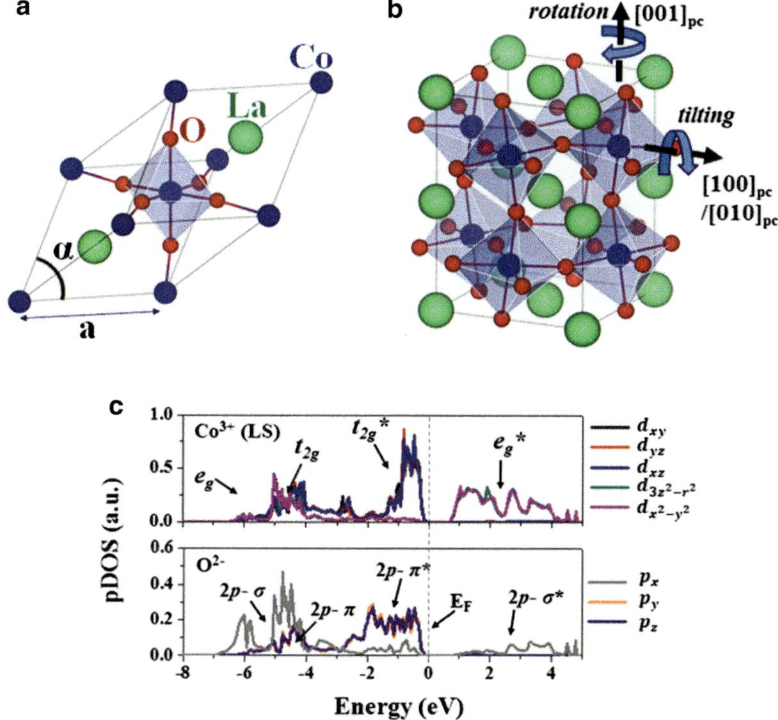

Fig. 1.9 (**a**) Rhombohedral unit cell of LaCoO$_3$ with the *large sphere* representing La, the *medium-sized sphere* representing Co, and the *small sphere* representing O. (**b**) $\sqrt{2} \times \sqrt{2} \times 2$ tetragonal supercell of LaCoO$_3$, showing tilting and rotation of CoO$_6$ octahedral network. (**c**) Projected density of states (pDOS) of the non-magnetic ground state of LaCoO$_3$ for Co^{3+} $3d$ orbitals (*up*) and the nearest neighbor O^{2-} $2p$ orbitals (*down*). The Fermi energy (*dashed vertical line*) is set to 0 eV. From [43]

concept introduced by Slater [39, 40]. Unlike the case of rare earths, where the band structure effects can be considered as perturbation, $3d$ metals show both correlated (local) and band (itinerant) behavior. This can be easily seen by comparing the level diagram in Fig. 1.8 with the actual local electronic structure of Co^{3+} ion in the ground state of LaCoO$_3$. LaCoO$_3$, or LCO, is a perovskite-type crystal. However, its crystal structure (the primitive cell is shown in Fig. 1.9a) is a little bit more complicated. In many perovskites, corner shared octahedra can rotate about various crystallographic axes (octahedral rotation and tilt) resulting in symmetry lowering and larger primitive cells. In the case of LCO the non-magnetic ground state is characterized by an antiphase octahedral rotation about the (111) axis, resulting in cell doubling along the (111) direction. Despite the distortion, the local structure of the $[CoO_6]^{9-}$ octahedron is practically intact. Due to crystal field splitting and Hund's exchange coupling being of the same order, Co^{3+} can access different spin states: low-spin (LS, $t_{2g}^6 e_g^0$, S = 0), intermediate-spin (IS, $t_{2g}^5 e_g^1$, S = 1),

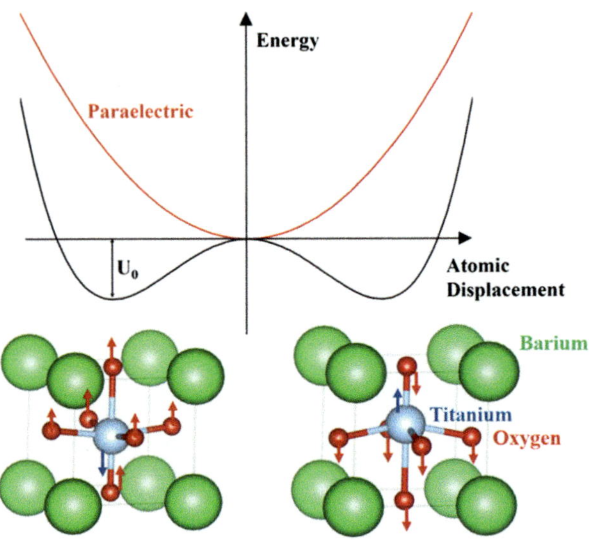

Fig. 1.10 Free energy as a function of atomic displacement in BTO. The order parameter η can be thought of as a relative displacement of Ti and oxygen sublattices

or high-spin (HS, $t_{2g}^4 e_g^2$, S = 2) [41–43]. In Fig. 1.9c we show the density of states projected on Co $3d$ and the nearest neighboring oxygen $2p$ orbitals. Note that instead of narrow localized states we observe broad (several eV wide) bands, yet the local picture is still useful!

Among the transition metal oxides there is a special group that is particularly relevant to the book. As many of these oxides are insulating, they can be polarized by applying an external electric field. In most cases, the polarization is relatively small. However in some materials such as BTO it can be extremely large, particularly at the right temperature. The phenomenon is known as ferroelectricity and materials exhibiting it are known as ferroelectrics. It is convenient to think of ferroelectrics in terms of a collection of microscopic dipoles assigned to each unit cell. In BTO one can think of displacing positively charged Ti ion (formal charge +4) with respect to the negatively charged oxygen octahedron (see Fig. 1.10). If all dipoles are aligned, the crystal develops a macroscopic spontaneous polarization. Crystals with spontaneous polarization are known as pyroelectric. On the other hand, different regions of a crystal may be aligned in opposite ways, and regions of uniform polarization are known as ferroelectric domains. If you start with a crystal with overall polarization equal to zero (domains with opposite polarization compensate each other) and apply a small electric field to it, the crystal will develop a small polarization. However, once the field is strong enough to cause domain switching, a very large polarization will develop as more and more dipoles are aligned. The polarization saturates when all microscopic dipoles point in one direction. If the field is now reduced, the crystal in general is unable to return to its original state and some residual or remnant polarization will remain even under zero applied field. It is of course, possible to make the polarization zero if one applies the field in the opposite direction. The value needed to achieve this is called the coercive field. If one keeps increasing the field in the opposite direction, everything repeats with the only difference that the polarization direction is now reversed. This hysteretic behavior is similar to that of a ferromagnet, and is responsible for the name of the phenomenon. It is important to stress that it is this hysteretic behavior rather than spontaneous polarization itself that makes a crystal ferroelectric.

The ferroelectric transition in BTO is associated with a second order structural phase transition, which is described within a single order parameter Ginsburg-Landau picture, assuming the order parameter η is proportional to polarization. In the absence of an electric field the thermodynamic potential has a characteristic double well shape shown in Fig. 1.10. Two minima correspond to two opposite orientations of polarization. In the presence of the electric field the thermodynamic potential can be written as [33]:

$$\Phi(T,\eta,E) = \Phi_0 + \frac{1}{2}\alpha(T - T_c)\eta^2 + \frac{1}{4}\beta\eta^4 - a\eta E$$

Here α and β are positive constants; the coefficient of the second order term changes sign at the Curie temperature T_c, resulting in the characteristic double well shape of the thermodynamic potential as a function of the order parameter (in the absence of a field). The equilibrium value of the order parameter η_0 in the absence of the electric field is obtained by simple differentiation, and is zero for $T > T_c$ and $\eta_0^2 = \sqrt{-\frac{\alpha(T-T_c)}{\beta}}$ below the Curie temperature. The equilibrium spontaneous polarization $P_s = -\frac{\partial\Phi}{\partial E}$ is then simply $a\eta_0$. In the presence of the field one can compute the susceptibility $\chi = \frac{\partial P}{\partial E} = a\frac{\partial\eta}{\partial E}$. The equilibrium value of the order parameter is determined simply by:

$$\frac{\partial\Phi}{\partial\eta} = \alpha(T - T_c)\eta + \beta\eta^3 - aE = 0$$

Taking the derivative with respect to the field we find

$$\frac{\partial\eta}{\partial E} = \frac{a}{\alpha(T - T_C) + 3\beta\eta_0^2}$$

This results in the Curie-Weiss dependence of the susceptibility above the transition temperature:

$$\chi = \frac{a^2}{\alpha(T - T_C)}$$

The Curie-Weiss constant is $C_{C-W} = 4\pi a^2/\alpha$.

Ferroelectrics are used in a wide spectrum of applications. In thin film form, they have been used for several years in rf devices and in nonvolatile memories. Components based on ferroelectric films are also being developed for various sensor and actuator applications and for tunable microwave circuits. An excellent review of applications of ferroelectrics in high frequency electronics and memories based on ferroelectric materials has been published by Setter et al. [44]. They have reviewed piezoelectric microsensors and microactuators, polar films in microwave electronics, polar ceramics in bulk acoustic wave devices, tunable microwave applications and ferroelectric field effect transistors (FeFETs). The second section deals with materials, structure (domains, in particular), and size effects. Another fundamental review of the recent progress in ferroelectric films has been published by Dawber et al. [45]. These authors discuss the physics relevant for the performance and failure of ferroelectric devices. They also provide a detailed account of the enormous progress made in the first-principles computational approach to understanding ferroelectrics. They also discuss in detail the important role that strain plays in determining the properties of epitaxial thin ferroelectric films.

In the last few years there has been a resurgence of interest in materials known as multiferroics [46]. These are materials that have two or more ferroic orders, for example ferromagnetism and ferroelectricity, simultaneously. Many of the currently studied multiferroic materials such as $BiFeO_3$, $BiMnO_3$, $PbVO_3$, $YMnO_3$, $TbMnO_3$ $TbMn_2O_5$ or $LuFe_2O_4$ are transition metal oxides. Interestingly, the original theoretical work on the magneto-electric effect was done on Cr_2O_3 [47]. Multiferroics can be classified according to the strength of coupling between the magnetic and electric orders. Type-1 multiferroics have weak coupling and ferroelectricity and magnetism are independent in origin. $BiFeO_3$ is the best known example of a Type-1 multiferroic. On the other hand, type-2 multiferroics have strong coupling between the two ferroic orders because one causes the other, i.e. magnetism causes ferroelectricity or ferroelectricity results in magnetism. Most of the rare-earth perovskite manganites such as $TbMnO_3$ are type-2 and have magnetically driven ferroelectricity. This strong coupling between the magnetic and electric properties results in a colossal magnetoelectric effect.

The attraction of multiferroic materials is in the electrical control of ferromagnetism and magnetic control of ferroelectricity at room temperature that may result in new kinds of device functionality. However, there are many challenges to applications of multiferroics. With type-1 multiferroics, where ferroelectric or magnetic properties occur at room temperature, the problem is the absence of magnetoelectric coupling. Although type-2 multiferroics have strong coupling between the magnetic and electric properties, the main drawback is that the polarization is about a thousand times too small for practical applications and occurs only at temperatures well below room temperature. By combining multiferroics with semiconductors in a monolithic fashion, applications in advanced memory technology, where for example magnetic data can be written electrically or even multi-state single bits, can be realized. An excellent review of using multiferroics for memory technology has been recently published by Thomas et al. [48].

1.3 Zintl Intermetallics

In Chap. 2, we shall discuss wetting between covalent semiconductors and ionically bonded transition metal oxides. Wetting is controlled largely by the interface energy, which in turn is controlled by the chemical bonds at the interface. In the case of a semiconductor-oxide interface there is a sharp boundary between the nature of the chemical bonding resulting in a very high energy cost of the interface. A way to alleviate this issue is to use transitional materials in which the bonding is intrinsically neither purely covalent nor ionic or metallic. In such cases the perturbation caused by the interface is not nearly as drastic, and the energy cost may be reduced substantially. One class of such materials is the Zintl compounds. As this class of materials is often outside the curriculum of typical physics and chemistry majors, we will now briefly introduce these unusual materials. For a more detailed treatment, readers are referred to the many excellent texts on the subject [49, 50].

Bonding in solids is governed by the laws of quantum mechanics while the interactions are Coulombic in nature. Three main types of strong chemical bonds are commonly discussed: covalent, metallic, and ionic. Covalent bonding can be explained as the interference between the overlapping atomic wave functions that form bonding and anti-bonding states. Covalent bonding has a strong directional character and a typical example is the sp^3 hybridization in semiconductors described above. Metallic bonding occurs when the valence electrons are highly delocalized and form a "gas" that permeates the entire volume of the crystal and is highly mobile. Metallic bonds are non-directional. Ionic bonding can be understood in terms of electrostatics. Elements with a large difference in electronegativity transfer their valence electrons from the electropositive atom (e.g. Na) to the electronegative atom (e.g. Cl) forming ions of opposite signs (Na^+ and Cl^-) that attract each other electrostatically (until they reach the regime of Pauli repulsion). It should be noted, however, that the boundaries between these bonding types are not sharp.

In 1929, German chemist Eduard Zintl studied a wide group of intermetallic compounds (solid phases that contain two or more metallic elements and optionally non-metallic components) [51]. He was especially interested in the transition area between the ionic and metallic compounds and began research on the compounds of alkali metals or alkaline earth metals with group IIIA–VIIA elements [52]. In addition to developing synthesis and structural analysis of these materials, Zintl discovered compounds that had not been observed at the time and displayed highly unusual structures and behavior. Although he studied intermetallics, some of them were exhibiting salt like properties such as a melting point higher than that of constituents, poor conductivity, and greater brittleness. In particular, he discovered structures, for example NaTl (see Fig. 1.11), for which typical electron counting rules applicable to salts would not apply anymore and so a new concept had to be introduced.

Zintl proposed that in these materials, the construction of the crystal structure was governed by a covalently-bonded framework of the negatively charged main

Fig. 1.11 Double diamond
crystal structure of NaTl
(*blue*: Na, *green*: Tl).
The covalent bonding
between the Tl$^-$ is indicated
by *yellow lines*. With
kind permission from
Springer Science+Business
Media: [56]

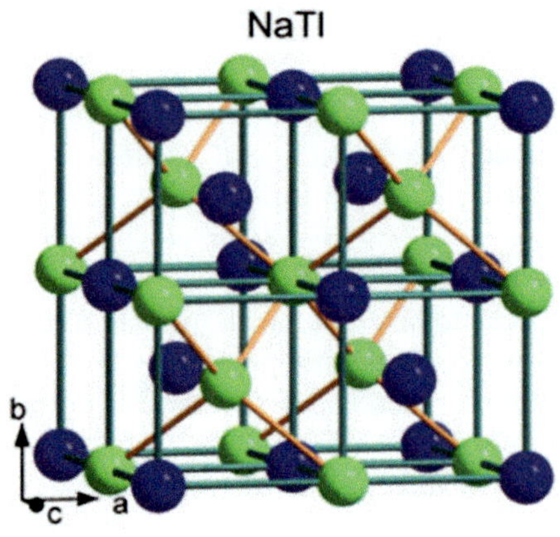

group metals with the positively charged alkali or alkaline earth elements occupying the voids in the framework. This theory was supported by the fact that the lattice constants in these compounds were, to first order, independent of the alkali or alkaline earth metal used.

Zintl also investigated which elements were forming anions with group IA or group IIA elements (Zintl hereby introduced the term polyanion). For the NaTl structure, he found that the electropositive atoms (group IA and group IIA) exhibited a volume contraction and therefore had to undergo some kind of electron transfer. He concluded that the electropositive metals were donating electrons to the main group metals, which were subsequently behaving as pseudoatoms (atoms with the equivalent valence electron configuration of a different atom) that determined the structure of the compound. In this sense, NaTl can be viewed as Na$^+$Tl$^-$, with the group IIIA Tl$^-$ atoms behaving as if they were group IVA elements and forming sp^3 bonds resulting in the diamond structure typical of group IVA materials. The Na$^+$ atoms, which occupy less space due to electron transfer, just fill up the interstitial spaces.

After Zintl's death in 1941, F. Laves proposed to call such materials Zintl phases and formulated the following rules [51]:

- Zintl phases crystallize in "nonmetallic", salt-like structures.
- Zintl phases are always those phases of the alloy system that include the greatest amount of electropositive metals.

These rules (especially the second one) soon proved to be unsatisfactory and subsequently, numerous attempts were made to redefine the concept of the Zintl phase. In particular, W. Klemm and E. Busmann stated: "In Zintl's idea, the formally negatively charged atoms that possess the same electron number as the neutral atoms of the nearest group elements, form polyanions with structures similar to the corresponding elements" [53]. This concept is sometimes referred to as the "Zintl-Klemm-Busmann" concept.

Another important refinement of the Zintl phase concept was made by W. Schaefer, H. Eisenmann and B. Mueller, who were studying compounds of the alkali/alkaline earth metals with the electronegative metals, metametals and semimetals of group IIIA–VA in 1973 [54]. They synthesized materials by fusing the elements together under an inert gas, direct reduction, solution and subsequent distillation in mercury or preparation in liquid ammonia, as used by Zintl. Through determination of the properties of these compounds, they were trying to obtain a new definition of Zintl phases, and a special focus was placed on the properties that could describe the underlying bonding in the materials to show the coexistence of ionic bonding and intermetallic phase. The amount of ionic bonding is indicated by the heat of formation, melting points, volume contraction on formation (especially a volume contraction of the alkali and alkaline earth elements would be an indication of a Zintl phase) and X-ray spectroscopy measurements. Due to the lack of information gathered on the specific materials in these areas, the authors were using X-ray crystal structure determination to investigate the compounds. The combination of the group IA and IIA elements with group VA elements and some elements of group IVA revealed structures typical for salts that had a high amount of ionic bonding, namely structures that have completely "isolated" group VA atoms (meaning they only have neighbors that are other kinds of atoms), fulfilling the conditions for maximally uniform charge distribution within the crystal. On the other hand the structures exhibited by the compounds of group IA and IIA elements with other elements of group IIIA–VA are partial lattices which are observed for single crystalline group IIIA–VA elements (see table in [54]).

Most of these structures can be explained by the Zintl-Klemm-Bussmann concept. The alkali or alkaline earth metals transfer electrons to the more noble components of the alloy, which form a partial lattice corresponding to their resulting outer electron configuration. However, there are structures in [54] that cannot be explained through this concept. The alloys in the CrB structure for example show planar zigzag chains typical for group IVA chains, whereas they should be in a helical structure typical for group VIA lattices. In spite of that, the (8-N)-rule, which states that elements of a main group N will show structures that allow 8-N nearest neighbors, is still maintained. This led the authors to the conclusion that the relation of pseudoatom lattices for Zintl phases should be omitted and instead the more general definition that elements have to obey the (8-N) rule in order to be Zintl phases has to be used. Even compounds with non-integral charges on the anions can be related to the next integral charge number. However, compounds were found by the authors that could not be explained through the extended Zintl-Klemm-Busmann concept proposed by Schaefer et al. [54]. These structures either have an extreme stoichiometry, which means they have significantly more elements of one type than the other, or they exhibit just a slight difference in electronegativity. This means that a transfer of electrons is very unlikely and explains why the extended Zintl-Busmann-Klemm concept is not applicable. Finally, it was clear that it is more convenient to "apply the term Zintl phases to intermetallic compounds which display a pronounced

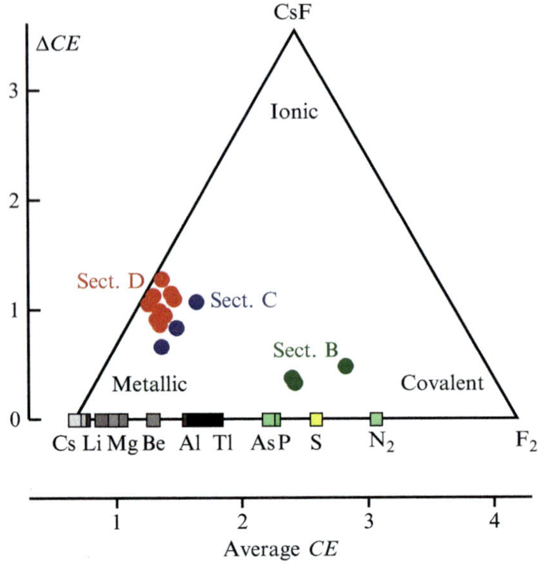

Fig. 1.12 Graph of the amount of the type of bonding in certain compounds. The *red, blue,* and *green circles* correspond to different types of compounds obeying the Zintl-Klemm-Busmann concept. It is clearly visible that Zintl phases combine different types of bonding. With kind permission from Springer Science+Business Media: [56]

heteropolar bonding contribution and in agreement with an ionic formulation in their anion partial lattices that obey the (8-N)-rule" [54]. One should keep in mind that apart from the ionic bonding, in Zintl phases, covalent bonding also plays an important role in the anionic partial lattice, which manifests in the validity of the (8-N)-rule. This has important implications for oxide-semiconductor epitaxy.

As research progressed, in order to delineate Zintl phases from intermetallics and insulators, three criteria were defined for Zintl phases [55]:

- A well-defined relationship exists between the chemical (lattice) and electronic configuration of the material. This is often referred to as satisfying electron counting rules (8-N-rule)
- The material is a semiconductor (sometimes one finds the requirement $E_{gap} <$ 2 eV), or at least, shows increasing electrical conductivity with increasing temperature.
- The material is either diamagnetic or exhibits temperature-independent (Pauli) paramagnetism.

All these criteria imply that Zintl phases have narrow homogeneity widths and electronic structure calculations show that for Zintl phases, the bonding states are fully occupied and separated from the empty, anti-bonding states (band gap). To date, the idea of Zintl phases as a transition between bonding types has remained and has been verified multiple times. In Fig. 1.12, the sum of configuration energies (a quantity directly related to electronegativity) is used as the *x*-axis and the difference in their configuration energies as the *y*-axis. This gives a separation of the regions of covalent, ionic and metallic bonding. As one can see, the Zintl phases referenced in the graph are between all bonding types, subsequently combining them.

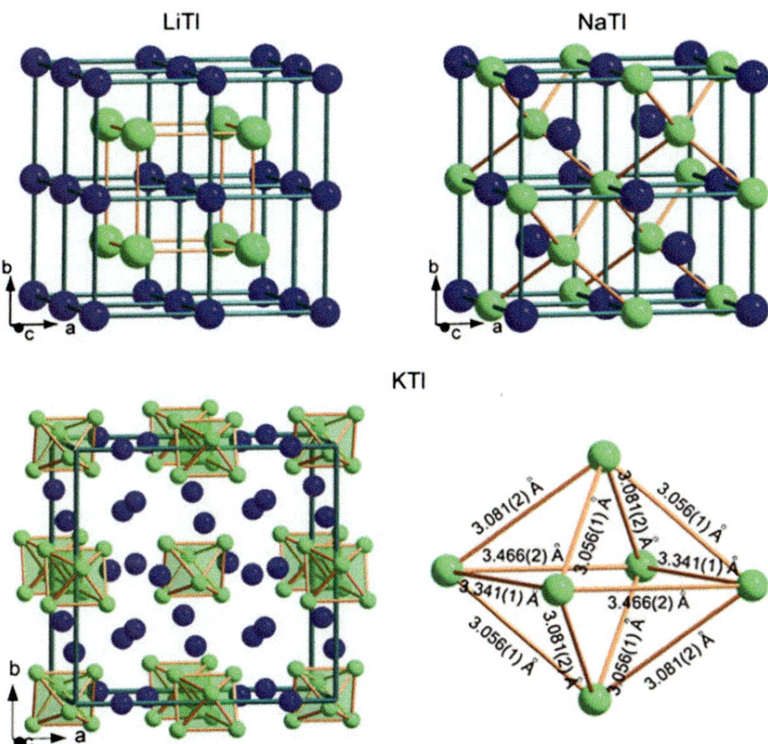

Fig. 1.13 Difference in the Zintl-phases LiM (M = Al, Ga, In) (NaTl-structure), LiTl (CsCl-structure) and KTl (no tetrel-like structure). With kind permission from Springer Science+Business Media: [56]

Current progress in the Zintl-Klemm formalism focuses especially on the role of the cations [56]. It is clear that the use of a certain cation is crucial for the formation of a certain kind of structure. This can be seen in the difference in the Zintl phases LiM (M = Al, Ga, In), which all crystallize in the NaTl structure, LiTl, which adopts the CsCl structure, and KTl, which has no structure typical for a tetrel (group IVA) element and forms distorted octahedra $[Tl_6]^{6-}$ (see Fig. 1.13). In the pseudoatom concept, all binary compounds should have the same structure. The difference can be explained by the over-simplification of the cations as mere electron donors, which again supports the use of the extended Zintl-Klemm concept that only addresses electron counting rules. However, the difference in the structures cannot be explained by only considering covalent bonding, which is the essential part of the extended Zintl-Klemm-concept where the cations do not contribute to any kind of bonding. One always has to consider both ionic and covalent bonding in Zintl phases.

In fact, recent studies have shown that the metallic bonding part also plays an important role [56–58]. Looking at the density of states of NaTl, the "classic" Zintl phase, one finds that near the Fermi level, the major contribution of the states arises from $6s$ and $6p$ orbitals of the Tl atoms. However, a significant contribution is also

Fig. 1.14 Density of states
and crystal orbital
Hamiltonian population
(COHP) of NaTl.
Reproduced with
permission from [53]

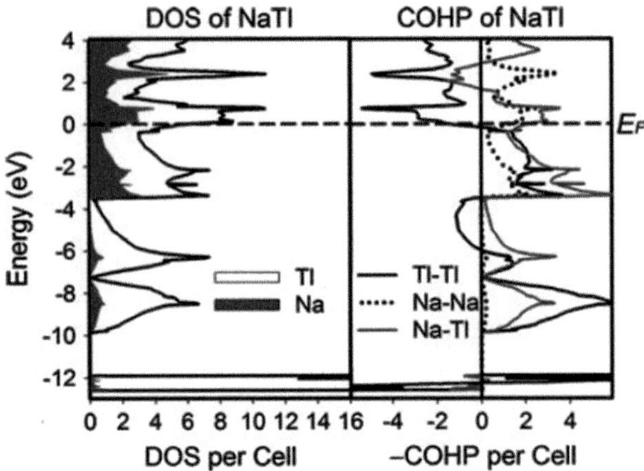

made by the Na orbitals (see Fig. 1.14), which suggests the purely ionic picture to
be inadequate, in which the atomic orbitals of the Na atoms should not interact at
all. This shows that although the beauty and simplicity of the Zintl-Klemm concept
lies in the observation of covalent bonding, one cannot forget about ionic and
metallic bonding when examining composition-structure relations in Zintl phases.

To emphasize this statement, one can for example, consider the binary compounds LiM, where M = Al, Ga, In, Tl (see above). Miller et al. found that in these
compounds a NaTl structure is favored by covalent bonding and the CsCl structure
is favored by ionic and metallic bonding [56]. These results can be obtained by DFT
calculations of the energy and separating it into Madelung (electrostatic) terms and
electronic (band structure) terms. The covalent interaction decreases as one moves
from LiAl to LiTl and eventually the ionic and metallic bonding outweighs the
covalent bonding in LiTl, which exhibits the CsCl-structure. This competition
between the bonding types can be found in many Zintl phases (see for example
[56] or [57]).

Although the Zintl-Klemm-concept is able to predict structures and properties of
specific intermetallics, there are examples where it fails to explain certain phenomena due to its simplicity. This shows us that the concept may still need further
refinement and one sometimes has to be careful when using the predictions of this
theory.

References

1. R.A. McKee, F.J. Walker, M.F. Chisholm, Phys. Rev. Lett. **81**, 3014 (1998)
2. X. Zhang, A.A. Demkov, H. Li, X. Hu, Y. Wei, J. Kulik, Phys. Rev. B **68**, 125323 (2003)
3. C.R. Ashman, C.J. Först, K. Schwarz, P.E. Blöchl, Phys. Rev. B **69**, 075309 (2004)
4. Z. Yu, Y. Liang, C. Overgaard, X. Hu, J. Curless, H. Li, Y. Wei, B. Craigo, D. Jordan,
   R. Droopad, J. Finder, K. Eisenbeiser, D. Marshall, K. Moore, J. Kulik, P. Fejes, Thin Solid
   Films **462**, 51 (2004)

5. Y. Liang, S. Gan, Y. Wei, R. Gregory, Phys. Status Solidi B **243**, 2098 (2006)
6. C. Rossel, B. Mereu, C. Marchiori, D. Caimi, M. Sousa, A. Guiller, H. Siegwart, R. Germann, J.-P. Locquet, J. Fompeyrine, D.J. Webb, C. Dieker, J.W. Seo, Appl. Phys. Lett. **89**, 053506 (2006)
7. J.W. Reiner, A. Posadas, M. Wang, M. Sidorov, Z. Krivokapic, F.J. Walker, T.P. Ma, C.H. Ahn, J. Appl. Phys. **105**, 124501 (2009)
8. V. Vaithnayathan, J. Lettieri, W. Tian, A. Sharan, A. Vasudevarao, Y.L. Li, A. Kochhar, H. Ma, J. Levy, P. Zschack, J.C. Woicik, L.Q. Chen, V. Gopalan, D.G. Schlom, J. Appl. Phys. **100**, 024108 (2006)
9. G. Niu, S. Yin, G. Saint-Girons, B. Gautier, P. Lecoeur, V. Pillard, G. Hollinger, B. Vilquin, Microelectron. Eng. **88**, 1232 (2011)
10. A.K. Pradhan, J.B. Dadson, D. Hunter, K. Zhang, S. Mohanty, E.M. Jackson, B. Lasley-Hunter, K. Lord, T.M. Williams, R.R. Rakhimov, J. Zhang, D.J. Sellmyer, K. Inaba, T. Hasegawa, S. Mathews, B. Joseph, B.R. Sekhar, U.N. Roy, Y. Cui, A. Burger, J. Appl. Phys. **100**, 033903 (2006)
11. J. Wang, H. Zheng, Z. Ma, S. Prasertchoung, M. Wuttig, R. Droopad, J. Yu, K. Eisenbeiser, R. Ramesh, Appl. Phys. Lett. **85**, 2574 (2004)
12. A. Posadas, M. Berg, H. Seo, A. de Lozanne, A.A. Demkov, D.J. Smith, A.P. Kirk, D. Zhernokletov, R.M. Wallace, Appl. Phys. Lett. **98**, 053104 (2011)
13. A. Posadas, M. Berg, H. Seo, D.J. Smith, A.P. Kirk, D. Zhernokletov, R.M. Wallace, A. de Lozanne, A.A. Demkov, Microelectron. Eng. **88**, 1444 (2011)
14. H. Seo, A.B. Posadas, C. Mitra, A.V. Kvit, J. Ramdani, A.A. Demkov, Phys. Rev. B **86**, 075301 (2012)
15. M.D. McDaniel, A. Posadas, T. Wang, A.A. Demkov, J.G. Ekerdt, Thin Solid Films **520**, 6525 (2012)
16. B.W. Wessels, J. Cryst. Growth **195**, 706 (1998)
17. D. Dimos, Annu. Rev. Mater. Sci. **25**, 273 (1995)
18. G. Eranna, B.C. Joshi, D.P. Runthala, R.P. Gupta, Crit. Rev. Solid State Mater. Sci. **29**, 111 (2010)
19. D.L. Polla, L.F. Francis, Annu. Rev. Mater. Sci. **28**, 563 (1998)
20. X. Chen, S. Shen, L. Guo, S. Mao, Chem. Rev. **110**, 6503 (2010)
21. S. Abel, T. Stoöferle, C. Marchiori, C. Rossel, M.D. Rossell, R. Erni, D. Caimi, M. Sousa, A. Chelnokov, B.J. Offrein, J. Fompeyrine, Nat. Commun. **4**, 1671 (2013)
22. A.A. Demkov, H. Seo, X. Zhang, J. Ramdani, Appl. Phys. Lett. **100**, 071602 (2012)
23. P.Y. Yu, M. Cardona, *Fundamentals of Semiconductors: Physics and Materials Properties*, 4th edn. (Springer, Berlin, 2010)
24. W.A. Harrison, *Electronic Structure and the Properties of Solids* (Dover, New York, 1989)
25. N.W. Ashcroft, N.D. Mermin, *Solid State Physics* (Cengage Learning, London, 1976) (Chap. 10)
26. P.G. Neudeck, Silicon carbide technology, in *The VLSI Handbook*, ed. by W.-K. Chen (CRC Press, Boca Raton, FL, 2006)
27. C.W. Bunn, Proc. Phys. Soc. Lond. **47**, 836 (1935)
28. Ü. Özgür, Y.I. Alivov, C. Liu, A. Teke, M.A. Reshchikov, S. Doğan, V. Avrutin, S.-J. Cho, H. Morkoç, J. Appl. Phys. **98**, 041301 (2005)
29. S. Maekawa, T. Tohyama, S.E. Barnes, S. Ishihara, W. Koshibae, G. Khaliullin, *Physics of Transition Metal Oxides*. Springer Series in Solid-State Sciences, vol. 144 (Springer, Berlin, 2004)
30. P.A. Cox, *Transition Metal Oxides: An Introduction to Their Electronic Structure and Properties*. The International Series of Monographs on Chemistry (Oxford University Press, Oxford, 1992)
31. V.E. Henrich, P.A. Cox, *The Surface Science of Metal Oxides* (Cambridge University Press, Cambridge, 1994)

32. M.E. Lines, A.M. Glass, *Principles and Applications of Ferroelectrics and Related Materials*.
    Oxford Classic Texts in the Physical Sciences (Oxford University Press, Oxford, 2001)
33. B.A. Strukov, A.P. Levanyuk, *Ferroelectric Phenomena in Crystals* (Springer, Berlin, 1998)
34. G.F. Dionne, *Magnetic Oxides* (Springer, New York, 1978)
35. J. Stöhr, H.C. Siegmann, *Magnetism* (Springer, Berlin, 2006)
36. D.G. Schlom, L.-Q. Chen, X. Pan, A. Schmehl, M.A. Zurbuchen, J. Am. Ceram. Soc. **91**, 2429
    (2008)
37. J.B. Goodneough, J.-S. Zhou, Chem. Mater. **10**, 2980 (1998)
38. J.H. van Vleck, Rev. Mod. Phys. **25**, 220 (1953)
39. J.C. Slater, Phys. Rev. **49**, 537 (1936)
40. J.C. Slater, Phys. Rev. **49**, 931 (1936)
41. P.M. Raccah, J.B. Goodenough, Phys. Rev. **155**, 932 (1967)
42. A. Podlesnyak, S. Streule, J. Mesot, M. Medarde, E. Pomjakushina, K. Conder, A. Tanaka,
    M.W. Haverkort, D.I. Khomskii, Phys. Rev. Lett. **97**, 247208 (2006)
43. H. Seo, A.B. Posadas, A.A. Demkov, Phys. Rev. B **86**, 014430 (2012)
44. N. Setter, D. Damjanovic, L. Eng, G. Fox, S. Gevorgian, S. Hong, A. Kingon, H. Kohlstedt,
    N.Y. Park, G.B. Stephenson, I. Stolitchnov, A.K. Taganstev, D.V. Taylor, T. Yamada,
    S. Streiffer, J. Appl. Phys. **100**, 051606 (2006)
45. M. Dawber, K.M. Rabe, J.F. Scott, Rev. Mod. Phys. **77**, 1083 (2005)
46. L.W. Martin, S.P. Crane, Y.-H. Chu, M.B. Holcomb, M. Gajek, M. Huijben, C.-H. Yang,
    N. Balke, R. Ramesh, J. Phys. Condens. Matter **20**, 434220 (2008)
47. I.E. Dzyaloshinskii, Sov. Phys. JETP **10**, 628 (1960)
48. R. Thomas, J.F. Scott, D.N. Bose, R.S. Katiyar, J. Phys. Condens. Matter **22**, 423201 (2010)
49. S.M. Kauzlarich, *Chemistry, Structure, and Bonding of Zintl Phases and Ions* (VCH Publishers
    Inc., New York, 1996)
50. T.F. Faessler, *Zintl Phases: Principles and Recent Developments* (Springer, Heidelberg, 2011)
51. G.J. Miller, Structure and bonding at the Zintl border, in *Chemistry, Structure and Bonding of
    Zintl Phases and Ions*, ed. by S.M. Kauzlarich (Wiley-VCH, New York, 1996)
52. E. Zintl, Z. Phys. Chem. **154**, 1 (1931)
53. W. Klemm, E. Busmann, Z. Anorg. Allg. Chem. **319**, 297 (1963)
54. H. Schaefer, B. Eisenmann, W. Mueller, Angew. Chem. Int. Ed. **12**, 694 (1973)
55. R. Nesper, Prog. Solid State Chem. **20**, 1 (1990)
56. G.J. Miller, M.W. Schmidt, F. Wang, T.S. You, Quantitative advances in the Zintl-klemm
    formalism, in *Zintl Phases: Principles and Recent Developments*, ed. by T.F. Faessler
    (Springer, Berlin, 2011)
57. F. Wang, G.J. Miller, Eur. J. Inorg. Chem. **26**, 3989 (2011)
58. M.H. Whangbo, L. Changhoon, J. Koehler, Eur. J. Inorg. Chem. **26**, 3841 (2011)

Chapter 2
Critical Issues in Oxide-Semiconductor Heteroepitaxy

In semiconductor/semiconductor heteroepitaxy, assuming that one is able to grow the correct phase of the material using the appropriate growth conditions, the two main challenges are the lattice and thermal mismatches between the substrate and the growing film [1]. Extensive work has been dedicated to address these difficulties including the lattice grading method [2], and the use of a compliant substrate for strain management [3]. The latter approach is based on a free, single crystal membrane that is sufficiently thin to deform elastically, thus allowing for total strain to be shared between the membrane and the heteroepitaxial layer grown upon it. These concepts have been utilized to reduce the defects in a variety of materials systems such as SiGe/SOI/Si [4], InGaAs/GaAs [5], and GaN/SOI [6] (SOI stands for silicon on insulator).

Thermal mismatch is an even a bigger problem in oxide-semiconductor integration because the difference in thermal expansion coefficients is greater. For example, the thermal expansion of Si is $2.6 \times 10^{-6} \, \mathrm{K}^{-1}$ and it is $8.8 \times 10^{-6} \, \mathrm{K}^{-1}$ in $SrTiO_3$ (STO). In other words at the growth temperature a semiconductor is slightly larger than what it is at room temperature, while the oxide is significantly larger, and thus one would expect large stresses to develop in the film upon cooling. As we shall see later in the book, this thermal mismatch has a real effect on the properties of thin oxide films grown on semiconductors at high temperature. On the one hand, one might exploit this difference. On the other hand, this makes low temperature deposition methods, such as atomic layer deposition (ALD) very attractive.

Luckily, nature gives us a break and lattice mismatch is a much less critical problem when depositing oxide films compared to semiconductor films. As semiconductors are mostly simple sp^3 covalently bonded materials, they are very sensitive to interatomic angles, and have a limited range of structural responses to lattice mismatch. Covalently bonded materials can only strain so much before they will relax to their normal lattice spacing, most commonly by forming edge dislocations that glide to the substrate-film interface. This concept is captured in the famous Matthews-Blakeslee model that relates the critical thickness of an epitaxial film to elastic strain, assuming that strain is relieved only through dislocation

A.A. Demkov and A.B. Posadas, *Integration of Functional Oxides with Semiconductors*, DOI 10.1007/978-1-4614-9320-4_2, © The Author(s) 2014

Fig. 2.1 Matthews-Blakeslee model for $Si_{1-x}Ge_x$ on Si. The *lower curve* is the equilibrium critical thickness from the model while the *upper curve* is a metastable condition calculated for growth at 500 °C. Copyright IOP Publishing. Reproduced from [9] by permission of IOP Publishing. All rights reserved

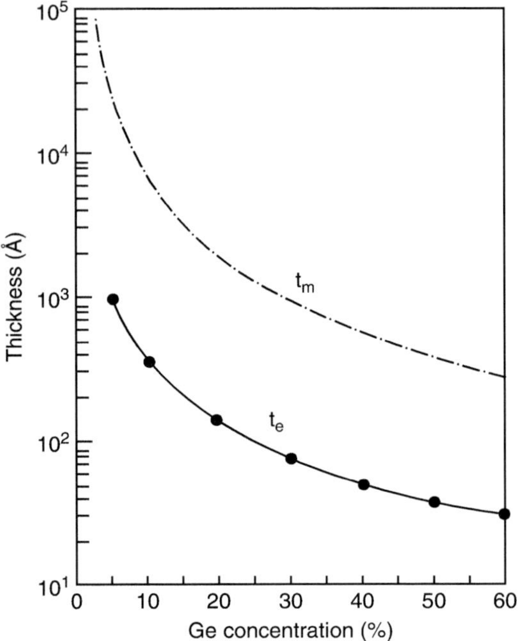

formation [7, 8]. The Matthews-Blakeslee equation (simplified for pure edge misfit dislocations) states that the critical thickness h_c can be expressed as:

$$h_c = \frac{b}{4\pi f(1+\nu)}\left[ln\left(\frac{h_c}{b}\right) + 1 \right]$$

Here f is the lattice mismatch; ν is the Poisson's ratio, and b is the Burgers vector of the misfit dislocation. The resulting curve for SiGe/Si is shown in Fig. 2.1 [9]. One can see that a strain of ~1 % (Ge content of 25 %) results in a critical thickness of ~10 nm and a strain of ~2 % (Ge content of 50 %) results in a critical thickness of ~4 nm.

Oxides are generally more tolerant of strain than semiconductors. Perovskite oxides being partly ionic are somewhat less sensitive to bond angle variation as long as the interatomic distances are maintained (Coulomb interaction depends mainly on the absolute distance between the charges). Also, perovskites have a much broader arsenal of responses at their disposal. Some are due to their more complicated crystal structure, and some to the peculiarity of transition metals. First, as we have discussed in Chap. 1, the octahedra can rotate and tilt which gives the oxide some freedom to change volume. Second, for certain transition metal ions, the octahedra can change its "stiffness" by changing the spin state of the transition metal ion, allowing the octahedra to distort. Third, lattice parameters of an oxide can often change by introducing oxygen vacancies into the crystal structure [10]. In other words, there are internal degrees of freedom that allow the material to lower its energy in response to strain [11]. As a result, it is not uncommon to epitaxially grow pseudomorphic oxide films with as much as several percent lattice mismatch to relatively large thicknesses exceeding the predicted critical thickness [12–14].

There are, however, three additional key problems unique to heteroepitaxy of perovskite oxides with covalent semiconductors. For high quality films the layer-by-layer or Frank-Van der Merwe growth is necessary. This is controlled by wetting at the oxide/semiconductor interface and is intimately related to the chemical bonding at the interface. Despite the fact that it is possible to match an oxide lattice to that of a semiconductor in the plane, there still is a problem of growing over a step edge, as the surface step height of the substrate is not necessarily matched by the out-of-plane inter-planar distance of the film. Last but not least, there is a symmetry difference between, for example, the diamond lattice of Si and the simple cubic lattice of a perovskite. This symmetry mismatch may result in twin and other domains, which could adversely affect the film properties. In the case of Si, the additional problem is oxidation and etching. At low pressure, oxygen etches Si owing to volatility of SiO, leaving craters on the surface [15, 16], while at higher pressure the formation of an amorphous SiO_2 layer destroys any possibility of epitaxial registry.

Here we will focus on the fundamental issues of oxide/semiconductor epitaxy, using $SrTiO_3$ on Si and GaAs as examples. However, these problems are universal and apply to all other systems discussed in this book, with the caveat that symmetry mismatch does indeed depend on the actual symmetry of the crystal and the types of domains possible on hexagonal substrates are different from those on a diamond or zincblende substrate.

2.1 Lattice Matching Oxides and Semiconductors

Looking at the diamond crystal structure of Si and ABO_3 perovskite structure of STO in Chap. 1, one is intrigued how exactly these two can be matched. The answer is given in Fig. 2.2. Si atoms at the surface are depicted with large spheres and smaller spheres correspond to atoms below the surface, with depth marked in units of lattice constant a. As can be seen from the figure, the surface unit cell of unreconstructed Si(001) is rotated $45°$ with respect to the conventional cubic cell of Si owing to the face-centered cubic (fcc) nature of the Si lattice. The lattice constant of a 1×1 surface cell is $a/\sqrt{2}$ or 3.84 Å, which is very close to 3.905 Å of cubic STO and results in 1.66 % compressive strain in a fully epitaxial oxide layer. This type of matching is often called a $45°$ rotation and is common to all perovskite on diamond (001) or zincblende (001) epitaxy. The critical thickness of STO on Si has been experimentally found to be approximately 4 nm [17].

Matching is of course different for the (111) orientation of cubic crystals [18, 19] or for hexagonal epitaxy [20]. For example, in Fig. 2.3 we illustrate the one-on-four lattice matching of cubic anti-bixbyite Gd_2O_3 on Si (111). Three stable phases of Gd_2O_3 can be found at ambient pressure. At room temperature, the cubic $Ia\bar{3}$ form is stable. It is followed by a monoclinic $C2/m$ phase at 1,500 K and hexagonal $P\bar{3}m1$

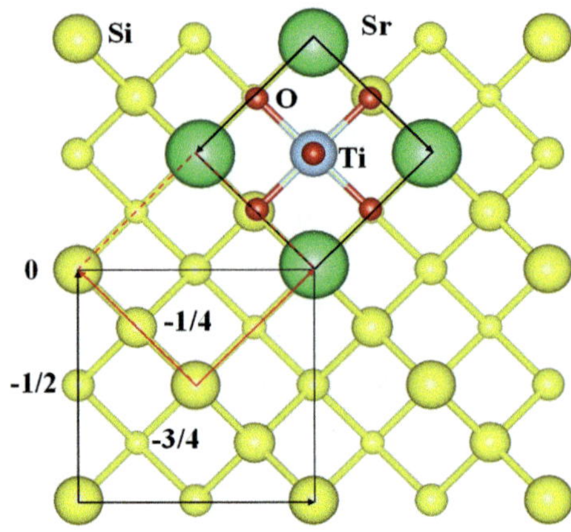

Fig. 2.2 Epitaxial matching of STO and Si (001). The 1 × 1 Si surface unit cell (*colored red*) is rotated 45° with respect to the bulk cubic cell (*colored black*) and matches the perovskite. The *numbers* refer to the vertical position with respect to the surface set at zero, in the units of Si lattice constant a = 5.43 Å

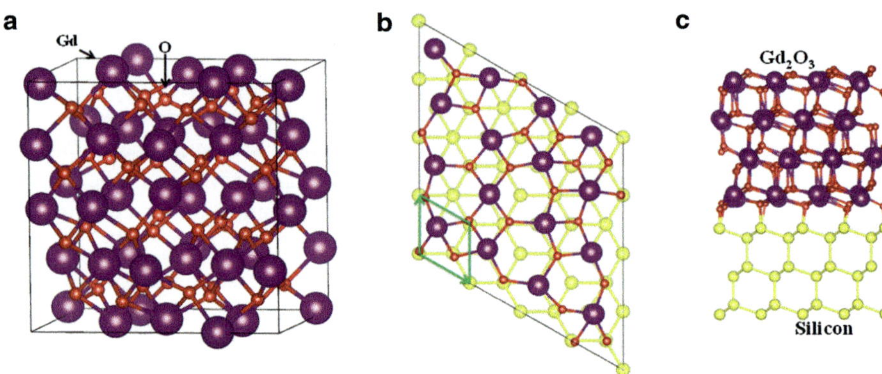

Fig. 2.3 Epitaxial matching of cubic Gd_2O_3 to Si (111). (**a**) Cubic unit cell of Gd_2O_3; (**b**) matching of one (111) unit cell of Gd_2O_3 to four unit cells of Si (111) (top view; Si unit cell is marked in *green*); (**c**) same matching, a side view along $(\bar{1}11)$

phase at 2,443 K. The ground state cubic phase of Gd_2O_3 is paramagnetic, but shows complex non-collinear antiferromagnetic behavior below 1.6 K [21, 22]. It is a large band gap (5.9 eV) insulator [23] with a medium dielectric constant $\varepsilon = 14$. The lattice constant of cubic Gd_2O_3 is 10.817 Å [24], and one unit cell of the (111) surface matches four unit cells of Si in the same orientation as shown in Fig. 2.3.

2.2   Wetting

The fundamental difficulty of perovskite/semiconductor epitaxy lies in thermodynamics. To achieve layer-by-layer growth, the film should wet the substrate. Wetting is controlled at the microscopic level by the interatomic forces. Knowing the surface energies of the substrate and film, and the energy of the interface

Table 2.1 Absolute surface energies $E_{surf}^{n \times m}$ and $\gamma^{n \times m}$ for various orientations and reconstructions

Orientation	Reconstruction	E_{surf} (eV/1 × 1 cell)			γ (J/m^2)		
		C	Si	Ge	C	Si	Ge
(111)	Unrelaxed	2.735	1.435	1.128	8.12	1.82	1.32
	Relaxed	2.165	1.372	1.116	6.43	1.74	1.30
	2 × 1 (right)	1.369	1.141	0.901	4.06	1.45	1.05
	2 × 1 (left)	1.369	1.136	0.893	4.06	1.44	1.04
	c(2 × 8)	2.346	1.109	0.865	6.96	1.41	1.01
	7 × 7	2.395	1.073	0.872	7.11	1.36	1.02
	H-covered	−2.760	−2.383	−2.249	−8.19	−3.03	−2.63
(110)	Unrelaxed	4.115	2.630	2.127	7.48	2.04	1.51
	Relaxed	3.264	2.190	1.661	5.93	1.70	1.17
	H-covered	−5.496	−4.644	−4.637	−9.99	−3.61	−3.32
(100)	Unrelaxed	3.780	2.174	1.691	9.72	2.39	1.71
	Relaxed	3.655	2.173	1.690	9.40	2.39	1.71
	2 × 1	2.222	1.321	1.035	5.71	1.45	1.05
	c(4 × 2)	2.222	1.285	0.985	5.71	1.41	1.00
	H-covered	−3.545	−4.853	−4.25	−9.11	−5.34	−4.56

Table taken from [25]

(γ_{sub}, γ_{film} and $\gamma_{interface}$, respectively), the condition of wetting can be simply expressed as:

$$\gamma_{sub} > \gamma_{film} + \gamma_{interface} \tag{2.1}$$

In other words, to achieve wetting the substrate should have high surface energy γ_{sub}, the film should have low surface energy γ_{film}, and the cost of having an interface $\gamma_{interface}$ should be low. Interestingly, it follows from this inequality that if material A (the film) wets material B (the substrate), then B is unlikely to wet A. In semiconductor/semiconductor epitaxy, the surface energies of the film and the substrate are often reasonably close. In Table 2.1 we list surface energies of common semiconductors for low index surfaces from [25]. More importantly, the nature of chemical bonding is only slightly modulated across the interface, staying predominantly covalent. This results in an interface energy that is relatively small. Consequently, achieving wetting is relatively easy, provided the surface energy of the growing film can be kept low under the growth conditions (sometimes a surfactant is required), and the main concern is the lattice mismatch resulting in too much elastic energy being stored in the film. In contrast, for semiconductor/perovskite epitaxy, none of this is generally true. In particular, the energy cost of going from an ionic oxide to a covalent semiconductor is rather high. One, therefore, has to be creative in designing template or wetting layers to reduce the normally high interfacial energy.

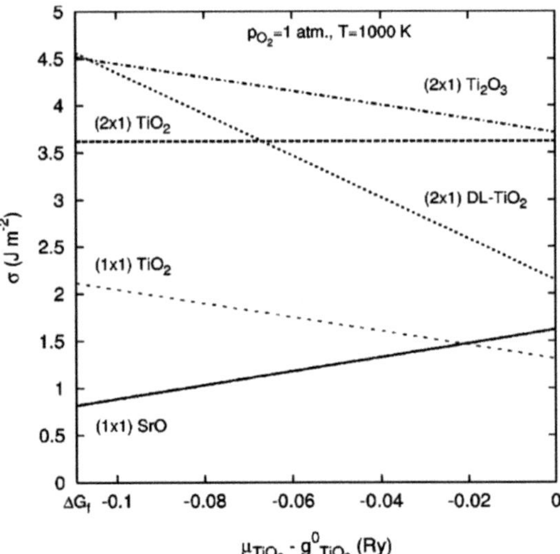

Fig. 2.4 Surface energies as a function of μ_{TiO_2} at $T = 1,000$ K and $p_0 = 1$ atm. Reprinted with permission from [28]. Copyright 2004 by the American Physical Society

Fortunately, the ABO_3 perovskite structure offers two (SrO and TiO_2) possible surface terminations and the surface energy is highly tunable [26–28]. Being a multicomponent system, the energy depends not only on the orientation and reconstruction, but also on the chemical environment as captured by the corresponding chemical potentials. In Fig. 2.4 we reproduce the surface energy diagram for STO from [28]. They considered 1×1 and 2×1 reconstructions of the (001) STO surface using first-principles DFT calculations. Surface energies were calculated as a function of TiO_2 chemical potential, oxygen partial pressure and temperature. The 1×1 unreconstructed surfaces were found to be energetically stable for many of the conditions considered. Under conditions of very low oxygen partial pressure, the 2×1 Ti_2O_3 reconstruction reported by Castell [29] was found stable. The graph corresponds to an oxygen pressure of 1 atm. and temperature of 1,000 K. Note the very wide range of surface energy from less than 1.0 to 4.5 J/m^2, and its sensitivity to the environment. The zero of chemical potential corresponds to TiO_2-rich environment.

Knowing surface energies, one can easily estimate what should be the energy of the interface to guarantee layer by layer growth. For example, for STO to wet Si, the surface energy of STO plus the energy of the interface should not exceed the surface energy of Si of ~1.7 J/m^2. With the STO surface energy ranging from 0.8 to almost 2.0 J/m^2 depending on the environment, this requires an interface with energy below 0.9 J/m^2 to achieve wetting [27]. This has been realized using a $SrSi_2$ template that has the stoichiometry of a bulk Zintl-Klemm intermetallic [30–33]. It is worth noting that this template also suppresses oxidation of Si below about 400 °C.

Recently, Demkov et al. explored theoretically the fundamental question of the bonding character change across the epitaxial interface between STO and GaAs

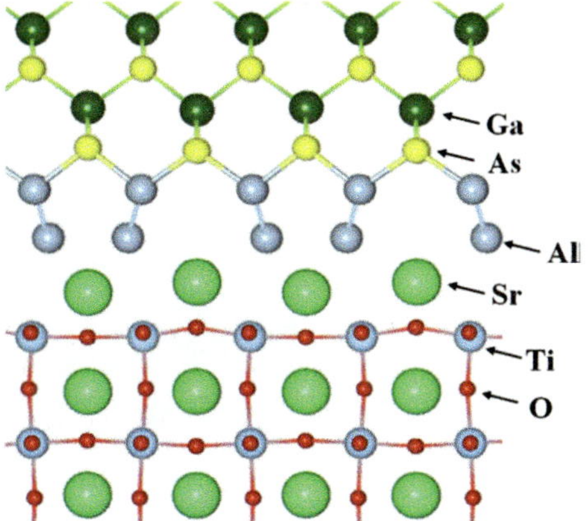

Fig. 2.5 Theoretical model of the STO/GaAs interface with a Zintl-Klemm $SrAl_2$ interlayer. Reprinted with permission from [34]. Copyright 2012, AIP Publishing LLC

using intermetallic Zintl-Klemm (Z-K) compounds as transition layers to ensure wetting [34]. The structure of cubic STO may be thought of as consisting of two types of alternating layers, a covalent TiO_2 layer and an ionic SrO layer. On the other hand, GaAs has zincblende structure, which is a manifestation of the sp^3 hybridization. Therefore, to form a high quality stable interface between a transition metal oxide material such as STO and an sp^3 covalent semiconductor such as GaAs, one has to change the fundamental nature of chemical bonding across the interface. If not addressed properly, this discontinuity in the chemical bonding results in a high interfacial energy γ_{int} of a few J/m^2. This high interfacial energy rather than the lattice mismatch is the main cause of 3D growth in perovskite/semiconductor epitaxy. Sr aluminides such as $SrAl_2$ offer a possible transition layer. Sr aluminide belongs to the Ae-Tr group of Zintl phases formed by triels and alkaline earths. The charge is transferred from the electropositive element Sr to the more electronegative element Al. Formally, Al^- has Si character, and forms structures characteristic to Si, i.e. diamond structure. For example, in the hypothetical cubic B32 (NaTl) structure Al atoms form a diamond-like four-connected network (see Chap. 1). In other words, the charge transfer from the electropositive to the electronegative species allows the latter one to assume the structural motif typical of Si, the next column element in the periodic table. It is precisely this property of Zintl compounds that can be exploited at the oxide/semiconductor interface.

In Fig. 2.5 we show the GaAs-STO interface proposed in [34]. The aluminide layer produced by replacing oxygen with Al in the SrO layer immediately following the TiO_2 surface plane serves as a transition from the d-orbital dominated bonding in the covalent octahedral Ti-O network to the tetrahedral network of AlAs. Note that AlAs is lattice matched to GaAs. The $SrAl_2$ interlayer separates STO from GaAs. GaAs is strained to match the STO lattice ($a_{theory} = 3.87$ Å). The Ga-As bond length in the bulk GaAs region ranges from 2.44 to 2.42 Å. At the interface the Al-As bond length is 2.42 Å, while the weaker Al-Al bonds in the Z-K layer are

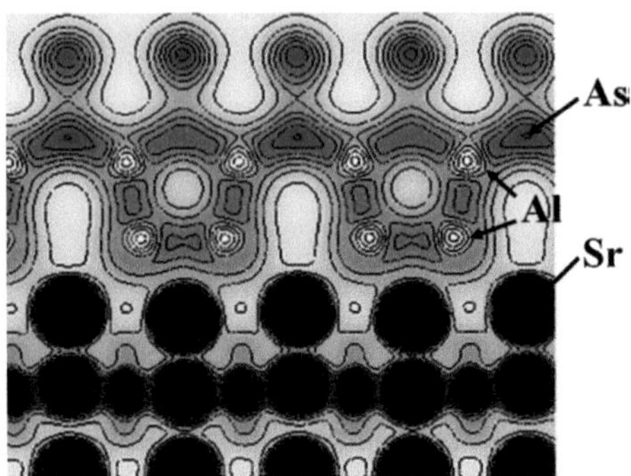

Fig. 2.6 Cross section of the charge distribution across the STO/Z-K/GaAs interface. The saturation level is set to 0.7 eÅ^{-3} (=12.6 % of the maximum density). The contours step interval is 0.05 eÅ^{-3}. Note the accumulation of charge between Al atoms representing the "covalent-like" Zintl-Klemm bonding. Reprinted with permission from [34]. Copyright 2012, AIP Publishing LLC

2.65 Å and 2.82 Å (to be compared to 2.82 Å in bulk SrAl$_2$). In Fig. 2.6, we show the charge density distribution in the plane containing Sr and Al atoms, with the contour plot overlaid. For clarity, the density saturation level is set to 0.7 eÅ^{-3} (12.6 % of the maximum charge density). Note the areas of relatively high electron density between the two Al atoms in the SrAl$_2$ interlayer. This pile up of charge is a Z-K bond between two metal atoms. The strength of these bonds is relatively low, as indicated by the low electron density.

Using the theoretical values for the surface energy of GaAs from [35], Demkov and co-workers assumed the average value of 1.0 J/m^2 representative of β2(2 × 4) reconstruction, which is stable in a wide range of As chemical potential. Then under Ti rich conditions the surface energy of STO is approximately 1.25 J/m^2 resulting in wetting of GaAs by STO as the mixed dimer (2 × 4) GaAs termination is stabilized. This is because under As and Ti rich conditions, the energy of the Zintl-based interface can be as low as 0.30 J/m^2. Indeed, Liang an co-workers have reported high quality epitaxial STO films on GaAs [36]. In addition, Demkov et al. computed the valence band offset at the GaAs/SrTiO$_3$ interface to be 2.50 eV in good agreement with recent experimental results [37]. Interestingly, the results of Demkov and co-workers also suggest a window for GaAs to wet STO which provides an explanation for the reported epitaxial growth of GaAs on STO, including a functional MESFET device [31].

Bulk properties of SrAl$_2$ were investigated theoretically by Slepko and Demkov [38]. They reported a density functional investigation of the orthorhombic (*Imma*) and cubic ($Fd\overline{3}m$) phases of this strontium aluminide. For the orthorhombic phase they calculated the work function and surface energy for (001), (010) and (100) oriented surfaces. The work function varies between 2.0 and 4.1 eV, and was shown to be determined by the predominant atomic species on the surface. Surface energy ranges from 0.32 to 1.84 J/m^2 were reported. More recently, Schlipf et al. have reported epitaxial growth by MBE of Zintl-phase SrAl$_4$ on the (001) oriented perovskite oxide LaAlO$_3$ using MBE [39]. Photoelectron spectroscopy measurements verified the Zintl-Klemm nature of the bonding in the material.

2.3 Kinetics Versus Thermodynamics: Chemical Reactivity

Even if the issues concerning lattice matching and wetting have somehow been resolved, the success of an epitaxial growth process is still dependent on an even more basic issue: thermodynamic stability of the film when in contact with the substrate at the growth conditions. For example, if the film reacts with the substrate while the film is growing and the reaction product is not lattice matched or does not wet the substrate then any chance for epitaxial growth is completely gone. This fundamental restriction severely limits the combinations of film and substrate materials that one can use to form epitaxial systems. Because many of the interesting functional oxides are ternary compounds, the relevant phase diagrams between them and semiconductors are often not yet completely mapped out adding to the difficulty of developing a process. It is for these reasons why there are very few epitaxial oxide on semiconductor systems that have been achieved to date. However, if we take advantage of the possibility of kinetic inhibition of some of these reactions between the substrate and the constituents of the oxide film, we may be able to work around some of these problems.

Let us look at the case of STO grown on Si by (see Chaps. 4 and 6 for details of the growth process). STO is an oxide where both metals are in their highest oxidation states. Depending on the arrival rates of the metals, there is a minimum oxygen partial pressure at which one is able to fully oxidize each metal. For a Ti metal flux of about one monolayer per minute, this pressure is experimentally found to be around 1–2×10^{-6} Torr [40]. Once formed, TiO_2 itself is stable against reduction down to oxygen partial pressures of $\sim 7 \times 10^{-9}$ Torr at 750 °C and $<1 \times 10^{-15}$ Torr at 500 °C, based on the heat of formation of TiO_2. This fact strongly hints that it is kinetics (the arrival rate of atoms), not thermodynamics that determines whether a material forms during MBE growth. Comparatively, Sr is much easier to oxidize than Ti. Sr needs a mere $\sim 1 \times 10^{-8}$ Torr oxygen partial pressure to form SrO at a rate of about one monolayer per minute [41]. Once formed, SrO is very difficult to reduce back to Sr metal unless one goes to temperatures above 1,000 °C under ultrahigh vacuum conditions. Therefore, in order to grow STO by MBE at one monolayer per minute, we need an oxygen pressure of about 2×10^{-6} Torr. Another critical growth parameter is the substrate temperature. The substrate temperature must be such that there is high surface diffusion but the bulk diffusion remains negligible. These criteria define a temperature window for layer-by-layer growth. For ionically bonded materials like SrO, the window is fairly wide and spans from about 1/9 to 1/3 the melting point [42]. For SrO with a melting point of $\sim 2,900$ K, this means an optimum growth temperature of 320–970 K (50–700 °C). For covalently bonded materials like TiO_2, the window is narrower ranging from about half the melting point to just below the melting point [42]. With a melting point of 2,130 K, this means an optimum growth temperature for TiO_2 of at least 1,065 K (~ 790 °C). As the two windows do not overlap, we take the midpoint of the gap and say 740 °C is the optimum STO growth temperature. Experimentally, however, because STO is not really SrO + TiO$_2$, the true growth window for layer-by-layer growth is more relaxed and flat STO can be grown at somewhat lower substrate temperature and oxygen partial pressure than in our simplified analysis [43], especially when co-depositing Sr and Ti where Ti oxidation is catalyzed by the presence of Sr, which is similar to the effect of using alkali metals to catalyze aluminum oxidation [44].

The growth conditions discussed above work very well when growing STO on STO substrates (homoepitaxy) but not necessarily on other substrates, particularly those that react readily with oxygen. We still have not yet addressed the issue of thermodynamic stability of the entire system during deposition. The bare Si substrate will rapidly form half a monolayer of amorphous SiO$_2$ when exposed for ~10 s to an oxygen pressure of 2×10^{-6} Torr at room temperature [45]. The substrate is not thermodynamically stable in the presence of oxygen but we need some minimum amount of oxygen to form the oxide film! This is the main problem that has prevented the development of epitaxially integrated oxides on silicon, and this is the problem to which the Zintl template provides a solution. While the detailed mechanism is still not clear at present, the half-monolayer Sr deposited on the Si(001) surface serves to protect the underlying Si from oxidation at modest temperatures and oxygen partial pressures. The Sr template has been found to be able to withstand conditions of up to ~400 °C and ~5 $\times 10^{-8}$ Torr O$_2$ for at least several minutes, keeping the oxygen on the surface and not allowing it to react with silicon [46]. The other key feature of the Sr Zintl template is that it preserves the surface lattice of silicon, allowing for epitaxy to occur.

We should note that even though the Sr Zintl template can withstand the presence of some oxygen at moderate temperatures, these conditions are still not optimal for layer-by-layer STO growth. The way around this is to kinetically limit Si oxidation in the presence of the template layer. We allow oxygen sufficient to oxidize a SrO layer but not to destroy the Zintl template into the growth chamber ($\sim 1 \times 10^{-8}$ Torr) at a relatively low temperature (~200 °C). We then deposit Sr under this low oxygen partial pressure, which becomes a SrO monolayer in contact with the Zintl template. As soon as this SrO layer is formed, we shut off Sr and open Ti while at the same time start increasing the oxygen pressure in order to form a partially oxidized TiO$_x$ layer in contact with the first SrO layer. We keep on increasing the oxygen partial pressure and alternately deposit SrO and TiO$_2$ to form a few unit cells of STO. At the end of this process, we ideally want to be at a partial pressure where the TiO$_2$ layer is fully oxidized, which is about 4×10^{-7} Torr when in contact with SrO. We have now formed a thin STO seed layer (two to ten unit cells is commonly used, see Chap. 6) on Si. However, because we used a very low growth temperature, the crystalline quality of this STO seed layer is quite poor and can even be amorphous if the stoichiometry is not perfectly matched. We cannot use a high growth temperature in the presence of oxygen as this can still result in Si oxidation. To improve the crystalline quality, we now remove all oxygen gas in the growth chamber and slowly heat up the STO seed layer to fully crystallize it. Full crystallization typically occurs at around 500 °C after a few minutes for stoichiometric samples. What we have done then to get around the issue of reactivity of the silicon with oxygen is to kinetically inhibit that process by lowering the substrate temperature when oxygen is present and utilizing the thermodynamic stability of the STO on the template layer to be able to fully crystallize the STO at higher temperature in the absence of oxygen. Normally, to grow crystalline oxides, one needs both a high temperature and high oxygen pressure. But because those conditions also lead to oxidation of the silicon, what we have done is to separate these two conditions in time. We first grow amorphous/ poorly crystallized STO with the right stoichiometry at low substrate temperature and sufficiently high oxygen pressure, preventing the formation of SiO_2. Then, we remove oxygen and increase the temperature to provide enough energy so the Sr, Ti, and O atoms find their proper places and form highly crystalline STO, with the oxygen in the STO film staying in STO. Once the crystalline STO seed layer has been formed this way, more STO or even other oxides can be deposited on top of it. Even under conditions of simultaneous high temperature and high oxygen pressure (up to a certain extent), the underlying STO seed layer is perfectly stable and will not be disrupted. Using the optimized conditions for layer-by-layer growth of STO on the STO seed layer grown on Si, some interfacial SiO_2 formation (~1–2 nm) does occur subsequently depending on the seed layer thickness, but this does not at all affect the initial STO seed layer quality. One last thing to note is that the seed layer is naturally oxygen deficient by virtue of the less than optimal initial oxygen pressure used to form it. This is usually healed by subsequent oxide deposition on the seed layer.

2.4 Twinning and Other Rotten Apples

Even if we resolve the wetting issue there still is a problem with the lattice symmetry mismatch between an oxide and semiconductor, in particular if you would like to grow a semiconductor on an oxide substrate. The difference in lattice symmetry brings up additional difficulties. For example, a zincblende material such as GaAs or a diamond-type material such as Si does not possess fourfold symmetry (i.e. the atomic positions are different when you rotate the cell by 90°) while a cubic perovskite oxide such as STO does. Therefore, even the two-dimensional nucleation of GaAs on STO would result in the formation of so-called twin boundaries. Simply put, the STO surface does not have anything to provide GaAs a directional preference. Thus, it can nucleate in one direction in one area of the terrace, and in an orthogonal direction somewhere else on the same terrace. When two such regions meet there will be a domain boundary. In Fig. 2.7 we show two different regions of GaAs on STO, where one region is rotated 90° with respect to another region.

For majority carrier devices (i.e. a transistor) some density of domains in GaAs can be tolerated. However, for minority carrier devices (especially light emitting diodes (LEDs) and lasers), domains are highly undesirable as they lead to reduction

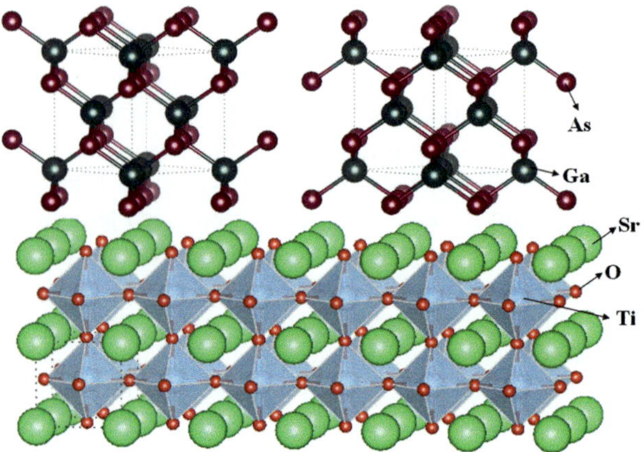

Fig. 2.7 Schematic of nucleation of two orthogonal domains of GaAs on STO

of minority carrier lifetime and non-radiative recombination processes. In order to avoid the formation of twin domains, one needs to provide a preferred direction on the oxide surface; in other words, break the fourfold symmetry of the cubic phase. One possibility is to use vicinal oxide surfaces where step edges may define a preferred nucleation direction. In GaAs epitaxy in Si, double steps are used to eliminate anti-phase domains. These are different type defects; the presence of a single step on a Si surface causes a Ga plane to meet an As plane along the step edge. Orthogonal domains we discuss here are of twin nature; the so-called diamond zig-zag chains meet at a 90° angle on the same terrace of the oxide surface. Though beneficial in that respect and also in reducing the threading dislocation density, the steps on a vicinal surface present a special challenge for oxide/semiconductor epitaxy as we will discuss in the next section. It will also be interesting to explore the role of step edges in selective chemisorption and whether strain can be used to drive a directional nucleation on the oxide surface.

Epitaxy is still possible in some cases even if the crystal symmetries are not totally compatible, for example, growing an orthorhombic material on either a cubic substrate or hexagonal substrate. In such cases, however, the material with lower symmetry grown on a substrate with higher symmetry will form orientation domains. These domains may or may not be detrimental depending on the specific application. One example is the growth of Gd_2O_3 on Si(100). Gd_2O_3 has a bixbyite structure (see Sect. 2.1) with a lattice constant of 10.82 Å, which is 0.4 % smaller than twice the Si lattice constant of 5.43 Å. However, because of interface energy considerations, Gd_2O_3 grows with the 110 orientation on Si(100). The 110 surface unit cell of bixbyite is rectangular-shaped and has twofold symmetry, with the long side having a lattice constant of 15.30 Å and the perpendicular side having a lattice constant of 10.82 Å. The longer side fits almost perfectly with four multiples of the Si(100) surface unit cell length of 3.84 Å (i.e. 4 × 3.84 = 15.36 ~ 15.30). However, this nice atomic matching of Si and O along one direction comes at a price of an incommensurate match between the shorter side of the bixbyite surface unit cell and the Si(100) surface unit cell. The epitaxial relationship is $Gd_2O_3[110]//Si[110]$

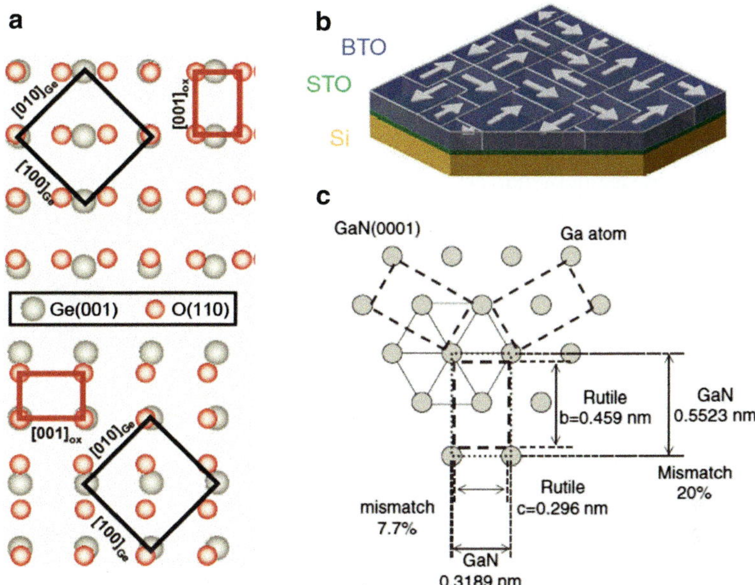

Fig. 2.8 (a) Two possible domains of (110) bixbyite on (100) Si/Ge. Copyright IOP Publishing. Reproduced from [47] by permission of IOP Publishing. All rights reserved; (b) domains in a-axis oriented BTO grown on Si. Reprinted by permission from Macmillan Publishers Ltd: [48], Copyright 2013; (c) three possible domains of (100) rutile TiO_2 on (0001) GaN. Reprinted from [49]. Copyright 2005, The Japan Society of Applied Physics

and $Gd_2O_3[100]//Si[-110]$. Lattice matching along one direction has been lost in order to reduce the interface energy. Because lattice matching of the long side of the rectangular bixbyite cell can occur in either of the two Si directions (it has square symmetry), there are then two possible domains as shown in Fig. 2.8a [47]. A similar but much more subtle domain structure occurs when growing a-axis oriented BTO on Si. Due to a combination of thermal expansion mismatch and strain relaxation, thick BTO films (>20 nm) on Si tend to crystallize such that their polar axis (the c-axis) lies in the plane of the film. One of the two a-axes also lies in plane and tries to match the Si surface unit cell. As with Gd_2O_3, since the Si has a surface with square symmetry, matching of the a-axis of BTO can freely occur in one of two perpendicular directions. This causes the c-axis of BTO to point randomly in-plane between the two orthogonal directions resulting in a ferroelectric domain structure like that shown in Fig. 2.8b [48]. A more pronounced example of orientation domains is when TiO_2 with tetragonal rutile structure is grown on wurtzite structure GaN. Rutile is observed to grow in the 100 direction, which has twofold symmetry, on the GaN(0001) surface, which has sixfold symmetry. The epitaxial relationship is $TiO_2[010]//GaN[10-10]$ and $TiO_2[001]//GaN[11-20]$. There are three possible orientations by which the 100 surface unit cell of rutile can match the atoms on the GaN(0001) surface and this results in three orientation domains as shown in Fig. 2.8c [49].

2.5 Step Edges

Typically, in semiconductor heteroepitaxy such as GaAs/AlAs or Si/SiGe the lattice mismatch in the vertical direction (normal to the interface) is exactly the same as in the lateral direction and is small. Therefore, surface steps present a major difficulty mainly if you grow a zincblende crystal like GaAs, on a diamond lattice such as that of Si. In this case you expect anti-phase domains (APD) running along the step edge [50]. A number of techniques have been proposed to battle this problem, including growth on highly vicinal surfaces with double height steps, to promote self-annihilation of APDs that results in APD-free GaAs on Si [51]. Unfortunately, step edges also cause problems in the case of oxide/semiconductor epitaxy, where matching two materials in the plane, does not in general, provide a corresponding match in the out-of-plane directions.

Consider, for example, the (100) surface of silicon. There are always step edges and terraces present on the surface, even for wafers cut as close as possible to the (100) orientation (nominally flat wafers). One can also cut and polish Si wafers with a particular miscut angle in a specific direction. In this case the surface looks like a staircase. As the miscut angle increases, the terrace width becomes smaller. And highly vicinal surfaces (large miscut angle, for example of 6°) are unstable towards step bunching. Step bunching results in wider terraces separated by higher steps or, in some cases, facets. As will be discussed later in the book, reconstruction of this surface results in formation of silicon dimers. Dimers are arranged in rows running along the (110) direction and separated from each other by so called troughs. The symmetry of the dimerized Si(001) 2×1 surface allows for two distinct types of surface step edges, distinguished by whether the dimerization direction on an upper terrace near a step is normal (Type A) or parallel (Type B) to the step edge [52]. For low miscut angles, the surface is characterized by single-height steps (S_A, S_B) alternating regions of 2×1 and 1×2 periodicity. The S_A single step is shown in Fig. 2.9a. The height of the steps is a quarter of the unit cell of silicon (5.43 Å), or 1.358 Å. This surface cannot have two S_A steps without an intervening S_B step [52–56]. However, at increasing miscut angles, double steps become energetically favored to keep terraces long [52–54]. In the lowest energy configuration (D_B) shown in Fig. 2.9b, dimer rows on all terraces run perpendicular to a step edge [52–56]. These single-domain miscut or *vicinal* Si (001) surfaces are used in semiconductor heteroepitaxy for control of antiphase domain growth and strain relief [57].

A miscut angle of 4° towards [110] is sufficient to produce a surface with only D_B steps [54, 55]. Comparing the reconstruction for nominally flat Si with that of the miscut wafer, one sees the nominally flat wafer is double-domain (2×1 and 1×2), while the 4°miscut wafer exhibits a single-domain reconstruction consistent with dimer rows running perpendicular to the step edge [54, 56]. Analysis of the splitting of the reflection high energy electron diffraction (RHEED) streaks allows for an estimate of terrace length [53–56]. A 4° miscut would produce terraces with a length of 3.86 nm and a step height of 2.71 Å.

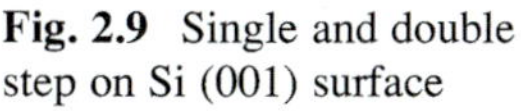

Fig. 2.9 Single and double step on Si (001) surface

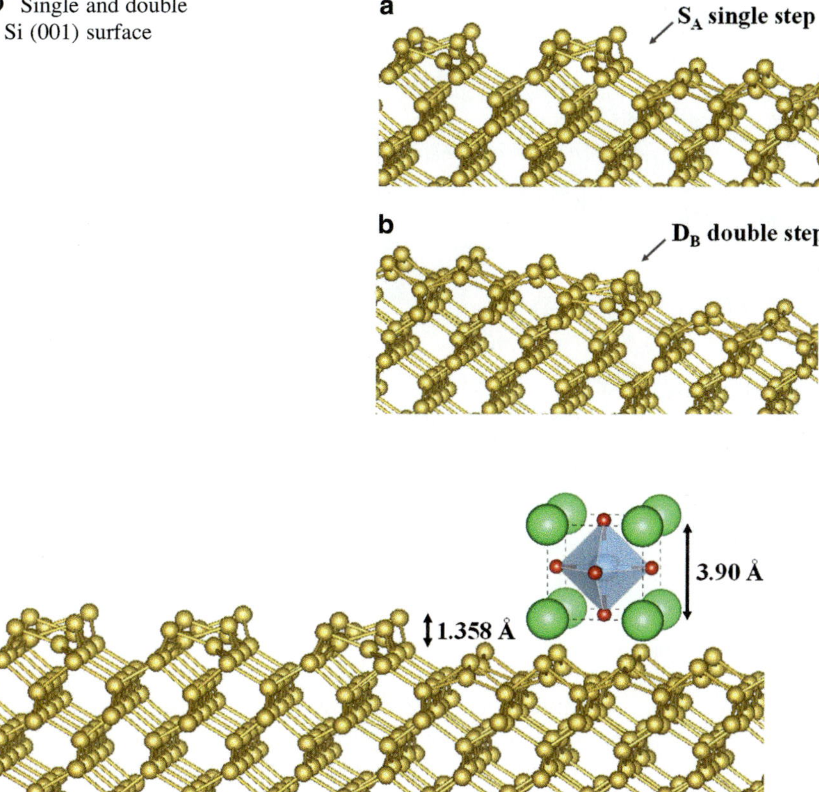

Fig. 2.10 Steps on the silicon (001) surface, a unit cell of ABO$_3$ perovskite oxide is shown to indicate the height difference

In Fig. 2.10 we show a single S$_A$ step on a Si (001) surface; also shown in the figure is a single unit cell of STO. Laterally the two are well matched (see Fig. 2.2). However, the STO unit cell is 3.9 Å tall, which does not match the step height on the silicon surface! Thus, even under the most ideal situation, STO grown on one silicon terrace may not match the STO grown on an adjacent terrace. This could lead to different domains of STO on the silicon surface with a density of domain walls of 10^{12} cm^{-2} (assuming a typical terrace width of 1,000 Å). As STO is highly ionic, such domain walls are most likely charged and may have adverse effects on materials properties.

The fundamental understanding of what happens as the oxide layers nucleated on different terraces meet at the edge is still largely missing. It is not clear whether the oxide layers grow continuously over step edges, form a line defect along the step edge, or a grain boundary forms along the step edge. Unfortunately, at present, there is still considerable debate about many widely observed grain boundary properties even in bulk perovskites.

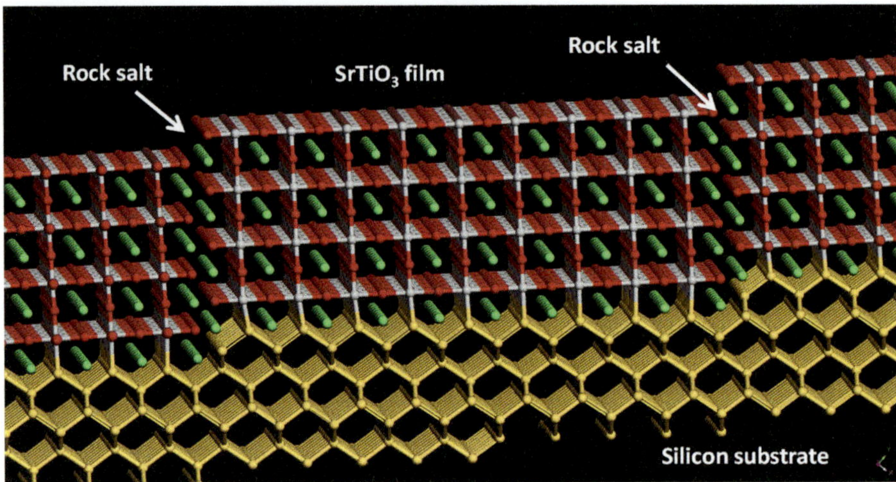

Fig. 2.11 Domain walls at step edges can be healed using the formation of quasi Ruddlesden-Popper layers

High-resolution transmission electron microscope studies and microanalysis results have suggested amorphous phases or cation interstitials to be the origin of the charge imbalance in the boundary plane [58]. More recently, Browning and Pennycook used the combination of Z-contrast imaging and electron-energy-loss spectroscopy (EELS) in the scanning transmission electron microscope (STEM) to study the correlation between the structural and the local electronic properties of STO grain boundaries [59]. They found that (001) tilt grain boundaries contain characteristic sequences of structural units that do not contain any intergranular grain boundary phases [60]. DFT calculations of these units now suggest that the behavior is more complicated than previously thought. In particular, Kim and co-workers found that it is energetically favorable for there to be an excess of oxygen vacancies in these units, and in the case of units centered on the Ti sublattice, a Ti excess [61]. Such non-stoichiometry leads to the formation of a highly doped n-type region at the boundary. Recently, Klie et al. have provided direct experimental evidence for the presence of the proposed excess of oxygen vacancies in the grain boundary plane that is independent of the cation arrangement [62].

Growth on a vicinal surface has been performed by Liang and co-workers, who sought to eliminate two-domain formation [63]. They used vicinal substrates with a nominal cutoff angle of 1.2° towards the (110) direction. However, the growth has proven challenging due to high surface reactivity caused by the high step density. A special case may be growth of STO on a 4° miscut Si wafer. Such a miscut towards the <110> direction, results in 3.86 nm wide terraces. That distance is close to approximately ten unit cells of STO. One possible way to heal the domain walls in this case could be the formation of quasi Ruddlesden-Popper planes along the step edge as shown in Fig. 2.11. The Ruddlesden–Popper (RP) type phases of general formula $A_{n+1}B_nO_{3n+1}$ or $AO(ABO_3)_n$ (where A is rare earth/alkaline

Fig. 2.12 The first three members of the Ruddlesden-Popper series of phases with formula $A_{n+1}B_nO_{3n+1}$

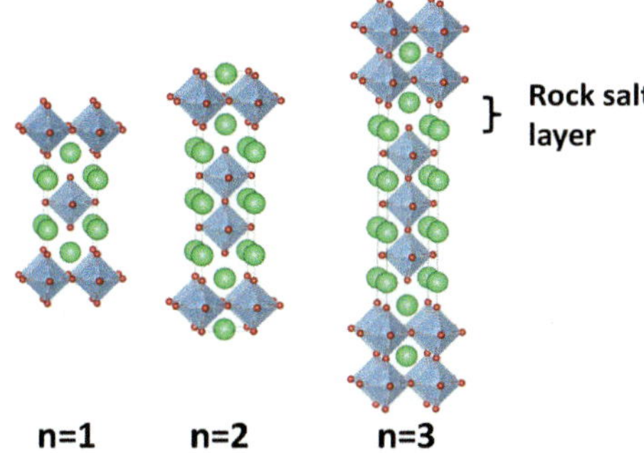

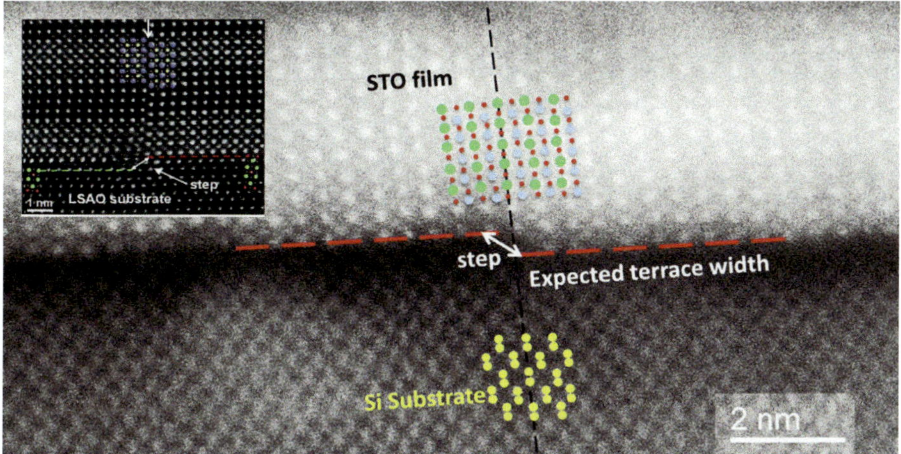

Fig. 2.13   Scanning transmission electron micrograph of STO grown on 4° miscut vicinal Si(100). (Image courtesy of D. J. Smith). *Inset*: a RP fault in LaNiO$_3$ grown across a step of the (La, Sr) AlO$_4$ substrate. Reprinted with permission from [67]. Copyright 2012, AIP Publishing LLC

earth ion, B is a transition metal ion) [64], crystallize with tetragonal or orthorhombic unit cell in the space group *I4/mmm* or *Fmmm*. The crystal structure of these phases can be described by the stacking of finite n layers of perovskites ABO$_3$ between rock salt AO layers along the crystallographic c direction. In Fig. 2.12 we show the first three members of this family. The stoichiometric ABO$_3$ can be viewed as a RP phase with $n = \infty$. RP phases for SrTiO$_3$ can be grown by MBE through precise control of the deposition process [65]. The inclusion of the rocksalt structure at the step edge allows for almost perfect matching of the STO on a Si terrace. The challenge is to stabilize the terrace size during the STO deposition. In Fig. 2.13 we show a STEM image of STO grown on a 4° miscut Si wafer [66]. A RP fault in a LaNiO$_3$ film grown across a step of an LSAT substrate from [67] is shown for comparison. The steps are clearly seen; however, the height appears to be larger than 2.71 Å and the terraces are significantly wider. This is most likely related to step bunching, often observed in vicinal Si (001) at elevated temperature in the presence of metals [68]. Surprisingly, the STO film appears to grow across the step uninterrupted.

2.6 The Role of the Interface

Assuming we have found a way to achieve monolithic integration of transition metal oxides and semiconductors, a natural question arises: where such hybrid structures may find useful applications. The answer to this question depends on whether it is the integrity of the interface itself or the top oxide layer that is of interest.

One of the benefits of the epitaxial interface is its low defectivity. The most celebrated oxide/semiconductor interface between silicon and silicon dioxide (SiO_2) as grown has only 10^{-10} electrically active defects per cm^2; that number can be reduced to 10^{-12} by a subsequent forming gas anneal that passivates dangling bonds at the interface. However, silicon dioxide is amorphous and dangling bonds at the interface appear at random. Theoretically, an epitaxial interface may be "defect free". One has to be careful with the terminology here, as there is always some equilibrium concentration of point defects controlled by their formation energy and temperature. The term "defect free" therefore implies equilibrium thermodynamic concentration of defects. Therefore an epitaxial oxide could be used instead of SiO_2 as a *gate dielectric*. This indeed was the original motivation of McKee et al. [30] when growing STO on Si. Another example would be using $YMnO_3$ as a ferroelectric gate for GaN [20]. Ironically, none of that came to be.

If one can find a way of growing a semiconductor layer epitaxially on the oxide substrate, and that oxide substrate may be integrated on the same or perhaps, a different semiconductor, one could have the epitaxial analogue of the *silicon on insulator* (SOI) structure. One example of this approach would be integration of Ge on Si using rare earth oxide buffers [18, 19]. An even more intriguing possibility is to use epitaxial oxide layers as buffers in integration of different semiconductors. Thus GaAs has been successfully integrated on Si (001) using an STO buffer [31], and more recently high quality GaN layers have been grown on Si(111) using the bixbyite form of Gd_2O_3 [69].

On the other hand, one can use the oxide layer epitaxially grown on a semiconductor as a *virtual substrate*. STO on Si would be a classic example of this approach. As high quality STO films can be grown on 200 mm Si wafers [70], this effectively opens the door for integrated oxide electronics as STO is a widely used substrate for growing ferroelectric, ferromagnetic and superconducting oxides [71, 72]. Last but not least is the opportunity to create novel oxide or hybrid heterostructures on semiconductor substrates. One such example would be photocatalytic structures integrated on Si (001) [73].

We are now at a point in time where the necessary technology is available to model, fabricate, and measure these functional oxides epitaxially integrated with semiconductors. In the next three chapters will briefly describe this necessary know-how before going into detail on actual epitaxial oxide on semiconductor systems.

References

1. J.Y. Tsao, *Materials Fundamentals of Molecular Beam Epitaxy* (Academic, San Diego, CA, 1993)
2. R.M. Sieg, S.A. Ringel, S.M. Ting, S.B. Samavedam, M. Curie, T. Lando, E.A. Fitzgerald, J. Vac. Sci. Technol. B **16**, 1471 (1998)
3. Y.H. Lo, Appl. Phys. Lett. **59**, 2311 (1991)
4. Z. Yang, J. Alperin, W.I. Wang, S.S. Iyer, T.S. Kuan, F. Semendy, J. Vac. Sci. Technol. B **16**, 1489 (1998)
5. C. Carter-Coman, A.S. Brown, N.M. Jokerst, D.E. Dawson, R. Bicknell-Tassius, Z.C. Feng, K.C. Rajkumar, G. Dagnall, J. Electron. Mater. **25**, 1044 (1996)
6. Z. Yang, F. Guarin, I.W. Tao, W.I. Wang, S.S. Iyer, J. Vac. Sci. Technol. B **13**, 789 (1995)
7. J.W. Matthews, A.E. Blakeslee, J. Cryst. Growth **27**, 118 (1974)
8. R. People, J.C. Bean, Appl. Phys. Lett. **47**, 322 (1985)
9. T.E. Whall, E.H.C. Parker, J. Phys. D **32**, 1187 (1999)
10. S.B. Adler, J. Am. Ceram. Soc. **84**, 2117 (2001)
11. H. Seo, A. Posadas, A.A. Demkov, Phys. Rev. B **86**, 014430 (2012)
12. L. Qiao, T.C. Droubay, T. Varga, M.E. Bowden, V. Shutthanandan, Z. Zhu, T.C. Kaspar, S.A. Chambers, Phys. Rev. B **83**, 085408 (2011)
13. V.V. Mehta, M. Liberati, F.J. Wong, R.V. Chopdekar, E. Arenholz, Y. Suzuki, J. Appl. Phys. **105**, 07E503 (2009)
14. L. Qiao, T.C. Droubay, M.E. Bowden, V. Shutthanandan, T.C. Kaspar, S.A. Chambers, Appl. Phys. Lett. **99**, 061904 (2011)
15. M.L. Yu, B.N. Eldridge, Phys. Rev. Lett. **58**, 1691 (1987)
16. N. Shimizu, Y. Tanishiro, K. Takayanagi, K. Yagi, Surf. Sci. **191**, 28 (1987)
17. J.C. Woicik, H. Li, P. Zschack, E. Karapetrova, P. Ryan, C.R. Ashman, C.S. Hellberg, Phys. Rev. B **73**, 024112 (2006)
18. H.J. Osten, A. Laha, M. Czernohorsky, E. Bugiel, R. Dargis, A. Fissel, Phys. Status Solidi A **205**, 695 (2008)
19. R. Dargis, A. Fissel, D. Schwendt, E. Bugiel, J. Krügener, T. Wietler, A. Laha, H.J. Osten, Vacuum **85**, 523 (2010)
20. A. Posadas, J.-B. Yau, C.H. Ahn, J. Han, S. Gariglio, K. Johnston, K.M. Rabe, J.B. Neaton, Appl. Phys. Lett. **87**, 171915 (2005)
21. H.R. Child, R.M. Moon, L.J. Raubenheimer, W.C. Koehler, J. Appl. Phys. **38**, 1381 (1967)
22. R.M. Moon, W.C. Koehler, Phys. Rev. B **11**, 1609 (1975)
23. M. Badylevich, S. Shamuilia, V.V. Afanas'ev, A. Stesmans, A. Laha, H.J. Osten, A. Fissel, Appl. Phys. Lett. **90**, 252101 (2007)
24. B.J. Kennedy, M. Avdeev, Aust. J. Chem. **64**, 119 (2011)
25. A.A. Stekolnikov, J. Furthmüller, F. Bechstedt, Phys. Rev. B **65**, 115318 (2002)
26. J. Padilla, D. Vanderbilt, Surf. Sci. **418**, 64 (1998)
27. X. Zhang, A.A. Demkov, H. Li, X. Hu, Y. Wei, J. Kulik, Phys. Rev. B **68**, 125323 (2003)
28. K. Johnston, M.R. Castell, A.T. Paxton, M.W. Finnis, Phys. Rev. B **70**, 085415 (2004)
29. M.R. Castell, Surf. Sci. **505**, 1 (2002)
30. R.A. McKee, F.J. Walker, M.F. Chisholm, Phys. Rev. Lett. **81**, 3014 (1998)
31. K. Eisenbeiser, R. Emrick, R. Droopad, Z. Yu, J. Finder, S. Rockwell, J. Holmes, C. Overgaard, W. Ooms, IEEE Electron. Device Lett. **23**, 300 (2002)
32. A.A. Demkov, X. Zhang, J. Appl. Phys. **103**, 103710 (2008)
33. E. Zintl, W. Dullenkopf, Z. Phys. Chem. B **16**, 195 (1932)
34. A.A. Demkov, H. Seo, X. Zhang, J. Ramdani, Appl. Phys. Lett. **100**, 071602 (2012)
35. K. Seino, W.G. Schmidt, F. Bechstedt, J. Bernholc, Surf. Sci. **507–510**, 406 (2002)
36. Y. Liang, J. Kulik, T.C. Eschrich, R. Droopad, Z. Yu, P. Maniar, Appl. Phys. Lett. **85**, 1217 (2004)
37. Y. Liang, J. Curles, M. McCready, Appl. Phys. Lett. **86**, 082905 (2005)

38. A. Slepko, A.A. Demkov, Phys. Rev. B **85**, 195462 (2012)
39. L. Schlipf, A. Slepko, A.B. Posadas, H. Seinige, A. Dhamdhere, M. Tsoi, D.J. Smith, A.A. Demkov, Phys. Rev. B **88**, 045314 (2013)
40. F.J. Walker, R.A. McKee, J. Cryst. Growth **116**, 235 (1992)
41. J. Lettieri, J.H. Haeni, D.G. Schlom, J. Vac. Sci. Technol. A **20**, 1332 (2002)
42. M.H. Yang, C.P. Flynn, Phys. Rev. Lett. **62**, 2476 (1989)
43. C.M. Brooks, L. Fitting Kourkoutis, T. Heeg, J. Schubert, D.A. Muller, D.G. Schlom, Appl. Phys. Lett. **94**, 162905 (2009)
44. N.A. Braaten, J.K. Grepstad, S. Raaen, S.L. Qui, Surf. Sci. **250**, 51 (1991)
45. E.G. Keim, L. Wolterbeek, A. Van Silfhout, Surf. Sci. **180**, 565 (1987)
46. Y. Liang, S. Gan, M. Engelhard, Appl. Phys. Lett. **79**, 3591 (2001)
47. A. Molle, C. Wiemer, M.D.N.K. Bhuiyan, G. Tallarida, M. Fanciulli, J. Phys. **100**, 042048 (2008)
48. S. Abel, T. Stöferle, C. Marchiori, C. Rossel, M.D. Rossell, R. Erni, D. Caimi, M. Sousa, A. Chelnokov, B.J. Offrein, J. Fompeyrine, Nat. Commun. **4**, 1671 (2013)
49. T. Hitosugi, Y. Hirose, J. Kasai, Y. Furubayashi, M. Ohtani, K. Nakajima, T. Chikyow, T. Shimada, T. Hasegawa, Jpn. J. Appl. Phys. **44**, L1503 (2005)
50. J.B. Posthill, J.C.L. Tarn, K. Das, T.P. Humphreys, N.R. Parikh, Appl. Phys. Lett. **53**, 1207 (1988)
51. K. Adomi, S. Strite, H. Morkoç, Appl. Phys. Lett. **56**, 469 (1990)
52. D. Chadi, Phys. Rev. Lett. **59**, 1691 (1987)
53. K. Sakamoto, T. Sakamoto, J. Electrochem. Soc. **136**, 2705 (1989)
54. E. Pehlke, J. Tersoff, Phys. Rev. Lett. **67**, 465 (1991)
55. D. Saloner, J.A. Martin, M.C. Tringides, D.E. Savage, C.E. Aumann, M.G. Lagally, J. Appl. Phys. **61**, 2884 (1987)
56. J. Reiner, K. Garrity, F.J. Walker, S. Ismail-Beigi, C.H. Ahn, Phys. Rev. Lett. **101**, 1 (2008)
57. F. Riesz, J. Vac. Sci. Technol. A **14**, 425 (1996)
58. Y.M. Chiang, T. Takagi, J. Am. Ceram. Soc. **73**, 3278 (1990)
59. N.D. Browning, S.J. Pennycook, J. Phys. D **29**, 1779 (1996)
60. M.M. McGibbon, N.D. Browning, M.F. Chrisholm, A.J. McGibbon, S.J. Pennycook, V. Ravikumar, V.P. Dravid, Science **266**, 102 (1994)
61. M. Kim, G. Duscher, N.D. Browning, K. Sohlberg, S.T. Pantelides, S.J. Pennycook, Phys. Rev. Lett. **86**, 4056 (2000)
62. R.F. Klie, M. Beleggia, Y. Zhu, J.P. Buban, N.D. Browning, Phys. Rev. B **68**, 214101 (2003)
63. Y. Liang, Y. Wei, X.M. Hu, Z. Yu, R. Droopad, H. Li, K. Moore, J. Appl. Phys. **96**, 3413 (2004)
64. S.N. Ruddlesden, P. Popper, Acta Crystallogr. **11**, 54 (1958)
65. J.H. Haeni, C.D. Theis, D.G. Schlom, W. Tian, X.Q. Pan, H. Chang, I. Takeuchi, X.-D. Xiang, Appl. Phys. Lett. **78**, 3292 (2001)
66. K. Kormondy, private communication
67. E. Detemple, Q.M. Ramasse, W. Sigle, G. Cristiani, H.-U. Habermeier, B. Keimer, P.A. van Aken, J. Appl. Phys. **112**, 013509 (2012)
68. H.-C. Jeong, E.D. Williams, Surf. Sci. Rep. **34**, 171 (1999)
69. F.E. Arkun, M. Lebby, R. Dargis, R. Roucka, R.S. Smith, A. Clark, ECS Trans. **50**, 1065 (2013)
70. X. Gu, D. Lubyshev, J. Batzel, J.M. Fastenau, W.K. Liu, R. Pelzel, J.F. Magana, Q. Ma, V.R. Rao, J. Vac. Sci. Technol. B **28**, C3A12 (2010)
71. C. Dubourdieu, J. Bruley, T.M. Arruda, A. Posadas, J. Jordan-Sweet, M.M. Frank, E. Cartier, D.J. Frank, S.V. Kalinin, A.A. Demkov, V. Narayanan, Nat. Nanotechnol. **8**, 748 (2013)
72. A.B. Posadas, C. Mitra, C. Lin, A. Dhamdhere, D.J. Smith, M. Tsoi, A.A. Demkov, Phys. Rev. B **87**, 144422 (2013)
73. T.Q. Ngo, A. Posadas, H. Seo, S. Hoang, M.D. McDaniel, D. Utess, D.H. Triyoso, C. Buddie Mullins, A.A. Demkov, J.G. Ekerdt, J. Appl. Phys. **114**, 084901 (2013)

Chapter 3
Predictive Engineering
of Semiconductor-Oxide Interfaces

Before discussing the density functional formalism used in most modern solid state calculations it is useful to put the problem into a broader context. This section is intended for beginning graduate students and can be omitted by the experts. The main difficulty of describing the solid state theoretically is its enormous complexity. A solid is comprised of electrons and nuclei interacting *via* Coulomb forces, so one has to describe correlated behavior of about 10^{23} strongly interacting particles! Clearly, this is an impossible task unless some simplifications are made. The first step is to separate light and fast electrons from slow and heavy nuclei. The original idea belongs to Max Born and Robert Oppenheimer (Max Born was born in Breslau, Germany in 1882, Robert Oppenheimer was born in New York in 1904) and was published in 1927 [1]. Note that Oppenheimer was only 23 years old when the paper came out. They suggested first to solve the purely electronic problem for some fixed configuration of nuclei $\overline{R}$:

$$\hat{H}_{el}\varphi_i\left(\vec{r}_1, \vec{r}_2, \dots, \overline{R}\right) = E_i^{el}\left(\overline{R}\right)\varphi_i\left(\vec{r}_1, \vec{r}_2, \dots, \overline{R}\right) \tag{3.1}$$

It is customary to include the nucleus-nucleus repulsion into the electronic Hamiltonian, so $\hat{H}_{el}$ is given by:

$$\hat{H}_{el} = \hat{T}_e + \hat{U}_{ee} + \hat{U}_{ep} + \hat{U}_{pp} \tag{3.2}$$

The first term is the electronic kinetic energy, and the other three describe electron-electron, electron-ion and ion-ion interactions, respectively. Once we solve this problem we have a complete set of functions to expand the total (electrons and nuclei) wave function of the system:

$$\Psi_s\left(\vec{r}_1, \vec{r}_2, \dots \vec{R}_1, \vec{R}_2 \dots\right) = \sum_i \chi_i(\overline{R})\varphi_i\left(\vec{r}_1, \vec{r}_2, \dots, \overline{R}\right) \tag{3.3}$$

A.A. Demkov and A.B. Posadas, *Integration of Functional Oxides with Semiconductors*, DOI 10.1007/978-1-4614-9320-4_3, © The Author(s) 2014

Of course, the complete set we are using is changing all the time as the nuclei move, and in each particular case one needs to specify which configuration is used. The wave-function (3.3) is the Born-Oppenheimer (BO) ansatz (if only the ground state of the electron system is included in the sum, the approach is known as the naïve BO wave function). One then inserts this expression into the Schrödinger equation for the entire system, and averages out fast electronic degrees of freedom. This is achieved by multiplying the whole expression by the conjugate of the electronic wave function, and integrating over all electronic coordinates.

$$[\hat{T}_p + E_j(\overline{R}) - E_s]\chi_{j,s}(\overline{R}) = -\sum_{\alpha,i} C_{\alpha,i}\chi_{i,s}(\overline{R}) \tag{3.4}$$

If we now neglect the terms involving derivatives with respect to nuclear positions (the right hand side of equation 3.4) with the exception of the nuclear kinetic energy, we end up having an effective Schrödinger-like equation for the coefficients χ which play the role of the nuclear wave functions:

$$[\hat{T}_p + E_j(\overline{R})]\chi_{js}(\overline{R}) = E_s\chi_{js}(\overline{R}) \tag{3.5}$$

The significance of this expression is that the potential energy of the nuclear motion is nothing more than the total electronic energy. The terms we have neglected in (3.4) to obtain (3.5) have operators of the following form acting on the nuclear wave functions:

$$C_{\alpha j} = -\left(\sum_i \frac{\hbar^2}{2m_p}2\langle\varphi_i|\nabla_\alpha|\varphi_i\rangle\nabla_\alpha + \sum_i \frac{\hbar^2}{2m_p}\langle\varphi_i|\nabla_\alpha^2|\varphi_i\rangle\right) \tag{3.6}$$

The matrix elements are taken between the electronic wave functions, but the operators themselves act on the nuclear coordinates. The first term can be thought of as an overlap between the electron wave function and the same function acted upon by a displacement operator, and if the electron wave function is localized this will be small. It can be shown that the second term scales as the ratio of the mass of electron to the mass of the proton (m_p) which is a very small number indeed (5.4462×10^{-4}). In other words, the mass difference and the localization of the electronic states suggest we can safely neglect the $C_{\alpha,j}$ in the right hand side. This is called the adiabatic approximation and $C_{\alpha j}$ is known as non-adiabaticity operator.

Unfortunately, we still do not know how to solve the many-electron Schrödinger equation. In principle, the problem can be solved directly using so-called Quantum Monte Carlo methods, but in practice approximations are needed. The Hartree-Fock theory is the simplest many electron theory which essentially treats electrons as independent (the dynamic electron-electron interaction is handled in electrostatic approximation), but takes into account the Pauli principle. Unfortunately, this approximation does not describe solids very well (mainly due to the total neglect of correlation effects). Density functional theory, which we will now describe, appears to do a better job.

3.1 Many Electron Problem and Density Functional Theory

The modern electronic structure theory of materials is based on density functional theory introduced by Walter Kohn (born in Vienna, Austria in 1923), Pierre Hohenberg (born in Paris, France in 1934) and Lu Sham (born in British Hong-Kong in 1938) in the mid-1960s [2, 3]. For this work Kohn shared the 1998 Nobel Prize with John Pople. The theory formulates the many-body problem of interacting electrons and ions in terms of a single variable, namely the electron density. The Hohenberg-Kohn theorem states that the electron density alone is necessary to find the ground state energy of a system of N electrons, and that the energy is a unique functional of the density [2]. Unfortunately, the precise form of that functional is not presently known. However, we do have reasonably good approximations, although the Hohenberg-Kohn theorem does not offer a specific method to compute the electron density. The solution for a slow varying density is given by the Kohn-Sham formalism [3], where an auxiliary system of non-interacting electrons in the effective potential is introduced, and the potential is chosen in such a way that the non-interacting system has exactly the same density as the system of interacting electrons in the ground state.

The Kohn-Sham (KS) equations below need to be solved iteratively until the self-consistent charge density is found:

$$\left[-\frac{1}{2}\nabla^2 + v_{eff}(r) \right] \varphi_i(r) = \varepsilon_i \varphi_i(r) \tag{3.7}$$

with the effective potential given by:

$$v_{eff}(r) = v(r) + \int \frac{n(r')}{|r - r'|} dr' + \frac{\delta E_{xc}[n]}{\delta n(r)} \tag{3.8}$$

where $v(r)$ is the external potential (e.g., due to ions) and $E_{xc}[n]$ is the exchange correlation energy functional. The exact form of this functional is not known and has to be approximated. The electron density is given by:

$$n(r) = \sum_{occ} |\varphi_i(r)|^2 \tag{3.9}$$

where the sum is over the N lowest occupied eigenstates. For a slowly varying density Kohn and Sham introduced the local density approximation (LDA):

$$E_{xc}[n] = \int \varepsilon_{xc}(n(r))n(r)dr \tag{3.10}$$

where $\varepsilon_{xc}[n]$ is the exchange and correlation energy per particle of a uniform electron gas of density n. It is important to keep in mind that it is the electron density that is the "output" of the KS equations. Strictly speaking, the eigenvalues of the KS equations $\{\varepsilon_i\}$ have no direct physical meaning; nevertheless they are often very useful when the single particle electronic spectra (band structures) are discussed. The reasons behind the tremendous success of the Kohn-Sham theory are easy to identify. By solving essentially a single electron equation not much different from that due to Hartree, but including the effects of exchange and correlation, one gets an upper estimate of the ground state energy of a many-body system. The theory is variational, and thus forces acting on the atoms can be calculated within the BO approximation (of course, as the Hamiltonian is approximate the ground state energy may not be correct). The equation however, is non-linear and an iterative solution is needed.

Typically, the KS equations are projected onto a particular functional basis set, and the resulting matrix problem is solved. In terms of the basis, when solving KS equations, one has two options. It is possible to discretize the equations in real space and solve them directly; these are so-called real space techniques [4]. These methods offer a number of advantages, including being basis free, free of costly fast Fourier transforms (FFT's), and easy to implement. They can handle charged systems better than plane waves, and the boundary conditions are easy to define for finite or partially periodic systems. Alternatively, one can choose a complete set of conventional functions. There are two major functional basis set types presently employed. For periodic systems plane waves offer an excellent expansion set which along with the fast Fourier transformations affords an easy to program computational scheme, the accuracy of which can be systematically improved by increasing the number of plane waves [5]. For systems with strong, localized potentials such as those of the first row elements, a large number of plane waves is necessary in the expansion, and calculations require the use of ultra-soft pseudopotentials (see below) to be feasible. The second choice is to use local orbitals such as e.g., atomic orbitals or any other spatially localized functions. Among the advantages of a localized basis set are a smaller number of basis functions, and sparsity of the resulting matrix due to the orbital's short range. The disadvantages are the complexity of multi-center integrals one needs, and absence of the systematic succession of approximations, since the set is typically either under-complete or over-complete. In both cases calculations are computer intensive.

3.2 Pseudopotential

Most likely the DFT-LDA approach would have been limited to small molecules if it were not for a pseudopotential method. Since only the valence electrons are involved in bonding, and these electrons see a weaker potential due to screening by the core electrons, one can substitute the full Coulomb potential due to ions $v(r)$ with a smooth pseudopotential. This effectively reduces the number of electrons one needs to consider to the valence electrons only. For example, only 4 instead of

14 electrons are needed for Si! The practical importance of this approximation should not be overlooked, as a typical diagonalization algorithm scales as N^3 with the size of the matrix, thus for silicon we get a factor of at least 42 for the speed-up (it would actually be a lot more owing to the basis being much larger for the core states)! The most elegant way to introduce a pseudopotential is due to Phillips and Kleinman [6]. Their construction is based on the use of a pseudo wave function for the valence states given by

$$\left|\varphi_v^{PS}\right\rangle = |\varphi_v\rangle + \sum_c \langle \varphi_c | \varphi_v^{PS} \rangle |\varphi_c\rangle. \tag{3.11}$$

In (3.11) the φ_v denotes the true valence wave function and the φ_c's the true core wave functions which are not known in practice and must be approximated. The sum runs over the core states. Applying the Hamiltonian of the system to $|\varphi_v\rangle$ leads to

$$\hat{H}\,|\varphi_v\rangle = E|\varphi_v\rangle = \hat{H}\left|\varphi_v^{PS}\right\rangle - \sum_c E_c |\varphi_c\rangle\langle\varphi_c|\varphi_v^{PS}\rangle \tag{3.12}$$

which can be rewritten into a Schrödinger-like equation

$$\left[T + V + \sum_c (E - E_c)|\varphi_c\rangle\langle\varphi_c|\right]\left|\varphi_v^{PS}\right\rangle = E\left|\varphi_v^{PS}\right\rangle \tag{3.13}$$

where the original potential term V is replaced by the Phillips-Kleinman pseudopotential

$$V^{PK} = V + \sum_c (E - E_c)|\varphi_c\rangle\langle\varphi_c| = V + V^R. \tag{3.14}$$

In (3.14) the term V is the original potential and V^R an additional contribution. V^R is repulsive as the core energies E_c are lower than the valence energies E. Moreover, as V^R depends on the core wave functions it vanishes outside a certain core region so that the Phillips-Kleinman pseudopotential becomes equal to the original potential and the pseudo wave function matches the true wave function (see Fig. 3.1). Obviously, as introduced, the Phillips-Kleinman potential is a non-local, non-Hermitian, energy dependent operator and thus is rather difficult to use. The significance of this development is therefore mainly conceptual. It suggests that a smooth function can describe the solution of the Schrödinger equation in the energy window of interest.

Today, pseudopotentials used in electronic structure calculations may be broadly divided in four classes: the hard norm-conserving pseudopotentials [7], soft pseudopotentials [8], Vanderbilt-type ultra-soft pseudopotentials [9], and projector augmented wave (PAW) pseudopotentials [10, 11]. The "softness" refers to how rapidly the potential changes in real space. The analogy comes from expanding a step function in a Fourier series; it takes a large number of plane waves to eliminate

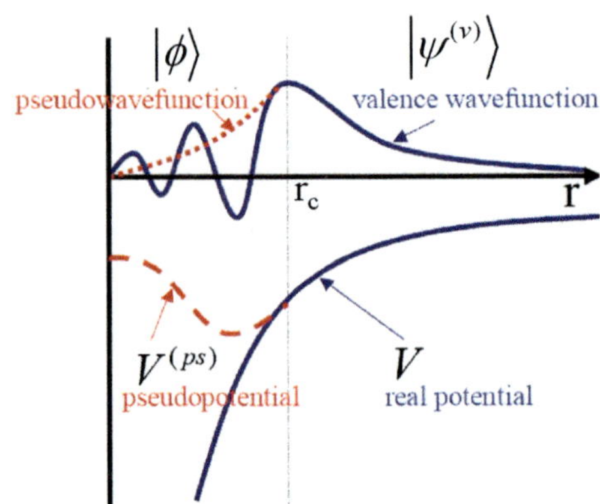

Fig. 3.1 A schematic of the pseudopotential and pseudo wave function. Both exactly match the real potential and wave function outside r_c

spurious oscillations at the step edge. On the other hand a "softer" function such as e.g., hyperbolic tangent can be expanded with greater ease. In general, hard pseudopotentials are more transferable. The choice of pseudopotential is in part dictated by the choice of a basis set used in a calculation. The use of local orbitals allows for a much harder pseudopotential. We will return to this point when discussing supercells. The pseudopotential is not uniquely defined, but this allows one to construct a pseudopotential that is simultaneously weaker and smoother than the original potential.

3.3   Energy Minimization and Molecular Dynamics

Once the solution of KS equations is found, the total energy in the LDA is given by:

$$E_{total} \approx \sum_i \varepsilon_i - \frac{1}{2}\iint \frac{n(r)n(r')}{|r-r'|}drdr' + \int n(r)\{\varepsilon_{xc}(n(r)) - \mu_{xc}(n(r))\}dr \quad (3.15)$$

where the exchange-correlation potential is given by $\mu_{xc} \equiv \frac{d}{dn}\{\varepsilon_{xc}(n(r))n(r)\}$. Now all ground state properties of the system can in principle be calculated. In particular, since we are using the Born-Oppenheimer approximation, the total energy of the electronic system which is a function of the ionic positions $\{\vec{R}_1, \ldots \vec{R}_i \ldots \vec{R}_N\}$, can be used as a an inter-atomic potential. Note that unlike potential functions used in classical molecular dynamics or molecular mechanics methods, the energy function $E_{total}\left(\vec{R}_1, \ldots \vec{R}_N^{\,N}\right)$ is not a sum of pair-wise interactions $\frac{1}{2}\sum_{i,j} V_{i,j}$ but a true many-body interaction energy computed quantum mechanically! One can easily calculate a force acting on any atom i in the direction α using the Hellmann-Feynman theorem $\left(\frac{\partial E}{\partial \lambda} = \langle\varphi(\lambda)|\frac{\partial H}{\partial \lambda}|\varphi(\lambda)\rangle\right)$

$$F_i^\alpha = -\frac{\partial E_{total}}{\partial R_i^\alpha}, \qquad \alpha = x, y, z. \tag{3.16}$$

At this point one can find the lowest energy atomic configuration by employing an energy minimization technique such as damped molecular dynamics or a conjugate gradient method. This process is then repeated until the forces on the atoms reach the desired level of convergence. Alternatively, a real molecular dynamics (MD) simulation can be launched. One has to keep in mind, however, that electronic "frequencies" $\frac{E_i - E_j}{\hbar}$ are much higher than a typical phonon frequency ω and for a stable simulation the time step needs to be a small fraction of the characteristic atomic period. The calculation then proceeds as follows: The KS energy is first calculated in a self-consistent manner for the initial atomic configuration, the Hellman-Feynman forces are evaluated, and atoms are moved to the next time step *via* some MD algorithm (Verlet, Gear, etc. [12]). At the new configuration the KS equations are solved again, and the procedure is repeated. Needless to say, these are very expensive calculations. They offer a significant advantage if a temperature dependence of a particular quantity is sought, since MD can be performed at finite temperature. For example, the Fourier transform of the velocity auto-correlation function gives the vibration spectrum, thus calculations performed at different temperature would give the temperature dependence of the phonon frequency.

3.4 Supercell/Slab Technique

As we have mentioned before, the plane wave method is particularly well suited for studying periodic systems. However, many systems of interest, and particularly interfaces and surfaces are manifestly non-periodic! Thus an artificial system with periodicity is created to simulate them. The geometry is often referred to as slab or supercell. We shall illustrate the idea for the case of a surface. Here one clearly deals with a system in which the periodicity in one direction (that perpendicular to the surface) is broken. To perform surface calculations with a plane wave basis set, a large simulation cell or a supercell is introduced in order to maintain artificial periodicity. A supercell contains a slab of bulk material (with many unit cells of the corresponding crystal) and a vacuum slab in the direction perpendicular to the surface as illustrated in Fig. 3.2 for the (101) surface of PtSi. Si (Pt) atoms are represented with yellow (blue) color. The [101] direction is along the long side of the supercell. In the two directions parallel to the surface the supercell has the usual bulk dimensions, and the periodic boundary conditions are used without any change. The periodic boundary condition in the direction normal to the surface is applied for the supercell dimension, rather than the physical crystal cell side. Thus the "universe" is filled with infinite parallel slabs of PtSi of certain thickness, separated by infinite parallel slabs of vacuum. It is crucial that the length of a

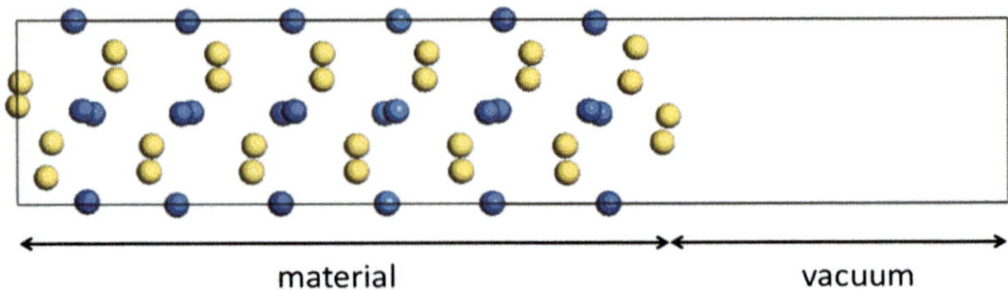

Fig. 3.2 Supercell used to simulate the (101) surface of PtSi. Si (Pt) atoms are represented with *yellow* (*blue*) color

supercell in the direction normal to the surface is large enough to eliminate any spurious interactions between the cells across the vacuum region. The thickness of a slab should be sufficient for bulk properties to be restored in the middle of it. The supercell obviously creates two surfaces, and it is advisable to use a symmetric termination of the slab to avoid an electric field forming due the potential differences of the two surfaces.

In principle, the larger the supercell chosen, the better it approximates the true surface (or rather a set of two identical surfaces). However, the calculation also becomes more demanding, as we shall now demonstrate. In the case of a periodic system we write the eigenfunctions $\psi_{n,k}(r)$ of the KS equations as Bloch functions:

$$\Psi_{n,k}(r) = u_{n,k}(r)e^{ikr} \tag{3.17}$$

where $u_{n,k}(r)$ is a lattice periodic function, n is the band index, and the wave vector k belongs to the first Brillouin zone (BZ). Since $u_{n,k(r)}$ is periodic, it can be expanded over the reciprocal lattice:

$$u_{n,k}(r) = \sum_{G'} \varphi_{n,k}\left(G'\right)e^{i\vec{G}'\vec{r}} \tag{3.18}$$

where G' are the reciprocal lattice vectors. This expansion goes to infinity! Note that we actually deal with two types of infinities here. One is due to the infinite periodic nature of the crystal and is captured by the wave vector k; the other comes from this expansion. For practical purposes the sum over G' is restricted to plane waves with kinetic energy below a given cutoff energy E_{cut}. Thus, defining the set $\Omega(G)$:

$$\Omega(G) = \left\{ \left|\frac{\hbar^2}{2m}\left|\vec{k} + \vec{G}\right|^2 \le E_{Cut}\right. \right\} \tag{3.19}$$

we obtain the following expansion of the Kohn-Sham wave functions:

$$\psi_{n,\vec{k}}(r) = \sum_{G \in \Omega(G)} \varphi_{n,k}(G)e^{i\left(\vec{G}+\vec{k}\right)\vec{r}} \tag{3.20}$$

The cutoff energy E_{cut} controls the numerical convergence and depends strongly on the elements which are present in the system under investigation. For example, first row elements with strong potentials require higher cutoff energy. Here we immediately see the weakness of the supercell method. In the direction normal to the surface, the reciprocal cell vectors $\left|\vec{G}_{\perp}\right|$ are very short due to a large length of the direct space cell (often many multiples of the physical cell lattice constant). Thus a very large number of plane waves is needed to reach convergence. This is the price one has to pay for the artificial periodicity. The introduction of ultra-soft pseudopotentials has made these calculations practical. The localized basis set would still have the advantage of being insensitive to the simulation cell size; however, the range of the orbitals should be sufficient to describe the vacuum decay.

3.5 Calculating Band Alignment and Dielectric Constants

Among the most useful applications of the DFT-LDA scheme, from the heterostructure development point of view, are calculations of the band disconti-nuity at the interface and of the dielectric constant. The discontinuity can be estimated using the reference potential method originally introduced by Kleinman [13]. Van de Walle and Martin proposed using the macroscopically averaged electrostatic potential as reference energy [14]. The method requires calculating a heterojunction AB in either slab (in this case you would have free surfaces) or supercell geometry to compute the average reference potential across the interface, and two additional bulk calculations to locate the valence band top (VBT) in materials A and B with respect to the average potential. For a supercell (or a slab) containing the interface, one calculates the average potential using the formula:

$$\overline{V}(z) = \frac{1}{d_1 d_2} \int_{z-d_1/2}^{z+d_1/2} dz' \int_{z'-d_2/2}^{z'+d_2/2} dz'' V\left(z''\right). \tag{3.21}$$

Where $V(z)$ is obtained by the xy-plane averaging (a simple $\frac{1}{(a_x \cdot a_y)} \iint_{cell} dxdy$ integration) of the electrostatic potential:

$$V(r) = -\sum_i \frac{Z_i e^2}{|r - R_i|} + e^2 \int \frac{n(r')}{|r - r'|} dr' \tag{3.22}$$

The parameters d_1 and d_2 are the inter-planar distances along the z direction (normal to the interface) in materials A and B, respectively. This produces a smooth

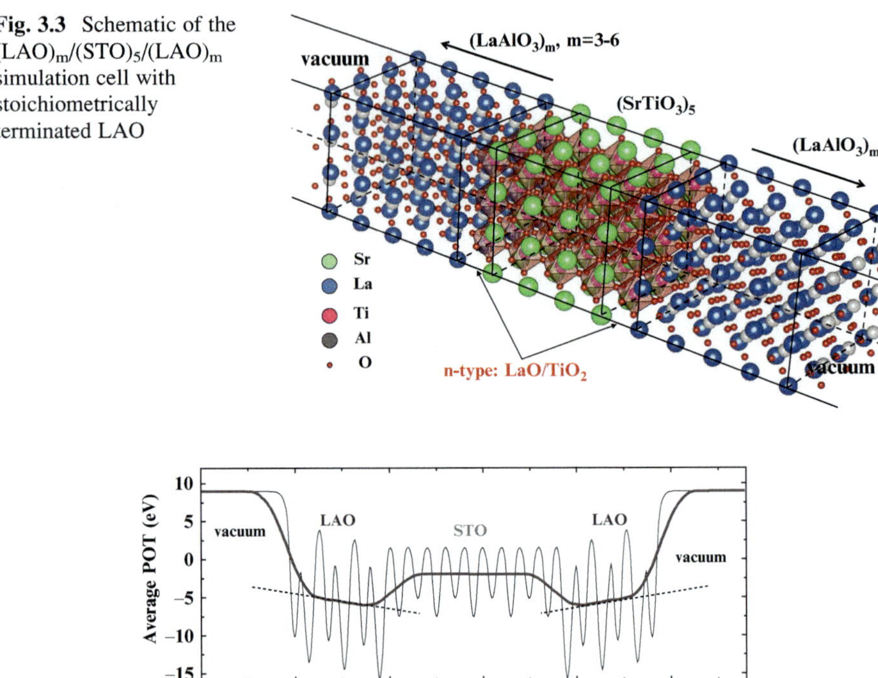

Fig. 3.3 Schematic of the (LAO)$_m$/(STO)$_5$/(LAO)$_m$ simulation cell with stoichiometrically terminated LAO

Fig. 3.4 The electrostatic potential (*black line*) across the (LAO)$_3$(STO)$_5$(LAO)$_3$ simulation slab

reference potential. Assuming that far away from the interface the potential reaches its bulk value one can place corresponding VBTs with respect to the average potential on both sides of the interface using the bulk reference, and thus determine the valence band offset. The conduction band offset has to be inferred using the experimental values of the band gaps, since those are seriously underestimated in the DFT-LDA calculations. The use of periodic boundary conditions creates certain difficulties when dealing with systems with intrinsic electric fields. In Fig. 3.3 we show a simulation cell used to calculate the electrostatic potential across the LaAlO$_3$/SrTiO$_3$ interface. As LaAlO$_3$ (LAO) is polar, one has to use a symmetric (LAO)$_m$/(STO)$_5$/(LAO)$_m$ heterostructure with vacuum termination to ensure the potential has the same value on both sides of the simulation cell. In Fig. 3.4 we show the plane-averaged electrostatic potential across this cell along with its macroscopic average shown as a thick red line. It is clear that in the vacuum region on both sides of the heterostructure the potential has the same value making the use of the periodic boundary conditions physically justified. Unfortunately, this technique requires doubling the size of the system to be calculated.

Calculating the dielectric constant/tensor from first principles is somewhat less straightforward due to the periodic boundary conditions used in most first principles codes. In brief, it is the absence of the surface in an infinite periodic solid that causes the problem. It is impossible to define the electrostatic dipole in a unique and physically meaningful way. Vanderbilt has shown that the change in electronic polarization is related to the polarization current and can be calculated using the geometric or Berry phase of the electrons [15]:

$$P_\alpha^{el} = \frac{i}{\Omega} \sum_{ki} \langle u_{ki} | \frac{\partial}{\partial k_\alpha} | u_{ki} \rangle, \tag{3.23}$$

where Ω is the unit cell volume, k is the Bloch vector, and u_{ki} is the cell periodic part of the Bloch wave function. Once the change in polarization with respect to a reference state of the system is determined, the Born effective charge tensors $Z_{m,\alpha\beta}^*$ (derivatives of the unit cell polarization in the direction α with respect to β displacements of the atomic sublattice m) can be evaluated. From these, the mode effective charge vectors $Z_{j,\alpha}^*$ can be computed and the dielectric constant is given by:

$$\varepsilon_{\alpha\beta} = \varepsilon_{\alpha\beta}^\infty + \frac{4\pi}{\Omega} \sum_j \frac{Z_{j\alpha}^* Z_{j\beta}^*}{\omega_j^2 - \omega^2} \tag{3.24}$$

The sum is over the phonon modes, Ω is the primitive cell volume. The electronic contribution $\varepsilon_{\alpha\beta}^\infty$ can be computed using the linear response theory. The values thus computed typically overestimate experiment by about 20 %, mainly due to the error in the band gap. A semi-empirical "scissor" correction is then used in which the conduction bands are moved up in energy by hand to match the experimental spectrum. To calculate the dielectric tensor in (3.24) one needs to know the vibrational frequencies of the system $\{\omega_j\}$ and displacement patterns corresponding to these modes. In other word one needs the phonon spectrum.

3.6   Phonon Calculations

The phonon spectrum of a solid can be computed in a relatively straightforward manner. Assuming the harmonic approximation, the total energy for such a system (3.15) can be expanded as

$$E = E_0 + \sum_{a,\kappa,\alpha} \sum_{b,\kappa',\beta} \frac{1}{2} \frac{\partial^2 E}{\partial v_{\kappa,\alpha}^a \partial v_{\kappa',\beta}^b} v_{\kappa,\alpha}^a v_{\kappa',\beta}^b \tag{3.25}$$

up to second order in atomic displacements $\left\{ \vec{v}_k^a \right\}$ from the equilibrium positions. The vectors $\left\{ \vec{v}_k^a \right\}$ represent a displacement of atom a in cell k ($\alpha, \beta = x, y, z$). The first order term of the expansion is zero as the system is assumed to be at the equilibrium configuration. The Hessian Θ (a matrix of second order energy

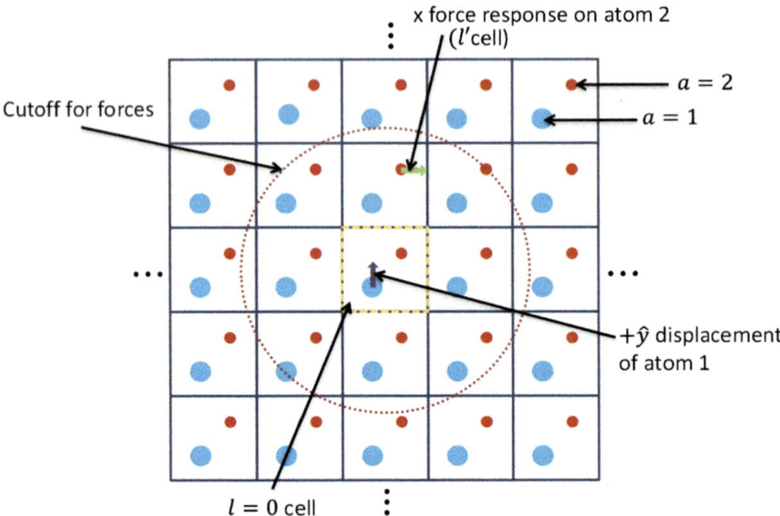

Fig. 3.5 Schematic of calculating the short range force constant matrix. The cut off radius is introduced, beyond which forces are considered to be negligible

derivatives) is known as the force constant matrix, and can be calculated from first principles:

$$\Theta_{\kappa a\alpha,\kappa' b\beta} = \frac{\partial^2 E}{\partial v^a_{\kappa,\alpha}\partial v^b_{\kappa',\beta}}. \tag{3.26}$$

In practice, one computes a numerical derivative of the Hellmann-Feynman force on atom **a** in cell **k** in the direction α due to a displacement of atom **b** in cell **k′** in the direction β as shown in Fig. 3.5. Formally, in a crystal this matrix has an infinite range and needs to be truncated for practical computations. If interatomic forces decay slowly with distance, special care needs to be taken when calculating the force constant matrix as will be described below.

Vibrational frequencies appear as the eigenvalues of the force constant matrix:

$$\sum_{b,\kappa',\beta}\left[\Theta_{\kappa a\alpha,\kappa' b\beta} - \delta_{kk'}\delta_{ab}\delta_{\alpha\beta}m_b\omega_i^2\right]\chi^i_{k'b\beta} = 0, \tag{3.27}$$

where m_b is the mass of atom type **b**. As the system is periodic, one can use the Bloch theorem and redefine the eigenvectors of the force constant matrix as follows

$$\chi^{i,\vec{\kappa}}_{lb\alpha} = \frac{e^{i,\vec{k}}_{a\alpha}}{\sqrt{Nm_b}}e^{i\vec{k}\left(\vec{R}_l+\vec{r}_b\right)}. \tag{3.28}$$

In (3.28), $\vec{k}$ is the wave vector in the first Brillouin zone, superscript i tracks the eigenmode, $\vec{R}_l$ is the Bravais lattice vector pointing to cell l, $\vec{r}_b$ is the position of atom b in that cell, and N is the number of primitive cells allowed by the periodic boundary conditions. We have also introduced the wave vector dependent polarization vectors $\vec{e}_a^{\,i,\vec{k}}$. The eigenvalue problem (3.25) can now be written as:

$$\sum_{b,l,\beta} \frac{\Theta_{la\alpha,l\,b\beta}}{\sqrt{m_a m_b}} e^{-i\vec{k}\left(\vec{R}_l+\vec{r}_a-\vec{R}_l-\vec{r}_b\right)} = \omega_i^2\left(\vec{k}\right) e_{a\alpha}^{i,\vec{k}} \tag{3.29}$$

The mass-normalized lattice Fourier transform of the real space force constant matrix Θ is known as the dynamical matrix:

$$D_{a\alpha,b\beta}^{\vec{k}} = \sum_l \frac{\Theta_{la\alpha,l\,b\beta}}{\sqrt{m_a m_b}} e^{-i\vec{k}\left(\vec{R}_l+\vec{r}_a-\vec{R}_l-\vec{r}_b\right)} \tag{3.30}$$

The square roots of the eigenvalues of the dynamical matrix $\omega_i^2\left(\vec{k}\right)$ give the desired phonon spectrum. Note that in practice, the infinite lattice sum in (3.28) is truncated, thus the dynamical matrix only includes the short range forces. The ionic interactions often present in transition metal oxides, result in long-range Coulomb forces between distant neighbors, which must be taken into account using the Madelung sum technique. Following Maradudin [16] this long-range correction to the dynamical matrix has the form

$$\vec{D}_{\alpha\beta}^{\,long}\left(\vec{k};a;b\right) = \frac{e^2}{V\varepsilon_0\varepsilon_\infty} \frac{\left[\vec{k}\,\overleftrightarrow{Z}^*(a)\right]_\alpha \left[\vec{k}\,\overleftrightarrow{Z}^*(b)\right]_\beta}{\left|\vec{k}\right|^2} \times \exp\left(-\frac{\left|\vec{k}\right|^2}{\rho^2}\right). \tag{3.31}$$

where $\overleftrightarrow{Z}^*(a)$ is the Born effective charge tensor of atom a, V the volume of the primitive cell, and ε_∞ the high frequency dielectric constant. The parameter ρ controls the range of the long-range correction. The total dynamical matrix is now the sum of (3.30) and (3.31).

The vibrational spectrum can be used to compute other thermodynamic properties of materials such as the vibrational free energy:

$$F_{vib} = rk_BT \int_0^\infty g(\omega)\ln\left[2\sinh\left(\frac{\hbar\omega}{2k_BT}\right)\right] d\omega, \tag{3.32}$$

where r is the number of degrees of freedom, and $g(\omega)$ is the phonon density of states. From this the heat capacity can be computed as:

$$C_v = \frac{1}{4k_BT^2} \int_0^{\infty} g(\omega) \frac{\hbar^2\omega^2}{\sinh^2\left(\frac{\hbar\omega}{2k_BT}\right)} d\omega. \tag{3.33}$$

Vibrational frequencies and specific heat can be compared with experiment such as neutron scattering, Raman or infra-red (IR) spectroscopies and calorimetry.

3.7 *Ab-Initio* Packages

Today many first principles codes are available. An example of a real space code is PARSEC [4]. VASP [17] and CASTEP [18] are plane wave codes. FIREBALL [19], SIESTA [20] and DMol [21] are local atomic orbital codes. The work horse of computational chemistry, GAUSSIAN, is a local orbital code using atomic orbitals expanded in terms of Gaussians to simplify multi-center integrations [22]. Linear response calculations can be performed with PWSCF [23] and *Abinit* [24]. Overall, DFT-LDA calculations give very accurate ground state properties such as e.g., structural parameters, elastic constants, and relative energies of different phases. The most serious drawback of the theory is its inability to describe the excited states, and thus to predict the band gap. Several methods have been developed to address this problem, such as the exact exchange method [25, 26], GW method [27], and Bethe-Salpeter method [28]. Unfortunately, all of these techniques require a significant increase in computational time. To learn more about the applications of the DFT-LDA formalism to high-k dielectrics we refer the reader to reference [29].

In particular, in the VASP code one can use a quasi-particle (QP) correction within the G_oW_o approximation [27]. The QP energies are obtained by solving the equation:

$$(T + V_{ext} + V_H)\psi_{nk}(\mathbf{r}) + \int d\mathbf{r}' \Sigma\left(\mathbf{r},\mathbf{r}' : E_{nk}\right)\psi_{nk}\left(\mathbf{r}'\right) = E_{nk}\psi_{nk}(\mathbf{r}), \tag{3.34}$$

where T is the kinetic energy of the electrons, V_{ext} is the external potential of the ions, and V_H is the Hartree potential. The energy dependent non-local self-energy operator Σ contains the exchange and correlation effects, and within Hedin's *GW* approximation [30] it takes the form:

$$\Sigma(\mathbf{r},\mathbf{r}' : \omega) = \frac{i}{4\pi} \int_{-\infty}^{\infty} e^{i\omega'\delta}G(\mathbf{r},\mathbf{r}',\omega + \omega')W(\mathbf{r},\mathbf{r}',\omega')d\omega', \tag{3.35}$$

where G is the Green's function and W is the screened Coulomb interaction. Then to first order the QP eigenvalues of (3.34) are obtained as:

$$E_{nk}^{QP} = \langle \psi_{nk} | T + V_{ext} + V_H + \Sigma | \psi_{nk} \rangle \tag{3.36}$$

In a non-self-consistent $G_0 W_0$ approximation, which has been shown to be a quite reasonable approximation [30, 31], the ψ_{nk} are chosen to be the Kohn-Sham LDA wave functions. In general, QP methods seem to work well for transition metal oxides [32].

3.8   Beyond the DFT-LDA

Despite its astounding success in materials theory, the failures of the DFT-LDA scheme are numerous, systematic, and well documented [33]. Many of these failures occur in transition metal (TM) oxides where the LDA, being a mean-field theory, fails to properly account for electron correlations (strictly speaking it is not possible to separate exchange and correlation in the LDA-DFT formalism). The physical reason for this failure is the relatively high degree of electron localization in the TM d-shells. Perdew and Zunger have shown that the self-interaction results in significant errors in single particle energy levels [34]. Self-interaction corrections (SIC) have been successfully implemented and used for calculations of TM oxides [35]. Unfortunately, SIC methods typically result in orbital dependent potentials. A very attractive scheme avoiding orbital-dependent potentials was suggested by Filippetti and Spaldin [36]. Another way to at least partially account for the electron correlation is the so-called LDA + U method [37]. Essentially, it amounts to solving a Hubbard problem within the unrestricted Hartree-Fock approximation for a chosen set of states. Lee and Pickett have successfully used it to describe magnetic ordering in Sr_2CoO_4 [38].

A somewhat different approach to fixing the shortcomings of the LDA came from quantum chemistry, and is known as the hybrid functional method. In a popular HSE formalism [39, 40], the exchange correlation functional is constructed from 25 % Hartree-Fock exchange (E_x) and 75 % of the generalized gradient approximation due to Perdew, Burke and Ernzerhof (PBE) [41]. In HSE the exact exchange is further decomposed into a long range and a short range part in real space. The range separation is determined by a parameter, μ, which is typically chosen as a distance at which the non-local long range interaction becomes negligible. The HSE exchange correlation functional is written as:

$$E_{xc}^{HSE} = \frac{1}{4} E_x^{sr,\mu} + \frac{3}{4} E_x^{PBE,sr,\mu} + E_x^{PBE,lr,\mu} + E_c^{PBE} \tag{3.37}$$

where the superscript sr and lr stand for short range and long range, respectively, and μ is the screening parameter mentioned earlier. The HSE method is implemented in the VASP code.

Many oxide materials demonstrate strongly-correlated behavior that results in exciting experimentally observed phenomena such as superconductivity or metal to insulator transitions. The dynamical mean field theory (DMFT) offers a possibility to investigate these regimes for which DFT-based methods do not work properly. The main physics, which DMFT captures, is the quantum fluctuations specified by the local Hamiltonian [42]. Compared with DFT, where charge density $n(r)$ is determined self-consistently, DMFT requires self-consistently determining the local Green's function $G\left(\vec{r}, \vec{r}; \omega\right)$, which includes information on the excitations [43]. Recently developed DFT + DMFT methodology [44, 45], which applies DMFT to "correlated orbitals" extracted from DFT and requires $n(r)$ to be the charge density used in the Kohn-Sham Hamiltonian (3.7 and 3.8), provides the most general approach in materials science.

Of course, this short description of DFT-based methods is not meant to be comprehensive, but rather to help the reader in navigating through the theoretical material in the following chapters. There are a number of excellent books on DFT that the interested reader may find useful [46–48].

References

1. M. Born, R. Oppenheimer, Ann. Phys. **84**, 458 (1927)
2. P. Hohenberg, W. Kohn, Phys. Rev. **136**, 864 (1964)
3. W. Kohn, L.J. Sham, Phys. Rev. **140**, 1133 (1965)
4. J.R. Chelikowsky, N. Troullier, Y. Saad, Phys. Rev. Lett. **72**, 1240 (1994)
5. M.C. Payne, M.P. Teter, D.C. Alan, T.A. Arias, J.D. Joannopoulos, Rev. Mod. Phys. **64**, 1045 (1992)
6. J.C. Phillips, L. Kleinman, Phys. Rev. **116**, 287 (1959)
7. D. Hamann, M. Schluter, C. Chiang, Phys. Rev. Lett. **43**, 1494 (1979)
8. N. Trulier, J.L. Martins, Phys. Rev. B **43**, 1993 (1991)
9. D. Vanderbilt, Phys. Rev. B **41**, 7892 (1990)
10. P.E. Blöchl, Phys. Rev. B **50**, 17953 (1994)
11. G. Kresse, J. Joubert, Phys. Rev. B **59**, 1758 (1999)
12. M.P. Allen, D.J. Tildesley, *Computer Simulation of Liquids* (Clarendon Press, New York, 1988)
13. D.M. Bylander, L. Kleinman, Phys. Rev. B **36**, 3229 (1987)
14. C.G. Van de Walle, R.M. Martin, Phys. Rev. B **39**, 1871 (1989)
15. R.D. King-Smith, D. Vanderbilt, Phys. Rev. B **47**, 1651 (1993)
16. G.K. Horton, A.A. Maradudin, *Dynamical Properties of Solids* (North-Holland, Amsterdam, 1974)
17. G. Kresse, J. Furtmuller, Phys. Rev. B **54**, 11169 (1996)
18. V. Milman, B. Winkler, J.A. White, C.J. Pickard, M.C. Payne, E.V. Akhmatskaya, R.H. Nobes, J. Quant. Chem. **77**, 895 (2000)
19. J.P. Lewis, K.R. Glaesemann, G.A. Voth, J. Fritsch, A.A. Demkov, J. Ortega, O.F. Sankey, Phys. Rev. B **64**, 195103 (2001)
20. J.M. Soler, E. Artacho, J.D. Gale, A. García, J. Junquera, P. Ordejón, D. Sánchez-Portal, J. Phys. Condens. Matter **14**, 2745 (2002)
21. B. Delley, J. Chem. Phys. **113**, 7756 (2000)
22. M.J. Frisch et al., *Gaussian 98* (Gaussian, Inc., Pittsburgh, PA, 1998)

23. S. Baroni, A. Dal Corso, S. de Gironcoli, P. Giannozzi, http://www.pwscf.org
24. X. Gonze, D.C. Allan, M.P. Teter, Phys. Rev. Lett. **68**, 3603 (1992). http://www.abinit.org
25. M. Städele, J.A. Majewski, P. Vogl, A. Görling, Phys. Rev. Lett. **79**, 2089 (1997)
26. M. Städele, M. Moukara, J.A. Majewski, P. Vogl, Phys. Rev. B **59**, 10031 (1999)
27. F. Aryasetiawan, O. Gunnarsson, Phys. Rev. Lett. **74**, 3221 (1995)
28. M. Rohlfing, S.G. Louie, Phys. Rev. B **62**, 4927 (2000)
29. A.A. Demkov, A. Navrotsky (eds.), *Materials Fundamentals of Gate Dielectrics* (Springer, Dordrecht, 2005)
30. L. Hedin, Phys. Rev. **139**, A796 (1965)
31. M.S. Hybertsen, S.G. Louie, Phys. Rev. B **32**, 7005 (1985)
32. S. Lany, Phys. Rev. B **87**, 085112 (2013)
33. R.O. Jones, O. Gunnarson, Rev. Mod. Phys. **61**, 689 (1989)
34. J.P. Perdew, A. Zunger, Phys. Rev. B **23**, 5048 (1981)
35. M. Arai, T. Fujiwara, Phys. Rev. B **51**, 1477 (1995)
36. A. Filippetti, N.A. Spaldin, Phys. Rev. B **67**, 125109 (2003)
37. V.I. Anisimov, P. Kuiper et al., Phys. Rev. B **50**, 8257 (1994)
38. K.W. Lee, W.E. Pickett, Phys. Rev. B **73**, 174428 (2006)
39. J. Heyd, G.E. Scuseria, M. Ernzerhof, J. Chem. Phys. **118**, 8207 (2003)
40. J. Heyd, G.E. Scuseria, M. Ernzerhof, J. Chem. Phys. **124**, 219906E (2006)
41. J.P. Perdew, K. Burke, M. Ernzerhof, Phys. Rev. Lett. **77**, 3865 (1996)
42. A. Georges, G. Kotliar, W. Krauth, M.J. Rozenberg, Rev. Mod. Phys. **68**, 13 (1996)
43. The relation between the local Green's function and charge density is given by $n(r) = \int dw/\pi\, f(\pi)G(r,r;\omega)$. DMFT includes the frequency dependence (dynamics) that DFT does not
44. K. Held, I.A. Nekrasov, G. Keller, V. Eyert, N. Bluemer, A.K. McMahan, R.T. Scalettar, T. Pruschke, V.I. Anisimov, D. Vollhardt, Phys. Status Solidi **243**, 2599 (2006)
45. G. Kotliar, S.Y. Savrasov, K. Haule, V.S. Oudovenko, O. Parcollet, C.A. Marianetti Rev, Mod. Phys. **78**, 865 (2006)
46. D. Sholl, J.A. Steckel, *Density Functional Theory: A Practical Introduction* (John Wiley and Sons, Hoboken, NJ, 2009)
47. R.G. Parr, W. Yang, *Density-Functional Theory of Atoms and Molecules* (Oxford University Press, Oxford, 1989)
48. R.M. Martin, *Electronic Structure: Basic Theory and Practical Methods* (Cambridge University press, Cambridge, 2004)

Chapter 4
Crystalline Functional Oxide Growth Methods

By necessity, the integration of semiconductors and functional oxides is by the use of thin film deposition methods that enable crystalline growth of the oxide on a semiconductor substrate. There are only a handful of techniques at present that have been shown to be capable of sufficient control of thickness, stoichiometry and oxidation conditions that will allow for the growth of single crystalline complex oxide layers on semiconductors. We very briefly describe these thin film deposition techniques, with particular emphasis on the growth of multi-component oxide materials. Readers that require a more detailed treatment are referred to the various books and review articles on thin film deposition techniques [1–4]. The growth of complex oxides is made difficult by the fact that the materials are composed of three or more elements. In order to be able to grow high quality thin films of these complex oxides, one must be able to either: (a) stoichiometrically transfer the composition of the source compound to the substrate, or (b) precisely control (with monolayer or better degree of precision) the fluxes of all the elements involved. The primary reason for this is due to the multi-component nature of a typical complex oxide and also because most of these complex oxides do not have line compositions [5]. This means that, unlike in the growth of compound semi-conductors, one cannot have an excess or overpressure of one element over another (with the exception of oxygen) without also resulting in an excess of that element in the film. In the case of growing complex oxides on semiconductors, even oxygen has to be controlled precisely since excess oxygen could end up oxidizing the underlying semiconductor substrate.

The ability to grow epitaxial layers of complex oxides onto semiconductors has developed only in the last 20 years. Tremendous progress in the thin film growth of multi-component oxides took off soon after the discovery of high-T_c superconduc-tors, for example, the development of the pulsed laser deposition technique [6] and the use of activated oxygen sources in molecular beam epitaxy [7]. Oxide deposi-tion methods are now capable of producing artificially layered materials that are comparable in crystalline quality to advanced semiconductor heterostructures [8, 9]. By adapting the concepts once limited only to the semiconductor field to oxide materials systems, new experimental platforms for integrating the two types

A.A. Demkov and A.B. Posadas, *Integration of Functional Oxides with Semiconductors*, DOI 10.1007/978-1-4614-9320-4_4, © The Author(s) 2014

of materials into a single structure with cooperative functionality are now starting to become routine. We are now at the point where it is possible to fabricate integrated sensor/transistor systems where the transistor function is directly coupled to an environmental stimulus, such as temperature, pressure, and electromagnetic fields. In this chapter, we will give a brief outline of five different thin film deposition methods that have been demonstrated to be capable of growing epitaxial oxide thin films on a semiconductor substrate. We will describe physical vapor deposition methods such as molecular beam epitaxy, pulsed laser deposition, and sputtering, as well as chemical vapor deposition methods, including metal-organic chemical vapor deposition and atomic layer deposition.

4.1 Molecular Beam Epitaxy

Molecular beam epitaxy (MBE) is a physical vapor deposition process involving the thermal evaporation of elemental sources under ultrahigh vacuum conditions [10–12]. The low base pressures of MBE systems (~10^{-10} Torr) coupled with the use of specialized evaporation sources known as effusion cells (Fig. 4.1) results in the evaporated material taking the form of atomic or molecular beams that are directed at the substrate. These beams can be quickly turned on and off by fast mechanical shutters (usually pneumatically actuated), enabling the growth of complicated multilayer structures that can have different compositions for its different layers. The flux of the atomic beam is controlled by the temperature of the crucible in the effusion cell and is directly related to the vapor pressure of the element to be evaporated. For the growth of oxides, a source of oxygen is also

Fig. 4.1 Commercial high temperature effusion cell made by DCA Instruments. Image taken from DCA website http://dca.co.uk/portfolio/high-temperature/

required. In most cases, this is either molecular oxygen, or an active oxygen species such as ozone or atomic oxygen from a plasma source. If one uses only molecular oxygen, the types of oxides one can grow are limited only to those that have metals that oxidize easily. For some oxides like $LaCoO_3$ and $LaNiO_3$, an activated oxygen source is needed to grow them. The use of activated oxygen such as ozone or atomic oxygen, however, necessitates the use of additional equipment. For simple oxides with the metal in its highest oxygen state, growth is relatively straightforward as one just uses a single metal flux in the presence of excess oxygen. For multi-cation oxides such as the perovskite oxides, an additional difficulty is to get the two metal fluxes to match in order to obtain the correct phase with the correct stoichiometry. As most of these complex oxides do not have very well-defined line compounds [5], a poorly calibrated metal flux often results in the formation of secondary phases; therefore, measurement of fluxes is a critical concern in MBE growth. Some functional oxides such as EuO (with Eu^{2+}) and $LaTiO_3$ (with Ti^{3+}) have cations that are not in their highest oxidation state. In these cases, one also has to be able to precisely control the amount of oxygen present to obtain the correct phase. For example, using more than 1×10^{-7} mbar of molecular oxygen with an Eu flux of ~8 Å/min results in the formation of paramagnetic Eu_2O_3 instead of ferromagnetic EuO [13].

Flux monitoring in MBE is usually performed by using one or more of three techniques: (1) using a quartz crystal microbalance; (2) using a nude ion gauge to measure beam-equivalent pressure; or (3) using atomic absorption spectroscopy. The method of using of an ion gauge to measure source fluxes involves mounting the gauge on a retractable arm and being able to move it to the substrate position. The change in pressure reading of the ion gauge when the source shutter is open is then recorded. This beam-equivalent pressure can be related to the flux through a simple equation [14]. Flux measurement using a quartz crystal microbalance utilizes the change of the oscillation frequency of a quartz crystal as its total mass changes from being deposited on by the evaporated materials [15]. The quartz crystal is mounted on an arm such that it can be moved to the position of the substrate during flux measurement. The accumulated mass in a given amount of time is calculated from the frequency shift and this number is then converted to a thickness from the density of the material being measured. The readings of quartz crystal microbalances can take a long time (over 30 min) to stabilize and are highly influenced by temperature. For this reason, most quartz crystal monitors are water-cooled. Both ion gauge beam equivalent pressure and quartz crystal microbalance flux measurements are typically accurate to about 5 %, which might not be sufficient for certain applications. A more accurate method of measuring flux is by means of atomic absorption spectroscopy [16]. In this technique, a beam of light is directed through the path of the evaporating material and a detector senses how much of the light has been absorbed. Because atoms absorb light at specific wavelengths, the absorption specific to a particular element can be monitored. This allows fluxes to be measured to an accuracy below 1 %. The major drawback is the more expensive instrumentation needed for an atomic absorption flux monitor.

One aspect of MBE that makes it better over other thin film deposition techniques is its ability to arbitrarily substitute cations (both species and amount) in the same growth run. For example, one can grow a continuously graded $Ba_{1-x}Sr_xTiO_3$ with MBE in a single growth while this would require multiple target changes and multiple runs with discrete steps using sputtering or pulsed laser deposition. MBE also allows for precise control of stoichiometry allowing one to study the detailed effects of various kinds of non-stoichiometry on the physical properties of these materials. A further advantage of MBE is that one has full control of the surface termination of the film because of its true atomic layer-by-layer capability compared with the unit cell block growth of pulsed laser deposition or sputtering. The low background pressure in MBE, even during growth in oxygen, allows for the use of in situ real-time characterization techniques. Reflection high energy electron diffraction (RHEED), including spectroscopic techniques based on the incident electrons in RHEED (e.g. x-ray fluorescence, Auger spectroscopy, electron energy loss spectroscopy, and cathodoluminescence) [17–20], are often used to "watch" the growth process in real time, providing information on the crystalline structure, lattice spacing, film thickness, surface roughness, and surface composition. Additional in situ characterization methods commonly used in MBE systems include pyrometry [21] and ellipsometry [22].

The main components of an oxide MBE system are the ultrahigh vacuum chamber with a base pressure around 10^{-10} Torr, the effusion cells that enable the highly directional evaporation of the source materials, oxygen gas source, substrate manipulator and heater, and in situ characterization tools (usually RHEED and flux measurement at a minimum). A schematic of a typical oxide MBE system is shown in Fig. 4.2. Complex oxide materials typically have a transition metal element as one of the components. Because of the low vapor pressure of many of the transition metals, it is challenging to evaporate these with sufficient flux and flux stability. Effusion cells are available that can be operated at temperatures close to 2,000 °C that can enable low but stable fluxes of some transition metals to be achieved. This is the case, for example, for titanium or vanadium. However, for some transition metals, the vapor pressure is still not sufficiently high at 2,000 °C to achieve a high enough flux, for example, niobium and ruthenium. For these elements, one usually utilizes an electron beam evaporator [23]. However, flux stability and run to run reproducibility is quite poor with an electron beam evaporator. Acceptable fluxes for MBE growth can be achieved once the vapor pressure of the material exceeds 10^{-3} to 10^{-2} Torr. The temperature needed to heat up an effusion cell to achieve the needed vapor pressure can be found from looking up vapor pressure vs. temperature tables and charts [24, 25]. When using effusion cells, another concern is compatibility between the material to be evaporated and the crucible material. The crucible material must not react with the element to be evaporated at the temperature of evaporation and, ideally, the material to be evaporated should not wet the crucible. Several tables of evaporation compatibility that are essential to practitioners of MBE are available online [26, 27].

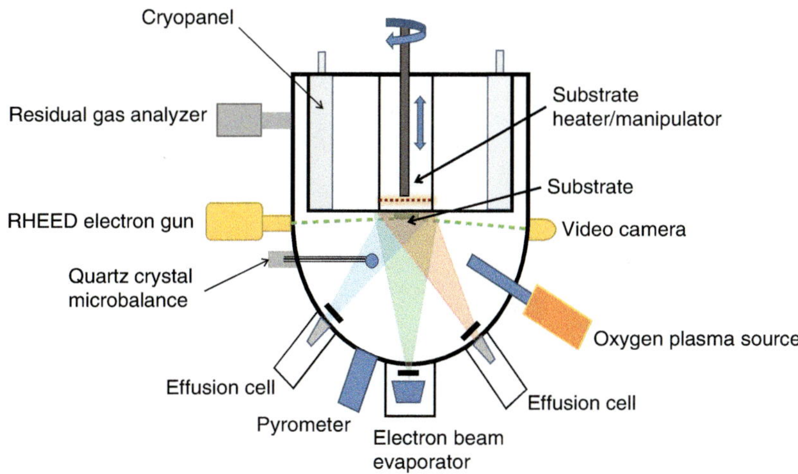

Fig. 4.2 Schematic of an oxide molecular beam epitaxy system

4.2 Pulsed Laser Deposition

Pulsed laser deposition (PLD), also known as laser ablation or laser MBE, is a method of depositing thin films by using a series of laser pulses to rapidly evaporate the source material, which is then transferred in the gas phase to the substrate [28–30]. The main benefit of PLD is that it is possible, under optimized conditions, to transfer the stoichiometry of the source material, known as the target, to the growing film, thus alleviating the need for precise control of individual evaporation rates of elements in a compound material. A schematic of a basic PLD system is shown in Fig. 4.3. The laser used for heating the target is typically situated outside the vacuum chamber and is often of the high-power KrF excimer variety. The laser is irradiated onto the target through a quartz lens. The evaporated atoms are hyperthermal because of the high energy of the laser and typically arrive at the substrate with kinetic energies ranging from 5 to 100 eV, allowing for sufficient surface diffusion at a lower substrate temperature in some cases, but could also result in re-sputtering of the growing film in others. In addition to the stoichiometric transfer of the source material to the substrate, another advantage of PLD is the ability to use a background gas during growth over a wide pressure range, from high vacuum ($\sim 10^{-7}$ Torr) to about 1 Torr. This is especially useful for growing oxides because being able to tune the oxygen partial pressure during growth over a wide pressure range provides another knob for growth process optimization. PLD, however, suffers from some drawbacks. The most significant is the ejection of microscopic particulates from the target that settle on the surface of the growing film, a phenomenon known as splashing [31]. Also, because of the highly directional plume of evaporated material produced by the laser (Fig. 4.4), PLD-grown films often suffer from composition and thickness uniformity

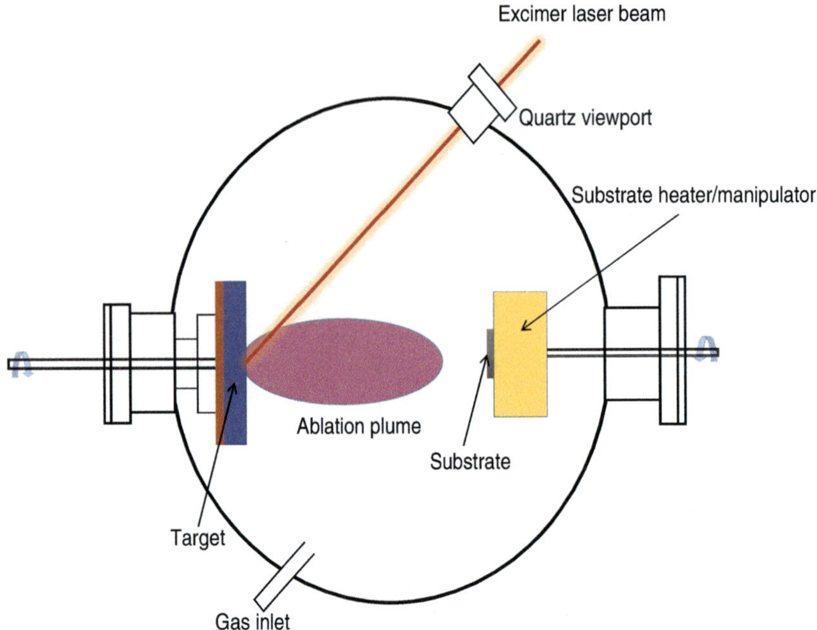

Fig. 4.3 Schematic of a basic pulsed laser deposition system

Fig. 4.4 Photograph of an ablation plume. Image taken from website of National Institute of Standards and Technology, Ceramics Division. http://www.ceramics.nist.gov/programs/thinfilms/pld.html

issues and are unsuited for deposition on large area substrates without additional modification of the deposition system.

The laser is the most crucial and expensive piece of hardware in a PLD system. The wavelength of the laser must be such that it would be strongly absorbed by the material to be evaporated but at the same time be able to deliver sufficiently high energy densities (>1 J/cm^2). To be able to use standard optical elements for focusing, the optimum wavelengths for PLD systems are in the 200–400 nm range. Due to these requirements, PLD systems are usually equipped with excimer lasers. For more on how excimer lasers work, the reader is referred to [32]. The most commonly used excimer lasers for PLD are based on either KrF (248 nm) or XeCl (308 nm). Between the laser and the growth chamber are optical elements that are used for focusing and steering the laser beam. For focusing and beam shaping, lenses that transmit UV light are needed. The most common lens materials are sapphire or UV-grade fused silica. Lenses are usually coupled with various apertures to minimize aberrations. The focused and collimated laser beam eventually passes through the laser port, which is where the laser enters the vacuum system of the growth chamber. The window is usually made of the same material as the lenses and should be optically flat and free of defects. The geometry of the growth chamber is an important element in PLD growth, specifically the relative positions of the beam focal point, the target, and the substrate. By necessity, the laser port and the target (which lies on the beam focal plane) are at some angle, usually around 45°. The PLD process is also quite sensitive to the distance between substrate and target so the substrate position needs to be adjustable in order to be able to optimize the growth for stoichiometric transfer. A rotating substrate stage is often used to improve deposition uniformity as well as to facilitate the use of RHEED. For epitaxial growth, it is also necessary to heat the substrate so the substrate manipulator also serves as a substrate heater. For the growth of oxides, the substrate heater must be oxygen-resistant yet still be capable of heating to at least 800 °C.

The final major component of a PLD system is the target holder/manipulator. The nature of the PLD process requires targets to be resurfaced periodically due to uneven target erosion and particulate buildup. For this reason, targets need to be relatively easily accessible and easily mounted/dismounted. To reduce uneven target erosion, many PLD systems employ a target rotation system or a laser scanning/rastering system. Targets commonly come in disc form and are mounted onto the target holder by mechanical clamping, bonding, or magnets. The target holder is normally water-cooled as the targets get quite hot from the laser. In some PLD systems, the targets are mounted in a multi-target carousel allowing several materials to be loaded at the same time. Multi-layer films of the different materials can then be easily grown by moving the carousel so that the laser hits the appropriate target.

Oxide thin film growth using PLD involves the control of many processing parameters that can strongly affect the composition and microstructure of the grown film [33]. The most crucial parameters for complex oxide growth are laser fluence and repetition rate. The fluence is a key parameter for retention of the target stoichiometry, which is the main benefit of using PLD. The repetition rate controls the degree of ionization and also the kinetic energy of the ejected particles of the target, which ultimately affect the morphology of the film as a result of particulate formation. Process parameters that are also important are the substrate temperature and the oxygen partial pressure, which control surface mobility and oxygen content of the growing film, respectively. The substrate to target distance is also important as it affects overall growth rate as well as composition in multi-component materials [34].

4.3   Sputter Deposition

Sputter deposition is a method of thin film deposition that involves the ejection of atoms from the surface of a solid source material due to a flux of highly energetic ions [35–37]. This process of knocking off atoms from a solid into the gas phase by means of kinetic energy transfer from incoming ions is known as sputtering [38]. Sputtering occurs when the incoming ions have a kinetic energy in the range of 50–1,000 eV. At higher kinetic energies (>50 keV), ion implantation occurs. The basic scheme of sputtering is shown in Fig. 4.5. Sputter deposition has several similarities to pulsed laser deposition. Similar to PLD, sputtering also uses source material in the form of targets, which are dense blocks of the material to be deposited in solid form. Sputtering is also able to stoichiometrically transfer the composition of source material to the substrate

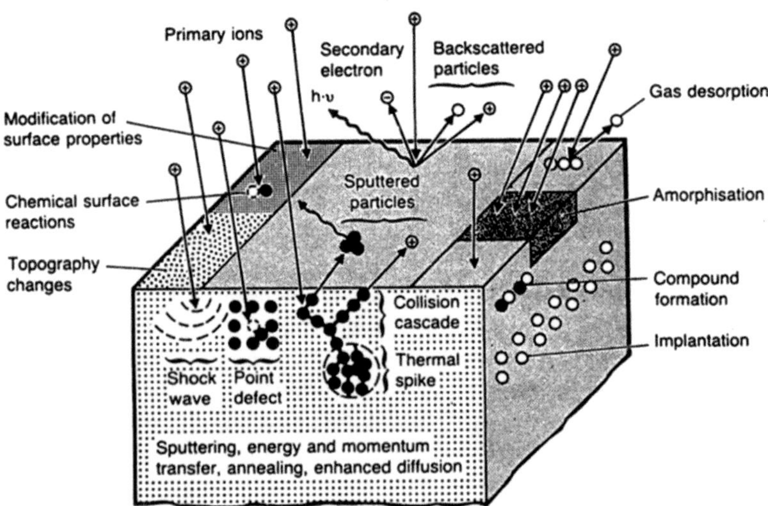

Fig. 4.5 Effects of ion bombardment on a solid surface. Image reprinted with permission from S.L. Rohde, "Sputter Deposition," in ASM Handbook, Volume 5: Surface Engineering, (ASM International, Materials Park, OH, 1994)

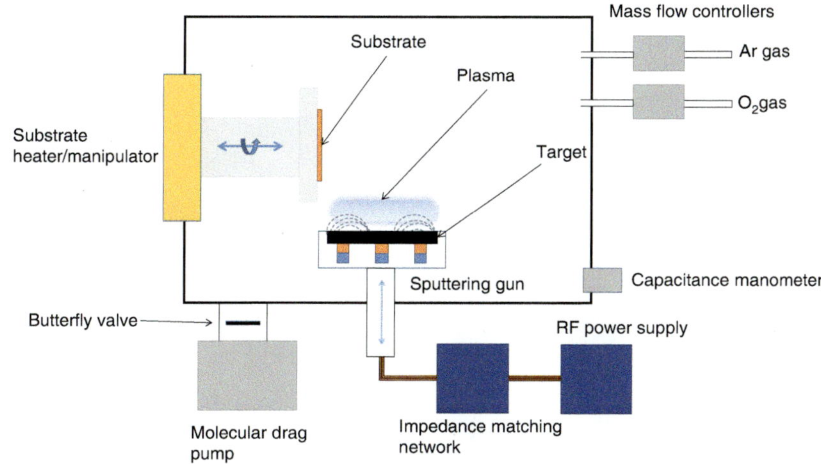

Fig. 4.6 Schematic of an off-axis RF magnetron sputtering system

making it very convenient for the growth of complex oxides. Sputtering, however, suffers from a limitation of not being able to utilize a wide range of pressures for the background gas. This is because the sputtering process itself requires a process gas that is ionized into a plasma that provides the energetic ions needed. The process gas used in complex oxide growth is typically a mixture of argon and oxygen and operated at pressures ranging from 1 to 1,000 mTorr. Sputtering is a less expensive way of achieving some of the advantages of PLD at the cost of being limited to relatively high pressures, preventing the use of many in situ characterization tools.

The key equipment required for sputter deposition systems, aside from the growth chamber itself, is the sputtering gun and its associated power supply. A basic schematic of a sputter deposition system is shown in Fig. 4.6. Because the creation of the plasma needed for sputtering to occur requires the application of large electric fields between the sputtering gun and the substrate, there is an obvious problem when trying to sputter insulating materials. Charge will build up at the surface of the insulating target and will quickly extinguish the applied field. To overcome this limitation, RF fields are employed for the sputtering of insulators [39]. RF power supplies are more expensive than DC power supplies and also require one to have an impedance matching tuning network. RF fields have frequencies that are high enough such that the heavy Ar ions in the plasma are unable to respond to the rapidly oscillating electric field. Most commercial sputtering systems use the frequency 13.56 MHz that is allowed by the United States Federal Communications Commission for arbitrary purposes. Because of the size (area) differential between the target (cathode) and the chamber walls (anode–ground), there is a higher electron concentration near the target and the Ar ions see an effective negative dc bias on the target (typically ~10 to 100 V). This self-bias is what accelerates the positive Ar ions to the target causing sputtering of the target material.

Fig. 4.7 Commercial
sputtering gun. Image
courtesy of Meivac, Inc.

In order to improve the efficiency of the sputtering process, several techniques have been developed to maintain the plasma near the surface of the target. The most commonly used method in modern sputtering systems for epitaxial oxide growth is the use of a planar magnetron configuration on the sputtering gun. Here, a cylindrical magnet with radially directed magnetic field lines is situated behind the target. The magnetic field lines are perpendicular to the electric field and this arrangement results in confinement of the plasma in a torus right in front of the target [40]. Figure 4.7 shows a commercial magnetron sputtering gun. The target is attached by mechanical clamping or magnets to the end of the sputtering gun, which acts as the cathode. A metallic target can be attached directly to the cathode but a target made of an insulating material requires a thermally and electrically conductive backing plate (usually copper). The target material itself is bonded using high temperature solder to the backing plate. The cathode region of the sputtering gun must be water-cooled to prevent the target from melting and also to protect the magnet inside the cathode. In early sputtering systems, the water used must be within a specific resistivity range to prevent the occurrence of electrochemical reactions inside the cathode assembly. More recent sputtering guns are now able to utilize a wider range of water resistivities.

There are also several process parameters that can be controlled in sputter deposition. As with all growth methods, the substrate temperature is important for crystalline, epitaxial growth. Parameters specific to sputtering are the forward power used for the plasma and the total pressure of the process gas (usually argon). These two parameters determine the growth rate and also the plasma composition and spatial extent which can be somewhat offset by substrate to target geometry. For oxide growth, the process gas is usually a mixture of argon and oxygen, with the oxygen needed to control the oxygen content of the growing film. One issue in sputtering is the different sputtering yields of different elements for a given kinetic energy of the incident species known as preferential sputtering [41]. Sputtering targets have to be conditioned for extended periods of time (several hours is typical) prior to use. By performing a pre-sputtering process, elements with high sputtering yield get ejected more than those with low sputtering yield. Eventually, the composition of the surface of the target changes so that it has a lot more of the low yield element than the high yield element. If done at the appropriate conditions, pre-sputtering can produce a surface target composition that compensates the sputtering yield difference and results in stoichiometric transfer of the bulk target composition.

Sputter deposition can also be performed in the so-called reactive sputtering scheme [42]. In this way, a metal oxide can be grown by sputtering elemental metal targets in a process gas containing oxygen to form the desired oxide composition. If the sputtering parameters and geometry are chosen correctly, the reaction can occur on or near the substrate surface, similar to what happens in chemical vapor deposition. Reactive sputtering is one method that can form suboxides of multivalent metals by controlling the amount of oxygen in the argon-oxygen process gas mixture.

Sputter deposition was first used for epitaxial complex oxide growth after the discovery of high-T_c superconductors [43]. For such multicomponent materials, a ceramic material with the same composition as the desired film is first synthesized and manufactured into a sputtering target. By using an oxygen-argon mixture for the sputtering gas, the deposited film is kept fully oxygenated, compensating for some inevitable volatile oxygen loss from the sputtered target. Sputtering was first performed using the so-called on-axis geometry where the substrate and target face each other. Similar to the splashing issues of PLD, on-axis sputtering also suffers from particulate deposition resulting in very rough film surfaces. A way around this was developed in 1990 by Eom et al. using a 90° off-axis geometry [44]. In this geometry, the sputtering gun axis is perpendicular to the substrate axis. This results in near complete elimination of particulates at the cost of a greatly decreased growth rate.

4.4 Chemical Vapor Deposition

Chemical vapor deposition (CVD) is a method of synthesizing solid thin films of a material from gas phase and surface chemical reactions of one or more precursors. CVD is a method known for its versatility as well as high growth rates. The structure and composition of the resulting thin film can be tailored by controlling the reaction chemistry as well as the deposition conditions. CVD is used for the deposition of a wide variety of films and coatings including dielectrics, metals, and epitaxial layers for microelectronics, hard coatings, and nanoparticles. For a more detailed treatment of the CVD process in general, the reader is referred to the several specialized texts on the topic [45–47]. The use of CVD for the deposition of epitaxial oxide thin films came to prominence after the discovery of high-T_c superconductors in 1986 [48]. This spurred significant technological development in the area of metal-organic CVD (MOCVD), which was also used for growing ferroelectric oxides [49].

Chemical reactions in CVD are typically endothermic and so energy must be supplied to maintain the reaction. Traditionally, this was done by heating the

Table 4.1 Commonly encountered variants of CVD and their basic descriptions

CVD variant name	Description
MOCVD (metal-organic CVD); also known as MOVPE (metal-organic vapor phase epitaxy)	Uses metal-organic compounds as precursors
VPE (vapor phase epitaxy)	Uses metal halides and hydrides as precursors
APCVD (atmospheric pressure CVD)	CVD process run at or near atmospheric pressure
LPCVD (low pressure CVD)	CVD process run in low vacuum conditions (~1 Torr)
PECVD (plasma enhanced CVD)	Uses a plasma to increase chemical reaction rates allowing for reduced growth temperature
PHCVD (photo-assisted CVD or photochemical vapor deposition)	Uses a UV radiation source to increase chemical reaction rates
CBE (chemical beam epitaxy); also known as MOMBE (metal-organic MBE)	Hybrid method combining the use of a volatilized metal-organic precursor with solid source elemental sources such as effusion cells
ALD (atomic layer deposition); also known as ALE (atomic layer epitaxy)	CVD process where the various precursors are dosed alternately rather than simultaneously and where the growth is self-limiting

substrate either inductively or resistively. Radiative heating has also been used. This thermal CVD can be disadvantageous when working with substrates and film materials that are not thermally stable at the high temperatures needed for the reaction. Two methods have been developed to bypass the need for high temperature by providing an alternative source of energy for the reaction. One is to use a plasma where electron bombardment of the precursors can initiate the formation of the necessary chemically active species (PECVD) [50]. A second method is to use UV radiation where photons get absorbed by the precursors and become chemically active (PHCVD) [51]. The CVD method is very versatile and has spawned many different specialized techniques, each with their own acronym. Table 4.1 lists some of the more commonly encountered CVD variants and their basic description.

A schematic of the basic parts of a CVD system is shown in Fig. 4.8. A CVD process involves having gaseous reactants admitted into the growth chamber (known as the reactor or reaction chamber) where a heated substrate is present. The gaseous reactants then undergo a chemical reaction at the various reaction zones near and on the surface of the substrate (Fig. 4.9) resulting in a solid material being deposited on the substrate and gaseous products that are driven out of the reactor. The major steps in a CVD process are precursor vaporization and transport; gas-phase reactions; mass transport to substrate; adsorption and surface diffusion; surface chemical reaction/nucleation; and desorption of by-products. The details of the gas flow dynamics and various chemical reactions in CVD have been reviewed extensively [52–54]. Figure 4.8 shows the three fundamental components of a CVD system: the gas delivery system, the reactor, and the exhaust system. The gas delivery system is normally custom designed for the specific set of precursors that one uses. Gaseous reactants are usually stored in gas bottles and the flow controlled by a pressure regulator and mass flow controller. For liquid or solid reactants, more complicated

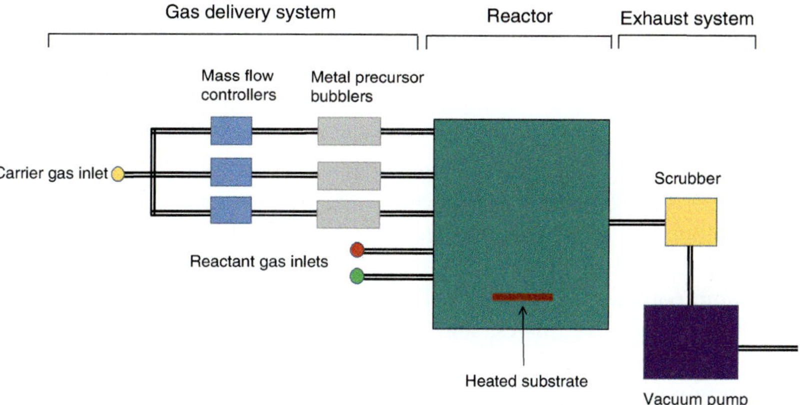

Fig. 4.8 Schematic diagram of a basic chemical vapor deposition system

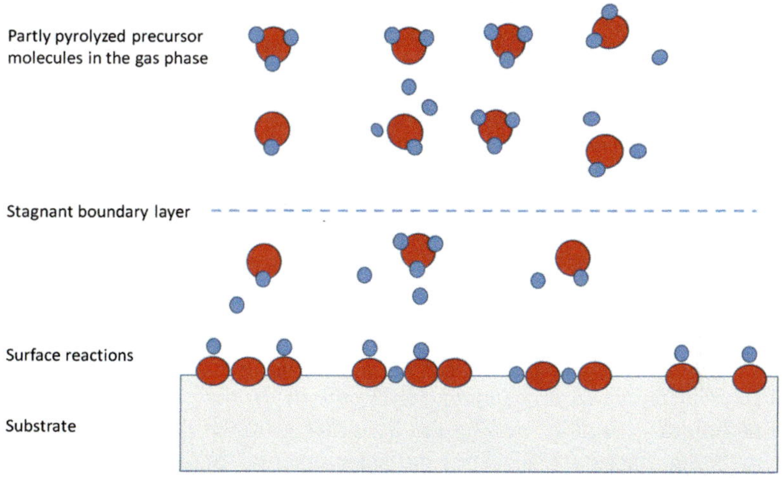

Fig. 4.9 The major reaction zones in chemical vapor deposition. Adapted from [1]

systems are needed. Typically, these reactants are heated to a sufficiently high temperature to evaporate the material, which is then picked up by an inert carrier gas flowing over or through the solid/liquid reactant. The reactor of a CVD system usually comes in two main types: a hot wall and a cold wall reactor. In a hot wall reactor, the heating elements are outside the reactor resulting in the substrate and reactor walls having the same temperature. In this case, deposition occurs not only on the substrate but also on the reactor walls. In a cold wall reactor, only the substrate is heated so there is no film deposition on the walls. The third and final component of a CVD system is the exhaust system, which includes the main vacuum pump, a means of controlling total pressure, and chemical scrubbers to remove or render inert the toxic, corrosive, or explosive by-products.

CVD precursor compounds are one of the most crucial elements in achieving a good CVD growth process. There are three characteristics of a good precursor for CVD. First, it must be sufficiently volatile (if not already gaseous) since vapor transport from the source to the substrate is necessary. Second, the precursor must decompose into the desired product at a reasonable temperature. Finally, the precursor should not undergo homogenous reactions (reactions with itself) in the gas phase. Over the last two decades, tremendous progress has taken place in the development of a wide variety of precursors for the growth of oxides, semiconductors, and metals. CVD precursors are usually hydrides, halides, or metal-organic compounds. Common ligands in metal-organic precursors include alkyls, alkoxides, β-diketonates, and amides. For a more detailed discussion of precursor chemistry, see [55]. For the growth of oxides, an oxidant is also usually needed. Common oxidants are molecular oxygen, ozone, or water vapor. Complex oxide growth using CVD is almost always of the MOCVD variety. MOCVD growth of ferroelectric, high-k dielectric, and superconducting oxides has been reported (see Sect. 4.7).

The advantages of CVD can be combined with MBE growth in a technique known as chemical beam epitaxy (CBE) or metal-organic molecular beam epitaxy (MOMBE). This technique was first utilized for the growth of III–V semiconductors where elemental Ga solid sources are combined with hydride As gaseous sources. With such a combination, adsorption-controlled growth of multi-component compounds is possible. Complex oxide growth using standard MBE is difficult because of the lack of line compositions so that precise flux matching between two metal fluxes as well as oxygen is needed. In the case of $SrTiO_3$, for example, elemental Ti has a very low vapor pressure even at temperatures exceeding 1,800 °C. This results in very slow growth rates of $SrTiO_3$ (<7 Å/min) and also being very prone to flux variations. By using a hybrid CVD-MBE method, one can then have a reaction-limited CVD-like growth mode rather than simply relying on arrival rates of atoms as in standard MBE growth. In the case of $SrTiO_3$, such a hybrid technique was pioneered by the Stemmer group at the University of California in Santa Barbara [56]. They utilize elemental Sr and the metal-organic compound titanium isopropoxide (TTIP) as the Ti source. At sufficiently high substrate temperatures, the TTIP molecules decompose at the substrate surface into TiO_2 + water + hydrocarbon fragments. The TiO_2 gets deposited while the volatile compounds get pumped away. By choosing the appropriate regime of TTIP/Sr flux ratio, one can reproducibly obtain a consistent Sr/Ti ratio in the STO film with very high growth rates and without compositional drift that can occur in MBE growth. In this surface reaction limited regime, the Sr/Ti ratio in the film is not affected by minor variations in the TTIP/Sr flux ratio. By using this MOMBE technique at a substrate temperature of 725 °C with an oxygen plasma at 5×10^{-6} Torr pressure, the Stemmer group has been able to deposit very high quality films of STO with lattice parameters exactly matching bulk crystals and XRD rocking curve width of $<0.01°$ at a growth rate of 34 Å/min [57]. A similar approach was used by Doolittle for the growth of $LiNbO_3$ from Li metal and $NbCl_5$ [5], and by King et al. for Y-Ba-Cu oxide using yttrium β-diketonate and elemental

Ba and Cu [58], where the Nb and Y sources used were CVD precursors due to the low volatility of the elemental materials. Hybrid CVD-MBE methods for other complex oxide systems are likely to be developed in the near future and may be key to manufacturability of these complex oxide materials.

4.5 Atomic Layer Deposition

Atomic layer deposition (ALD) is a special variant of the chemical vapor deposition process. The ALD technique involves alternating self-limited surface reactions of a precursor chemical with the substrate. For each ALD subcycle, a precursor is "pulsed" and carried to the surface of the substrate by a carrier gas. After the surface reaction saturates, the excess precursor and reaction by-products are purged by an inert gas, usually the same gas as the carrier gas. For the growth of metal oxides, a typical ALD cycle consists of alternating pulses of a metal precursor and a non-metal precursor (also known as the reactant). The growth process consists of many repetitions of this ALD cycle until the desired film thickness is achieved. Since the surface reactions usually saturate below a coverage of one monolayer, ALD allows for some degree of atomic layer control of the deposition similar to MBE. The basic principle of ALD is shown schematically in Fig. 4.10 where

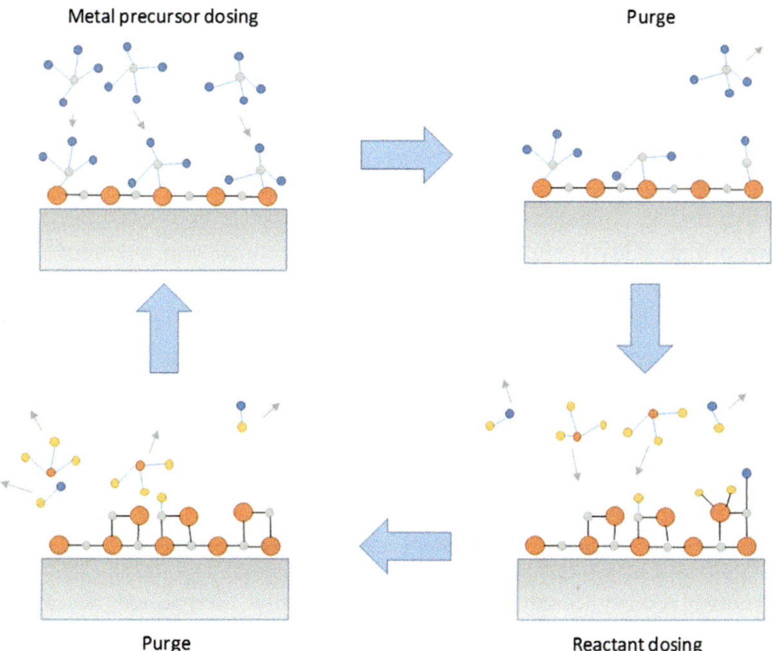

Fig. 4.10 Schematic diagram of a typical atomic layer deposition cycle

a cycle of the growth of a binary metal oxide is illustrated. The substrate is first exposed to the metal precursor in Step 1 until the surface reaction saturates. The excess reactants are then purged using an inert gas in Step 2. In Step 3, the substrate is then exposed to the non-metal precursor or reactant, which forms the metal oxide on the surface. Finally, in Step 4, another purge is performed driving off reaction by-products and excess reactant. The surface is ready for the next ALD cycle, which is repeated as needed. ALD was originally developed in Finland in the late 1970s as a means of obtaining highly uniform, pinhole-free coatings for electroluminescent displays [59]. Because of its very low growth rate, it was not until the last decade that ALD gained popularity as a deposition method due to its capability of depositing high-k dielectric materials in high aspect ratio geometries (conformal growth) without any pinholes [60]. ALD is now utilized for the deposition of materials in a wide variety of applications including micro-electromechanical, catalytic, magnetic, and optical applications. Details of the ALD process have been extensively reviewed by specialists in the field [61–63].

The unique capabilities of ALD are derived primarily from the chemistry of the precursors used. It is not sufficient that the precursors are delivered in alternating fashion. Only those precursors that result in self-limiting growth through rapidly saturating surface reactions are suitable for ALD growth. Self-limiting growth means that the same amount of material is deposited irrespective of the precursor dosing time, at least above a certain critical precursor dose. ALD precursors are generally chemically similar to those used for CVD so that these chemicals need to be volatile and should not etch or dissolve into the substrate. However, there are some additional requirements in order for a precursor to be suited for ALD. One requirement is that the precursor should not spontaneously decompose thermally at the growth temperature. This is important for the self-limiting nature of the process. Further, because of the alternating nature of precursor delivery, there is no danger of gas-phase reactions occurring between the precursors and so more reactive versions of precursor chemicals can be used. An illustration of self-limiting growth is shown in Fig. 4.11. The growth rate per cycle becomes constant above a certain precursor pulse time. Precursor decomposition or etching of the film will cause deviations in the growth rate as shown in Fig. 4.11. Hence, in a true ALD process, the film thickness is controlled only by the number of deposition cycles at a given temperature. One advantage of ALD over MBE then is that there is no need to precisely control the source fluxes but one can still obtain atomic level composition control with good reproducibility. A disadvantage, however, is that there is some unavoidable impurity incorporation from the precursor molecules and the carrier gas in ALD-grown films [62].

Most ALD reactors are of the continuous flow type where an inert gas (usually argon) is flowed at a fixed flow rate. The inert gas serves as both the carrier gas and the purge gas and is commonly flowed such that the reactor pressure is kept approximately constant in the range of 1–10 Torr. A typical ALD reactor consists of two main parts: the reaction chamber, and the precursor source manifold. ALD reaction chambers are of two general types. One is called the cross-flow and the

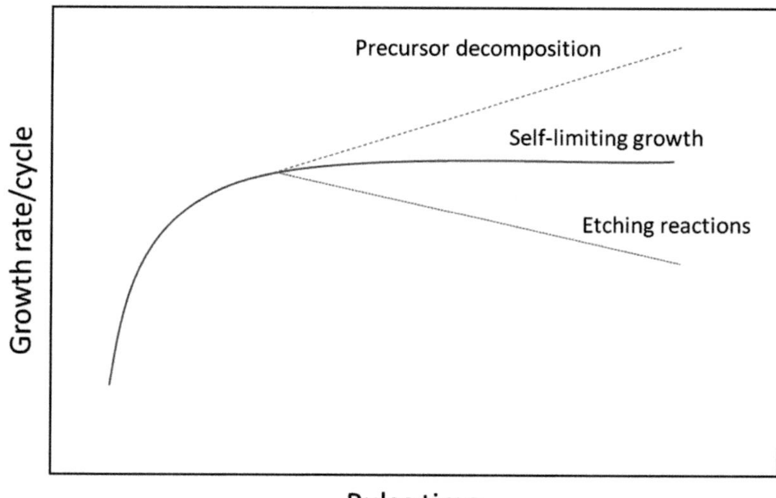

Fig. 4.11 An illustration of self-limiting growth. Beyond a certain dosing time, the film growth rate should saturate as a function of dosing time. Precursor decomposition or etching of the film by the precursor is likely occurring if the growth rate does not saturate. Adapted from [45]

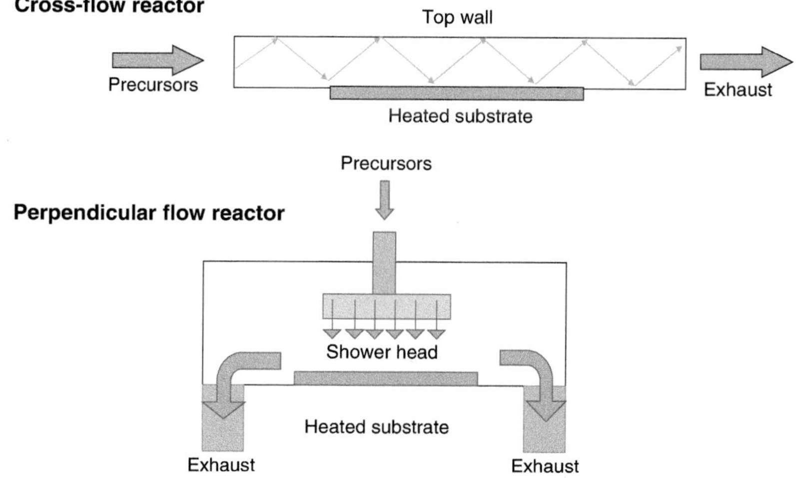

Fig. 4.12 Two main types of reactor geometries used for ALD. Adapted from [45]

other is the perpendicular-flow or showerhead type (Fig. 4.12). While cross-flow reactors benefit from being able to be purged relatively quickly, it is sensitive to flow non-idealities such as possible premature precursor decomposition and by-product re-adsorption, especially at sharp edges, leading to thickness and

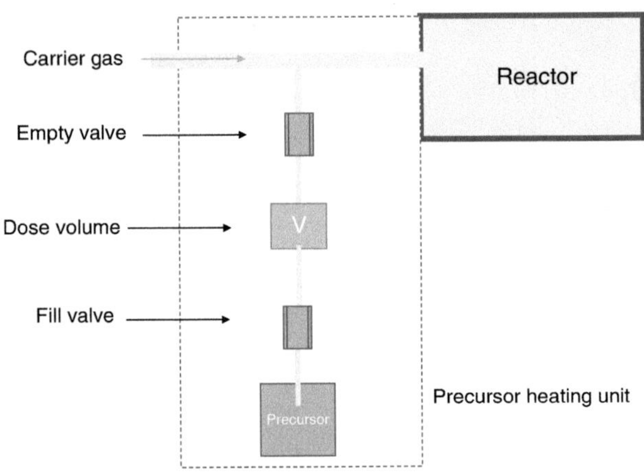

Fig. 4.13 Schematic of a method of introducing fixed doses of precursor into the reaction chamber. By opening the fill and empty valves sequentially, only a preset volume of precursor is introduced with each cycle. Adapted from [45]

composition non-uniformities [61]. ALD reaction chambers are also frequently operated in the hot wall mode, which prevents precursor condensation. The chamber wall is often the same temperature as the substrate hence resulting in film deposition on the walls as well. This means that periodic cleaning of the interior walls of an ALD reactor is necessary. The other key part of an ALD reactor is the precursor source manifold. Precursor sources for ALD are of two main types: mechanically valved high vapor pressure sources, and inert gas-valved low vapor pressure sources [61]. We will describe only the more common high vapor pressure source in this section. Precursors are contained in a vessel located outside the reaction chamber. The precursor vessel is connected to the carrier gas flow through a series of mechanical valves (Fig. 4.13) that allow a simple control of dosing. To prevent condensation of the precursor, the entire assembly must be heated. In a typical high vapor pressure source, the vapor pressure in the precursor container is higher than the pressure in the carrier gas lines. When the valves are pulsed open, the precursor vapor is simply drawn into the carrier gas and carried to the reaction chamber. Another method of transferring precursor material into the carrier gas is through the use of bubblers. In this case, the carrier gas is forced to go through or over the precursor compound. The use of bubblers allows for somewhat lower vapor pressure materials to still be used with mechanically actuated valves and without excessive heating of the precursors or gas lines.

ALD precursors are the key component that allow for self-limiting growth. This means that precursors must not thermally decompose but rapidly react with each other on the surface of the substrate or growing film. They must also be highly volatile to allow for delivery to the substrate with high fluxes as the dosing time is usually limited by vapor phase transport. The most common metal precursors for ALD come in seven major categories: halides, β-diketonate complexes, alkoxides,

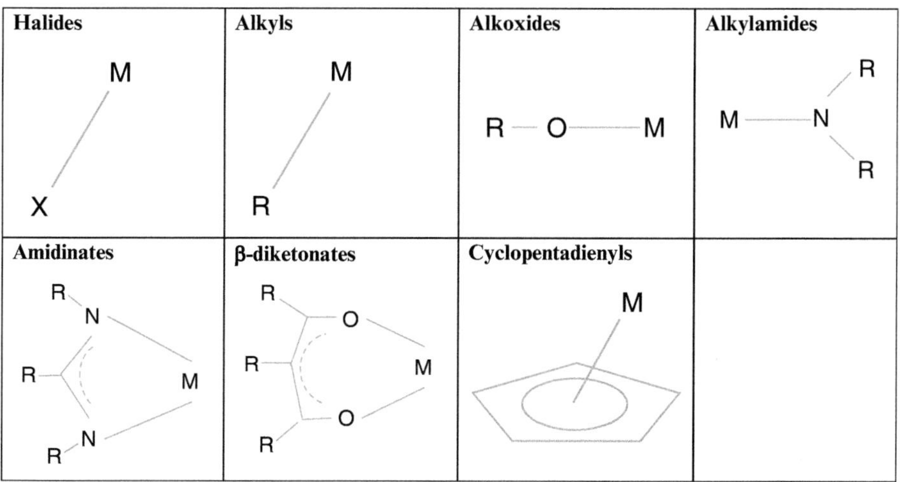

Fig. **4.14** The main classes of metal precursors used for ALD growth. *Note*: X = halogen,
R = alkyl group, M = metal

alkylamides, amidinates, metal alkyls and cyclopentadienyls. Figure 4.14 shows
the molecular structures of these common ALD precursor groups. Detailed
chemistries of the various ALD metal precursors are reviewed in [55]. For the
oxygen precursor, water is by far the most commonly used. For reactions needing a
stronger form of the oxidant, ozone and oxygen plasma are often used as well
[62]. In ALD growth, it is typical to first characterize the growth characteristics of a
single metal precursor in terms of film growth rate as a function of precursor dosing
time, purge time, and growth temperature.

The substrate temperature is one of the critical growth parameters in ALD.
The use of a too low temperature could result in either condensation of the precursor
or in the slowing down of the surface reaction. The use of a too high temperature on
the other hand could lead to spontaneous decomposition of the precursor. Finding
the temperature range where self-limiting growth occurs is critical. Tests to measure
growth rate as a function of dose time are also important. The growth rate at fixed
temperature should saturate as a function of dose time above a certain critical value.
If the growth rate continues to rise with temperature, precursor decomposition is
likely occurring and growth involves a CVD component. For some precursors, this is
unavoidable and one must carefully deal with the precursor decomposition.

ALD growth provides a way of depositing oxide thin films that are highly
uniform and conformal while also being pinhole free, making this technique
excellent for dielectric and ferroelectric layers. Because of its self-limiting nature,
growth is highly repeatable and readily scalable. The submonolayer saturation also
allows one to have atomic layer control and be able to grow artificial multilayer
materials without interdiffusion due to its low thermal budget.

4.6   The Growth of SrTiO₃ Thin Films

To give the reader a better understanding of the differences between the various methods and their relevant growth parameters, we discuss examples of the deposition of epitaxial $SrTiO_3$ (STO) thin films using each of the five deposition techniques discussed above. The examples are meant to highlight the differences in the relevant growth parameters for each of the methods and are not meant to show the best reported films for each method.

Brooks et al. reported on homoepitaxial growth of STO using MBE [64]. Growth was done at a substrate temperature of 650 °C under a background molecular oxygen pressure of 5×10^{-7} Torr. Some films were also grown using oxygen with a small amount of ozone (~10 %) although it was found that the presence of ozone did not yield any noticeable effect on the film crystalline quality. Sr was supplied from elemental Sr in an effusion cell while Ti was supplied from a Ti-Ball source. The STO growth was done using alternate monolayer dosing of SrO and TiO_2 to a total thickness of 250 unit cells (~100 nm). Shutter open times were based on initial measurements using a quartz crystal microbalance and optimized using RHEED intensity oscillations. The total growth rate was 6 Å/min corresponding to a shutter open time of 20 s. By carefully controlling the Sr/Ti flux ratio using analysis of the RHEED oscillation fine structure, the bulk lattice constant was achieved in the deposited film.

Lee and Koinuma reported the epitaxial growth of STO on Si using a TiN buffer layer by means of PLD [65]. Both the TiN and the STO layers were grown by PLD in a chamber with a base pressure of 2×10^{-7} Torr. For STO deposition, a commercial target was utilized. The surface of the target had to be ground by emery paper after each deposition as the laser causes the target surface to become conductive due to loss of oxygen. During the STO growth, the substrate was maintained at a temperature of 650 °C and the laser fluence used was 1–2 J/cm^2 (pulse width of 30 ns and repeat rate of 5 Hz). The ambient oxygen pressure was varied from high vacuum (~10^{-5} Torr) to 150 mTorr. When no additional oxygen was used, STO films were crystalline when deposited at a temperature above 400 °C with improved crystallinity at 550 °C. The x-ray diffraction rocking curve widths were found to be about 0.9°. Performing the deposition under very high oxygen ambient of 150 mTorr showed poorer film crystallinity with rocking curve width rising to 1.7°. X-ray pole figures confirmed that the film was epitaxial.

Deposition of epitaxial STO by RF magnetron sputtering was reported by Wang et al. on $LaAlO_3$ single crystal substrate [66]. The sputter deposition system had a base pressure of 4×10^{-7} Torr and was equipped with a sputtering gun with an off-axis geometry. The gun was placed 50 mm away from the substrate surface plane and 50 mm away from the substrate normal. For the process gas, a mixture of argon to oxygen in the ratio 3:1 was used with the total pressure maintained at 100 mTorr. The growth was done at a substrate temperature of 650–850 °C and a RF forward power of 80 W. The growth rate was determined to be 5.6 Å/min. After deposition, the chamber was backfilled with 0.5 atm O_2 while the sample was cooled at 4 °C/min.

The target used had a composition of Sr$_{1.1}$Ti$_{0.9}$O$_x$ to compensate for Sr loss during the sputtering process since Sr metal is somewhat volatile. Tests using a stoichiometric target showed 20 % Sr-deficiency and poor crystallinity in the deposited films. Films grown using the Sr-enriched target were stoichiometric and had x-ray diffraction rocking curve widths of 0.03° (same as the single crystal substrate).

Epitaxial STO was deposited by conventional MOCVD by a group at Northwestern University using a fluorinated Sr precursor that overcame the low volatility and thermal instability issues of the more commonly used CVD Sr precursor compound [67]. The precursors used were Ti(OPr^i)$_4$ (titanium isopropoxide) and Sr(hfac)$_2$(tetraglyme) (strontium hexafluoroacetylacetonate tetraglyme). The metal precursors were heated to 44 and 105 °C for the Ti and Sr precursors, respectively, and carried to the reactor using argon gas. Wet oxygen (oxygen bubbled through deionized water) was used as the oxidant. The oxygen flow rate was set at 50 sccm with the total argon + oxygen flow rate being 120 sccm. The reactor was maintained at a total pressure of 4.0 Torr during the deposition. The STO film was grown on a single crystal LaAlO$_3$ (001) substrate that was heated to 810 °C with growth rates ranging from 40 to 120 Å/min. The width of the rocking curves measured using x-ray diffraction was found to be as low as 0.36° indicating good crystalline quality. An alternative MOCVD approach was reported by Dubourdieu et al. using a so-called single-source precursor where a single compound contains both Sr and Ti in a 1:1 ratio [68]. The precursor they used was Sr$_2$Ti$_2$(OPr^i)$_8$(thd)$_4$ (thd = tetramethylheptanedionate). Using liquid injection MOCVD (with octane as the solvent), they were able to deposit crystalline STO at temperatures as low as 500 °C. Because the films were grown on Si, the STO films were not epitaxial but were textured polycrystalline (no in-plane epitaxy). The flow rates used were 300 sccm for oxygen and 300 sccm for the argon carrier gas with a working pressure in the range of 2–10 Torr.

The ALD growth of STO was first reported by Vehkamaki et al. in 2001 using glass substrates [69]. They used strontium tri-isopropyl cyclopentadienyl and titanium isopropoxide as the metal precursors, and water as the oxidant. While the Ti precursor behaved in self-limiting fashion, it was found that the Sr precursor underwent partial thermal decomposition. The Sr composition depended not only on the growth temperature but also on the Sr precursor dosing time. In order to prevent thermal decomposition of the Ti precursor, growth temperatures were limited to 325 °C or lower. Films grown with Sr/Ti precursors cycle ratios of slightly less than 1:1 (6:7 or 5:6) showed Sr/Ti ratios in the deposited film of 1:1. Pulse times for water and the Ti precursor were fixed at 0.6 s each. The Sr precursor pulse time needed to produce optimal stoichiometry was 0.2 s. The films became crystalline (textured polycrystalline) after a 500 °C air anneal although thick (>200 nm) films and films dosed with more water showed crystallinity as grown. Epitaxial films of STO on Si substrates were reported by McDaniel et al. in 2013 using an ultrathin MBE-grown STO buffer [70]. The precursors and carrier gas used were the same as Vehkamaki et al. but with longer pulse and purge times.

Pulse times were 2 s for Sr, 1 s for Ti, and 1 s for water, while purge times were 10–15 s long. Cycle ratios of 1:1 yielded stoichiometric films at the 250 °C growth temperature. The ALD reactor was maintained at a pressure of 1 Torr during the growth. The films were epitaxial as grown with x-ray diffraction rocking curve widths of 0.34°.

4.7   Survey of Complex Oxides Grown by Various Deposition Methods

In Table 4.2, we provide a survey of the different complex oxide materials systems grown by the five deposition techniques discussed above that have been reported in the literature. The table is limited to single phase materials based on the perovskite structure and that have been reported to be epitaxial or highly-oriented crystalline thin films.

Table 4.2 Survey of crystalline perovskite oxides that have been grown by each of the five different growth methods discussed

Material	MBE	PLD	Sputtering	MOCVD	ALD
$SrTiO_3$	✓	✓	✓	✓	✓
$BaTiO_3/(Ba,Sr)TiO_3$	✓	✓	✓	✓	✓
$CaTiO_3$	✓	✓	✓	✓	
$PbTiO_3/Pb(Zr,Ti)O_3$	✓	✓	✓	✓	✓
$GdTiO_3$	✓				
$Bi_4Ti_3O_{12}$	✓	✓	✓	✓	✓
$RScO_3$		✓	✓	✓	✓
$LaAlO_3$	✓	✓	✓	✓	✓
Vanadates	✓	✓			
$LiNbO_3$	✓	✓	✓	✓	✓
$Pb(Mg,Nb)O_3$		✓	✓	✓	
$SrBi_2Ta_2O_9$		✓	✓	✓	✓
Tantalates	✓	✓	✓	✓	
Bismuthates	✓	✓			
Chromates	✓	✓	✓		
$LaMnO_3/(La,A)MnO_3$	✓	✓	✓	✓	✓
Hexagonal $RMnO_3$	✓	✓	✓	✓	✓
Ferrates	✓	✓			✓
$BiFeO_3$	✓	✓	✓	✓	
MFe_2O_4	✓	✓	✓	✓	
$SrRuO_3$	✓	✓	✓	✓	
$LaCoO_3/(La,Sr)CoO_3$	✓	✓	✓	✓	✓
$RNiO_3$	✓	✓	✓	✓	✓
$YBa_2Cu_3O_x/Bi_2Sr_2Ca_2Cu_3O_x$	✓	✓	✓	✓	

Note: A = alkaline earth metal, M = transition metal, R = rare earth metal

References

1. D.L. Smith, *Thin-Film Deposition: Principles and Practice* (McGraw-Hill, New York, 1995)
2. P.M. Martin (ed.), *Handbook of Deposition Technologies for Films and Coatings: Science, Applications and Technology* (Elsevier, Amsterdam, 2010)
3. K. Seshan (ed.), *Handbook of Thin-Film Deposition Processes and Techniques: Principles, Methods, Equipment and Applications* (Noyes Publications/William Andrew Pub, Norwich, NY, 2002)
4. M. Ohring, *Materials Science of Thin Films: Deposition and Structure* (Academic Press, San Diego, CA, 2002)
5. W.A. Doolittle, A.G. Carver, W. Henderson, J. Vac. Sci. Technol. B **23**, 1272 (2005)
6. D. Dijkkamp, T. Venkatesan, X.D. Wu, S.A. Shaheen, N. Jisrawi, Y.H. Min-Lee, W.L. McLean, M. Croft, Appl. Phys. Lett. **51**, 619 (1987)
7. D.D. Berkley, B.R. Johnson, N. Anand, K.M. Beauchamp, L.E. Conroy, A.M. Goldman, J. Maps, K. Mauersberger, M.L. Mecartney, J. Morton, M. Tuominen, Y.-J. Zhang, IEEE Trans. Magn. **25**, 2522 (1989)
8. D.G. Schlom, J.H. Haeni, J. Lettieri, C.D. Theis, W. Tian, J.C. Jiang, X.Q. Pan, Mater. Sci. Eng. B **87**, 282 (2001)
9. A. Ohtomo, D.A. Muller, J.L. Grazul, H.Y. Hwang, Nature **419**, 378 (2002)
10. M. Henini (ed.), *Molecular Beam Epitaxy: From Research to Mass Production* (Elsevier, Oxford, UK, 2013)
11. R.F.C. Farrow (ed.), *Molecular Beam Epitaxy: Applications to Key Materials* (Noyes Publications, Park Ridge, NJ, 1995)
12. E. Parker, *The Technology and Physics of Molecular Beam Epitaxy* (Plenum Press, New York, 1985)
13. R. Sutarto, S. Altendorf, B. Coloru, M. Moretti Sala, T. Haupricht, C. Chang, Z. Hu, C. Schüßler-Langeheine, N. Hollmann, H. Kierspel, H. Hsieh, H.-J. Lin, C. Chen, L. Tjeng, Phys. Rev. B **79**, 205318 (2009)
14. C.E.C. Wood, D. Desimone, K. Singer, G.W. Wicks, J. Appl. Phys. **53**, 4230 (1982)
15. G.J. Davies, D. Williams, III-V MBE Growth Systems, in *The Technology and Physics of Molecular Beam Epitaxy*, ed. by Parker. (Plenum Press, New York, 1985)
16. Kometani, Wiegmann, *JVST* **12**, 933 (1975)
17. S. Hasegawa, S. Ino, Y. Yamamoto, H. Daimon, Jpn. J. Appl. Phys. **24**, L387 (1985)
18. P.G. Staib, J. Vac. Sci. Technol. B **29**, 03C125 (2011)
19. P. Staib, W. Tappe, J.P. Contour, J. Cryst. Growth **201–202**, 45 (1999)
20. "In situ cathodoluminescence," SVT Associates Application Note. No. 1101 (1998), http://www.svta.com/uploads/documents/CL_Ap_Note_1101.pdf
21. "kSA Bandit blackbody temperature measurement," k-Space Associates Technology Overview Note (1998), http://www.k-space.com/wp-content/uploads/kSA_BandiT_Blackbody.pdf
22. UVISEL In-Situ Spectroscopic Ellipsometer, Product Overview. Horiba Scientific, http://www.horiba.com/us/en/scientific/products/ellipsometers/in-situ-and-in-line/uvisel-in-situ/uvisel-in-situ-spectroscopic-ellipsometer-3718/
23. D.M. Mattox, *Handbook of Physical Vapor Deposition Processing* (Noyes Publications, Westwood, NJ, 1998), pp. 273–275
24. R.E. Honig, D.A. Kramer, RCA Review **30**, 285 (1969)
25. "Vapor Pressure Charts," International Union for Vacuum Science, Technique and Applications, http://iuvsta.org/iuvsta2/index.php?id=643
26. "Thin Film Evaporation Guide," Vacuum Engineering and Materials Co., Inc., http://www.vem-co.com/sites/default/files/pdfs/VEM_Thin_Film_Evaporation_Guide.pdf
27. "Deposition Techniques," Kurt J. Lesker Company, http://www.lesker.com/newweb/deposition_materials/MaterialDeposition.cfm?pgid=0
28. D.B. Chrisey, G.K. Hubler (eds.), *Pulsed Laser Deposition of Thin Films* (Wiley, New York, 1994)

29. R. Eason (ed.), *Pulsed Laser Deposition of Thin Films: Applications-Led Growth of Functional Materials* (Wiley-Interscience, Hoboken, NJ, 2007)
30. M.N.R. Ashfold, F. Claeyssens, G.M. Fuge, S.J. Henley, Chem. Soc. Rev. **33**, 23 (2004)
31. D.B. Chrisey, G.K. Hubler (eds.), *Pulsed Laser Deposition of Thin Films* (Wiley, New York, 1994). Chap. 6, pp. 167–198
32. D.B. Chrisey, G.K. Hubler (eds.), *Pulsed Laser Deposition of Thin Films* (Wiley, New York, 1994). Sec. 2.2, pp. 24–38
33. D.B. Chrisey, G.K. Hubler (eds.), *Pulsed Laser Deposition of Thin Films* (Wiley, New York, 1994). Chap. 9, pp. 225–264
34. M.G. Norton, C.B. Carter, Physica C **172**, 47 (1990)
35. K. Wasa, M. Kitabatake, H. Adachi, *Thin Film Materials Technology: Sputtering of Compound Materials* (William-Andrew, New York, 2004)
36. D.M. Mattox, *Handbook of Physical Vapor Deposition Processing* (Noyes Publications, Westwood, NJ, 1998). Chap. 6, pp. 315–377
37. S.L. Rohde, in *Sputter Deposition*, ed. by ASM International Handbook Committee. ASM Handbook, vol. 5: Surface Engineering (ASM International, Materials Park, OH 1994)
38. P. Sigmund, Sputtering by Ion Bombardment: Theoretical Concepts, in *Topics in Applied Physics*, vol. 47: Sputtering by Particle Bombardment I (Springer-Verlag, Berlin, 1981)
39. J.H. Keller, W.B. Pennebaker, IBM J. Res. Dev. **23**, 3 (1979)
40. J.A. Thornton, J. Vac. Sci. Technol. **15**, 171 (1978)
41. M. Stepanova, S.K. Dew, J. Vac. Sci. Technol. A **19**, 2805 (2001)
42. J. Musil, P. Baroch, J. Vlček, K.H. Nam, J.G. Han, Thin Solid Films **475**, 208 (2005)
43. W. Gawalek, W. Michalke, H. Bruchlos, T. Eick, R. Hergt, G. Schmidt, Phys. Status Solidi (a) **109**, 503 (1988)
44. C.B. Eom, J.Z. Sun, B.M. Lairson, S.K. Streiffer, A.F. Marshall, K. Yamamoto, S.M. Anlage, J.C. Bravman, T.H. Geballe, S.S. Laderman, R.C. Taber, R.D. Jacowitz, Physica C **171**, 354 (1990)
45. A.C. Jones, M. Hitchman (eds.), *Chemical Vapor Deposition: Precursors, Processes and Applications* (RSC Publishing, Cambridge, 2009)
46. G.B. Stringfellow, *Organometallic Vapor Phase Epitaxy*, 2nd edn. (Academic Press, New York, 1999)
47. M.L. Hitchman, K.F. Jensen, *Chemical Vapor Deposition* (Academic Press, New York, 1993)
48. A.D. Berry, D.K. Gaskill, R.T. Holm, E.J. Cukauskas, R. Kaplan, R.L. Henry, Appl. Phys. Lett. **52**, 1743 (1988)
49. G.J.M. Dormans, P.J. van Veldhoven, M. de Keijser, J. Cryst. Growth **123**, 537 (1992)
50. A.C. Jones, M. Hitchman (eds.), *Chemical Vapor Deposition: Precursors, Processes and Applications* (RSC Publishing, Cambridge, 2009). Chap. 12
51. A.C. Jones, M. Hitchman (eds.), *Chemical Vapor Deposition: Precursors, Processes and Applications* (RSC Publishing, Cambridge, 2009). Chap. 11
52. A.C. Jones, M. Hitchman (eds.), *Chemical Vapor Deposition: Precursors, Processes and Applications* (RSC Publishing, Cambridge, 2009). Chap. 3
53. G.B. Stringfellow, in *Crystal Growth of Electronic Materials*, ed. by E. Kaldis (Elsevier, Amsterdam, 1985)
54. P.M. Martin (ed.), *Handbook of Deposition Technologies for Films and Coatings: Science, Applications and Technology* (Elsevier, Amsterdam, 2010). Chap. 7
55. A.C. Jones, M. Hitchman (eds.), *Chemical Vapor Deposition: Precursors, Processes and Applications* (RSC Publishing, Cambridge, 2009). Chap. 5
56. B. Jalan, P. Moetakef, S. Stemmer, Appl. Phys. Lett. **95**, 032906 (2009)
57. B. Jalan, R. Engel-Herbert, N.J. Wright, S. Stemmer, J. Vac. Sci. Technol. A **27**, 461 (2009)
58. L.L.H. King, K.Y. Hsieh, D.J. Lichtenwalner, A.I. Kingon, Appl. Phys. Lett. **59**, 3045 (1991)
59. T. Suntola, J. Hyvarinen, Annu. Rev. Mater. Sci. **15**, 177 (1985)
60. M.T. Bohr, R.S. Chau, T. Ghani, K. Mistry, IEEE Spectrum **44**, 29 (2007)

61. M. Ritala, M. Leskela, in *Handbook of Thin Film Materials*, vol. 1, ed. by H.S. Nalwa (Academic Press, New York, 2002), pp. 103–159
62. A.C. Jones, M. Hitchman (eds.), *Chemical Vapor Deposition: Precursors, Processes and Applications* (RSC Publishing, Cambridge, 2009). Chap. 4
63. C.H.L. Goodman, M.V. Pessa, J. Appl. Phys. **60**, R65 (1986)
64. C.M. Brooks, L.F. Kourkoutis, T. Heeg, J. Schubert, D.A. Muller, D.G. Schlom, Appl. Phys. Lett. **94**, 162905 (2009)
65. M.B. Lee, H. Koinuma, J. Appl. Phys. **81**, 2358 (1997)
66. X. Wang, U. Helmersson, L.D. Madsen, I.P. Ivanov, P. Münger, S. Rudner, B. Hjörvarsson, J.-E. Sundgren, J. Vac. Sci. Technol. A **17**, 564 (1999)
67. S.R. Gilbert, B.W. Wessels, D.B. Studebaker, T.J. Marks, Appl. Phys. Lett. **66**, 3298 (1995)
68. C. Dubourdieu, H. Roussel, C. Jimenez, M. Audier, J.P. Sénateur, S. Lhostis, L. Auvray, F. Ducroquet, B.J. O'Sullivan, P.K. Hurley, S. Rushworth, L. Hubert-Pfalzgraf, Mater. Sci. Eng. B **118**, 105 (2005)
69. M. Vehkamäki, T. Hänninen, M. Ritala, M. Leskelä, T. Sajavaara, E. Rauhala, J. Keinonen, Chem. Vap. Depos. **7**, 75 (2001)
70. M.D. McDaniel, A. Posadas, T.Q. Ngo, A. Dhamdhere, D.J. Smith, A.A. Demkov, J.G. Ekerdt, J. Vac. Sci. Technol. A **31**, 01A136 (2013)

Chapter 5
Thin Oxide Film Characterization Methods

In this chapter, we give a very brief overview of some of the more commonly used techniques for routinely characterizing the structural, electronic, and chemical properties of epitaxial thin films. This chapter is primarily designed to introduce thin film growth practitioners to the most common characterization methods available and what types of information can be obtained from them. More in-depth treatments of each technique can be found in specific references mentioned within each section of this chapter. Table 5.1 gives a list of the methods discussed in this chapter, summarizing the information that can be obtained with each technique, as well as the limitations of the technique. For a more detailed description of these and other thin film characterization techniques, see the texts by Czichos [1], Martin [2], and Woodruff and Delchar [3]. We do not include in this chapter techniques that look at specific physical properties such as mechanical, electrical/thermal transport, and magnetic properties. A good review of such techniques can also be found in [1].

5.1 Electron Spectroscopies

In the various methods classified as electron spectroscopy, the energy and/or momentum of electrons ejected from a material, as a result of being excited by either photons or an electron beam, is analyzed [4]. There are three major categories of electron spectroscopy methods: x-ray photoelectron spectroscopy (XPS), ultraviolet photoelectron spectroscopy (UPS) including angle-resolved photoemission spectroscopy (ARPES), and Auger electron spectroscopy (AES). In XPS and UPS (Fig. 5.1a), high energy photons are used as the excitation source, which can be either soft x-rays (~1–2 keV) or vacuum or extreme ultraviolet radiation (<50 nm wavelength). In AES, on the other hand (Fig. 5.1b), a high energy (~3–20 keV) electron beam is used to excite electrons in the sample. XPS and AES are both

A.A. Demkov and A.B. Posadas, *Integration of Functional Oxides with Semiconductors*, DOI 10.1007/978-1-4614-9320-4_5, © The Author(s) 2014

Table 5.1 Summary of the more commonly used thin film characterization techniques

Technique	Probing beam	Detected quantity	Information obtained	Limitations/notes
XPS	X-rays	Photoelectrons	Composition, chemical state, thickness	Limited to outer 10 nm sampling depth; depth profiling possible
UPS	UV photons	Photoelectrons	Work function, valence band spectrum, surface states	Limited to a few nm sampling depth
ARPES	UV photons	Photoelectrons	Band structure; surface states	In-plane momentum-resolved UPS
AES	Electrons (~5 keV)	Secondary electrons	Composition, chemical state	Limited to outer 10 nm sampling depth; can be done with very high spatial resolution (sub-100 nm)
XRD	X-rays	Scattered x-rays	Film crystal structure and phase; film orientation/texture; film strain	Bulk sensitive; not suitable for amorphous or ultrathin (<4 nm) films
XRR	X-rays	Reflected x-rays	Film thickness, surface roughness, density	Limited to smooth films with thickness ranging from 2 to 200 nm; complicated information instruction for multiple layers
SEM	Electrons (~10 keV)	Secondary electrons	Imaging of microstructure; elemental mapping	Electron beam can damage surface; elemental analysis is only semi-quantitative
TEM	Electrons (~100 keV)	Transmitted electrons (both elastic and inelastic)	Heterostructure layer structure; crystalline structure; elemental mapping; chemical state (EELS); atomic-resolution	Complicated sample preparation that can introduce artifacts; image contrast strongly dependent on imaging conditions
LEED	Electrons (~100 eV)	Diffracted electrons (backscattered)	Surface crystal structure; surface reconstructions	Surface sensitive; atomic position information possible (LEED I–V)
RHEED	Electrons (~10 keV)	Diffracted electrons (forward scattered)	Surface crystal structure; surface reconstructions	Real-time measurement during growth possible; requires high vacuum (not suitable for CVD, sputtering); more complicated interpretation compared to LEED
PL	Photons (optical to UV)	Secondary photons	Luminescence properties; overall sample quality (defect density); concentration of optically active defects (e.g. oxygen vacancies); bandgap	Not suitable for materials that do not show luminescence; cannot detect electronic states that relax non-radiatively

	Probe	Signal detected	Information obtained	Comments
Raman	Photons (near IR to UV)	Photons (inelastically scattered)	Vibronic spectrum; chemical bond strength; phase identification; disorder analysis	Sample luminescence can interfere; signal is very weak; high spatial resolution possible
Ellipsometry	Photons (near IR to UV)	Photons (elastically scattered)	Thickness; density; dielectric constant; refractive index	Complicated analysis for thick multilayers; real time in situ measurement possible; strongly model dependent analysis
FTIR	Photons (IR)	Reflected photons	Vibronic spectrum; film thickness; adsorbate identification	Very weak signal
ISS	Ions (He at 1 keV)	Scattered ions	Surface composition; surface termination	Very surface sensitive; semi-quantitative only
MEIS	Ions (He at 100 keV)	Scattered ions	Depth-resolved elemental composition; some structural information possible	Modeling difficult for light elements on heavy substrates; limited to ~100 nm sampling depth
RBS	Ions (He at 1 MeV)	Backscattered ions	Elemental composition as a function of thickness	Poor depth resolution; absolute compositions possible; destructive
SIMS	Ions (Ar at 20 keV)	Secondary ions	Depth-profile of elemental composition with high spatial resolution	High sensitivity to concentration but no information on chemical state
STM	Sharp tip in close proximity to surface	Tunneling current	Surface roughness; atomically-resolved density of states of surface	Requires ultrahigh vacuum and very smooth and clean surface for atomic resolution; does not work on insulating substrates; tip artifacts possible
AFM	Sharp tip in close proximity to surface	Force gradient between tip and surface	Surface roughness; mapping of surface potential, magnetic domains, ferroelectric domains	Can work with insulating substrates; atomic resolution possible under certain conditions; tip can be modified to sense other forces such as magnetic

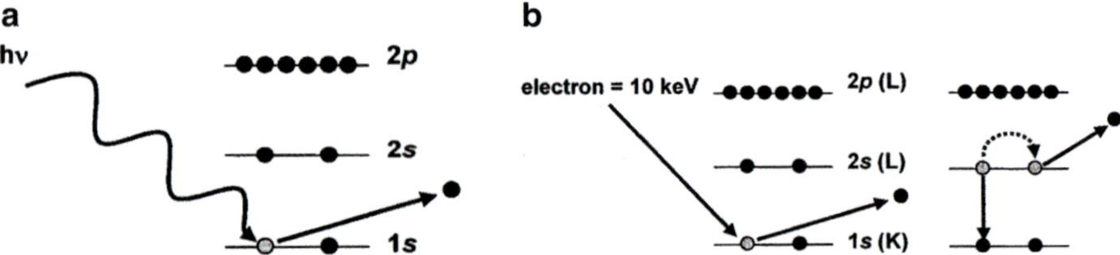

Fig. 5.1 (**a**) Basic principle of photoemission. A photon (x-ray photons in the case of XPS) is absorbed by a 1s core electron. If the photon has sufficient energy, that core electron will be photoejected from the atom with the excess energy transformed into kinetic energy of the photo-electron. If the x-ray energy is known, the binding energy of the core electron can be determined by measuring the kinetic energy of the photoelectron. (**b**) Basic principle of the Auger relaxation process. A core hole (K-shell) is filled by an electron from a higher level orbital (L-shell). The excess energy is used to eject a third electron (usually one with the same principal quantum number as the relaxing electron). The kinetic energy of the Auger electron is independent of the excitation energy. The ejected electron shown is known as a KLL Auger electron

Fig. 5.2 X-ray photoelectron spectrum of the Nb 3d core level of a mixed niobium oxide thin film. There are two sets of peaks, with each set consisting of a spin-orbit split pair ($3d_{5/2}$ and $3d_{3/2}$) separated by 2.7 eV. One set comes from NbO (203.8 and 206.5 eV binding energy) while another set comes from NbO_2 (206.8 and 209.5 eV binding energy). XPS is useful for identifying oxidation states and quantifying their relative amounts

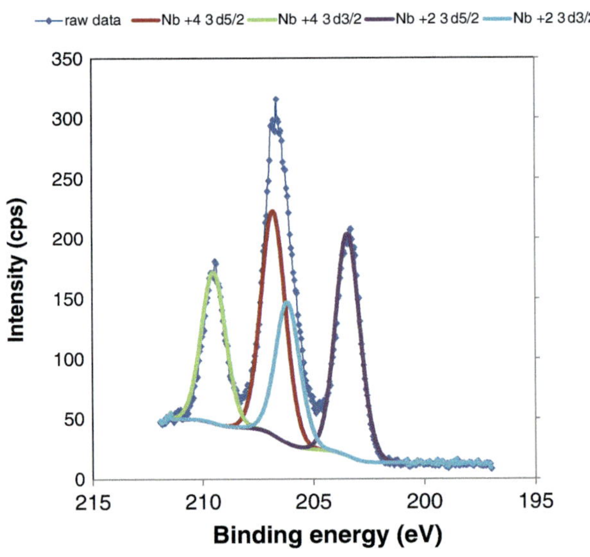

routinely used to determine chemical composition and chemical state of the sample. Each element produces a characteristic spectrum with features appearing at specific energies with a fixed relative intensity. This is then used to determine what chemical species are present in the sample. By using theoretical or empirical atomic sensitivity factors, XPS and AES can also be used for quantitative analysis of the chemical composition (Figs. 5.2 and 5.3). In both techniques, all elements except hydrogen and helium can be analyzed [5]. All electron spectroscopy methods rely on the detection of an electron ejected from the solid. Because electrons scatter very strongly, only electrons near the surface manage to escape the solid and reach the

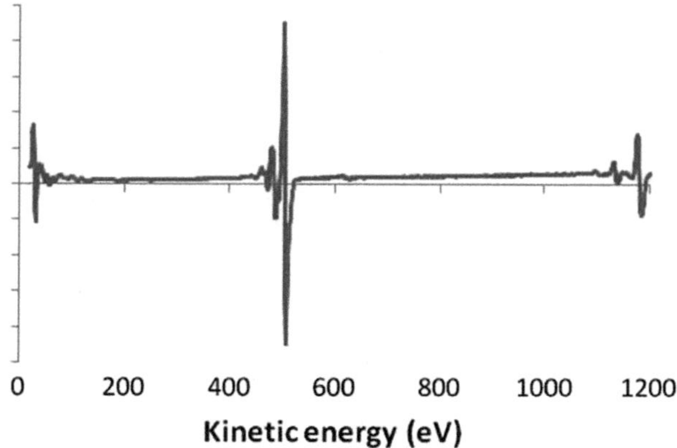

Fig. 5.3 Auger electron spectrum of MgO thin film grown on a GaN/sapphire substrate. The O KLL feature is located around 508 eV kinetic energy while the Mg KLL and LVV features are at 1,185 eV and 43 eV, respectively. Auger spectroscopy is a quick method of determining the surface composition of thin film materials

detector. The effective sampling depth follows an exponential decay curve with the characteristic length being closely related to the inelastic mean free path of the outgoing electron. This mean free path is somewhat universal, depending primarily on the kinetic energy of the outgoing electron and only weakly on the specific material. Typical inelastic mean free paths range from less than 10 Å for a kinetic energy of 200 eV and about 30 Å for a kinetic energy of 1.5 keV [6, 7]. This makes XPS and AES surface sensitive techniques. In the case of UPS, the photons only interact with the valence electrons and are mainly used to study the effects of surface contamination and adsorbates on the electronic structure of the sample. It is also utilized as a means of measuring the work function of a material by determining the kinetic energy cutoff at which electrons are no longer able to escape the solid.

When using UPS, one can gain additional information about the band structure of the sample by using angle-resolved photoemission (ARPES). In ARPES, the sample is rotated through both the polar and azimuthal directions while energy distribution curves are taken (Fig. 5.4). The two angles can be reconstructed into the two-dimensional coordinates in reciprocal space, allowing a 2D band mapping of the sample. By tuning the photon energy, one can also tune the value of k_z, the wavevector normal to the surface, allowing for 3D band structures to be measured as well. ARPES is often used to study the nature of surface electronic states in well-ordered materials and to map out the band structure of a material [8].

By utilizing an ion gun to progressively sputter away the top layers of a sample, one can use XPS to determine the composition and chemical state as a function of depth. This method is known as depth profiling and is routinely available with most modern commercial instruments. The disadvantage of this technique is that it is destructive by necessity and it is also possible for the energetic ions from the ion gun to interact with the sample resulting in a change in the chemical state. For sensitive samples and if non-destructive measurements are needed, an alternative to

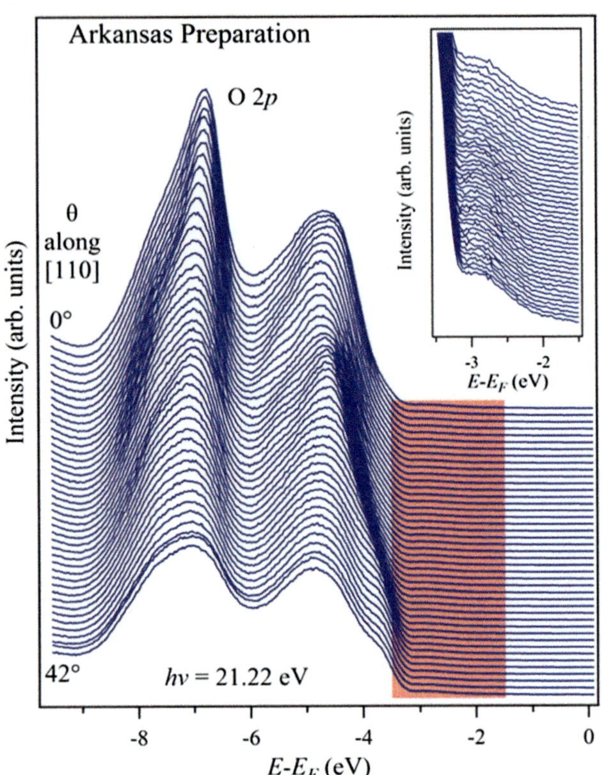

Fig. 5.4 Angle-resolved photoemission energy distribution curves for a series of emission angles θ along the [110] direction of a TiO_2-terminated $SrTiO_3$ single crystal surface. A close-up of the *shaded region* is shown in the *inset* and shows a mid-gap state about 800 meV above the top of the valence band that has very little dispersion, which has been attributed to a Ti-H state. From [11]

depth profiling is to use angle-resolved XPS (ARXPS). This is only suitable for depth profiling of relatively thin layers, about 8–10 nm. By collecting XPS data as a function of emission angle, the composition profile of a sample can be determined from how the various XPS peak areas change with angle [9].

An electron spectroscopy technique (XPS, UPS, AES) has three main elements—the excitation source, the electron energy analyzer (including the detector), and the sample stage. All of these are housed in an ultrahigh vacuum chamber. In many analysis chambers, there are additional equipment for further sample characterization (e.g. LEED) and for sample cleaning (heating stage, sputtering gun). We will discuss the excitation source and analyzer in more detail in the following (Fig. 5.5).

For XPS, the excitation source is an x-ray anode. The emitted x-rays are often monochromated for high resolution work. In laboratory XPS systems, there are two x-ray lines that are commonly used: Mg Kα (1,253.6 eV) and Al Kα (1,486.6 eV). The natural line widths of these lines without monochromation is about 0.7 eV for Mg Kα and 0.85 eV for Al Kα, which is usually sufficient for chemical species identification. However, for high resolution work such as deconvolution of core levels and line shape analysis, even narrower line widths of the photon source are needed. Diffraction from a curved quartz crystal surface is commonly utilized to suppress the Kα_2 line of Al as well as the bremsstrahlung background and satellites,

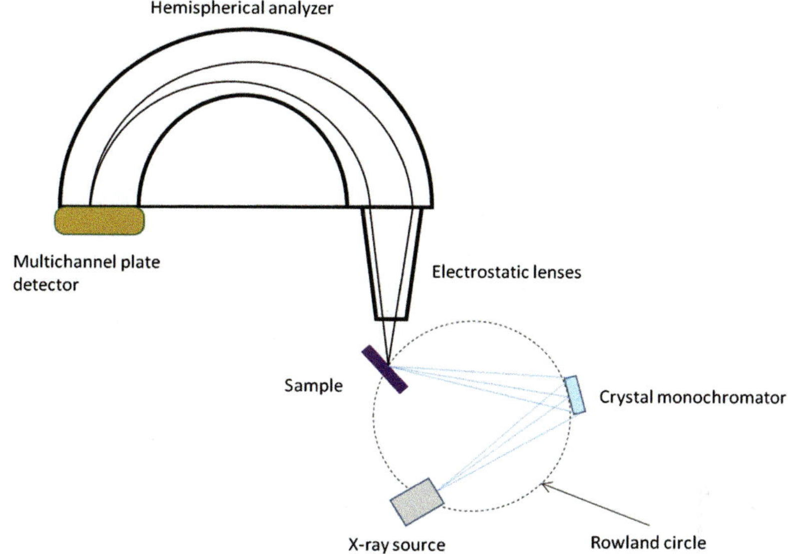

Fig. 5.5 Basic schematic of an x-ray photoelectron spectroscopy system showing fundamental components including the x-ray source, quartz crystal monochromator, sample holder/manipulator, electrostatic lenses, hemispherical analyzer, and multichannel plate detector

yielding monochromated line widths on the order of 0.2 eV for Al [10]. The x-ray anode and monochromator are sometimes separated from the main analysis chamber by a thin (1–2 μm) Al foil. This thin foil provides vacuum isolation between the x-ray anode chamber, which has higher background pressure from the anode, and the analysis chamber, while allowing x-ray photons to go through. The x-ray anode is typically operated at a power of a few hundred watts, resulting in a photon flux on the order of 10^{12} per second for typical source to sample geometries in laboratory XPS systems.

For UPS, the most commonly used photon source is a rare gas discharge lamp (Ar, Ne, or He). The most commonly used lines are He I (21.22 eV), He II (40.82 eV), Ne I (16.85 eV), Ne II (26.9 eV), Ar I (11.83 eV), and Ar II (13.48 eV) [10]. All these lines have satellites and it is often useful to have a UV grating that allows one to select a particular line without its satellite for easier data analysis. The natural line widths of these lines are only a few meV making them very useful for detailed valence band analysis and band structure measurements using ARPES. Because of the high pressure needed (~1 Torr) to sustain the gas discharge, a combination of low conductance paths (capillaries) and multi-stage differential pumping is often employed to create the pressure gradient between the UV source and the analysis chamber. Using electron cyclotron resonance generated He plasma with a retractable capillary allowing for a working distance of ~1 cm, a photon flux of ~10^{16} per second has been achieved by VG Scienta in their VUV 5000 source [12]. This high photon flux in combination with 2D detectors described below, allow for the recording of spectra in a matter of seconds.

For AES, the excitation source is a beam of electrons. Electrons are generated by an electron gun. The electrons can be scanned and focused on specific areas of the sample to a width of 10–20 nm allowing for elemental mapping of the surface. The most common electron gun used in AES is a tungsten cathode filament in the shape of a hair pin that is resistively heated, similar to that used in LEED or RHEED. However, the lateral resolution of an electron beam from this type of electron gun is poor. For elemental mapping, LaB_6 cathodes or field emission guns similar to those used in advanced TEM machines are utilized instead. The electron guns have electrostatic or both electrostatic and magnetic lenses to provide the ability for beam steering and beam shaping [13].

The electron energy analyzer is the second key element of an electron spectroscopy analysis system (Fig. 5.5). The analyzer determines the number of electrons of a given kinetic energy leaving the sample. The analyzer filters out electrons with energies outside the window of interest and directs them to the detector, which determines the electron count rate. The energy window is then scanned through the energies of interest. There are two common methods for selecting the kinetic energy of electrons resulting in two electron analyzer geometries—the concentric hemispherical analyzer (CHA) and the cylindrical mirror analyzer (CMA) [14]. The CMA is most often used in AES systems since that geometry allows for the collection of electrons on the same side as the primary electron beam. In most cases, the primary electron gun is located on the central axis of the CMA. The inner cylinder is usually grounded and the outer cylinder is at some negative potential that causes electrons of a particular kinetic energy to be directed and focused onto the detector at the backside of the CMA. For better resolution, a double pass CMA is sometimes used. The CHA is more commonly found in XPS/UPS systems. It works on a similar principle where the inner hemisphere and outer hemisphere are at different negative potentials, with the magnitude of the outer hemisphere potential being larger. The median equipotential line between the spheres is aligned to the path of incoming photoelectrons that have been accelerated/decelerated to a particular pass energy. CHAs generally have higher energy resolution than CMAs making them the analyzer of choice for ARPES, even though the transmittance is lower than a CMA. After passing through the electron energy analyzer, the selected electrons are directed to an electron detector, usually a simple channeltron or a microchannel plate detector (MCP). These basically count the electrons that are able to pass through the analyzer thereby providing a count rate for a particular kinetic energy [3, 14].

5.2   X-ray Scattering

There are two techniques based on the scattering of incident x-rays that are routinely used for thin film analysis: x-ray diffraction (XRD) and x-ray reflection (XRR). X-ray diffraction is mainly used to determine the lattice constants, degree of misorientation, level of strain, and presence of secondary phases in thin films. X-ray reflection, on the other hand, is used to determine film thickness and surface/

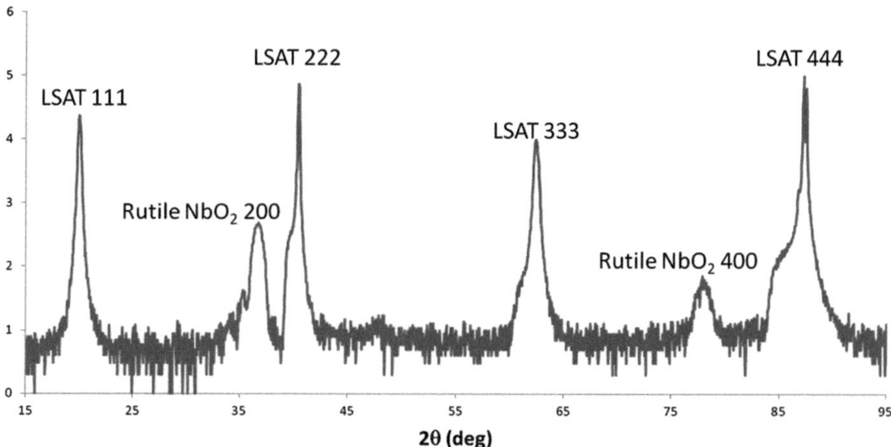

Fig. 5.6 Symmetric $2\theta - \theta$ scan of rutile NbO_2 on $La_{0.2}Sr_{0.8}Al_{0.6}Ta_{0.4}O_3$ (LSAT) substrate. Only the $h00$ peaks from the film are visible, indicative of a single crystallographic orientation

interface roughness. A good review of x-ray diffraction techniques as applied to epitaxial films can be found in [15]. In XRD/XRR, a beam of x-rays (most commonly Cu Kα radiation) is directed at a well-aligned sample. The angle between the sample and the x-ray source is then scanned through, with the detector angle locked to be twice the sample-source angle. This is known as the symmetric $2\theta - \theta$ scan. In such a scan, peaks corresponding to the lattice spacings perpendicular to the sample surface are observed in the scan (Fig. 5.6). A single phase, single-crystalline film should only have peaks from the substrate and from one orientation of the film material. The presence of extra peaks signifies the presence of secondary phases and orientations that are normally considered detrimental for epitaxial layers. A second type of scan routinely performed on epitaxial thin films is the rocking curve scan (Fig. 5.7). A rocking curve is performed by fixing the detector at an angle 2θ corresponding to a Bragg peak in the film. The sample is then scanned about the ω axis around a value ω = θ. This produces a peak with a full width at half maximum that is often used as a measure of the overall degree of crystalline order in the film. Very highly ordered films should produce a rocking curve width that is limited by the substrate. For thin oxide films grown on single crystal oxide substrates as measured by laboratory XRD systems, the rocking curve width of the film and substrate is around 0.05° for a high quality film. Non-epitaxial but highly oriented films tend to have rocking curve widths of several degrees. In order to definitely show epitaxial growth, one has to perform an azimuthal or φ scan, with the sample and detector set to an angular position corresponding to an off-normal Bragg peak (Fig. 5.8). For perovksite materials, a commonly used off-normal peak is the {103} or the {113} reflection. An epitaxial film should exhibit peaks at only specific φ angles that are symmetry related to similar off-normal peaks from the substrate. Oriented polycrystalline films show continuous rings with or without intensity modulation about the φ axis.

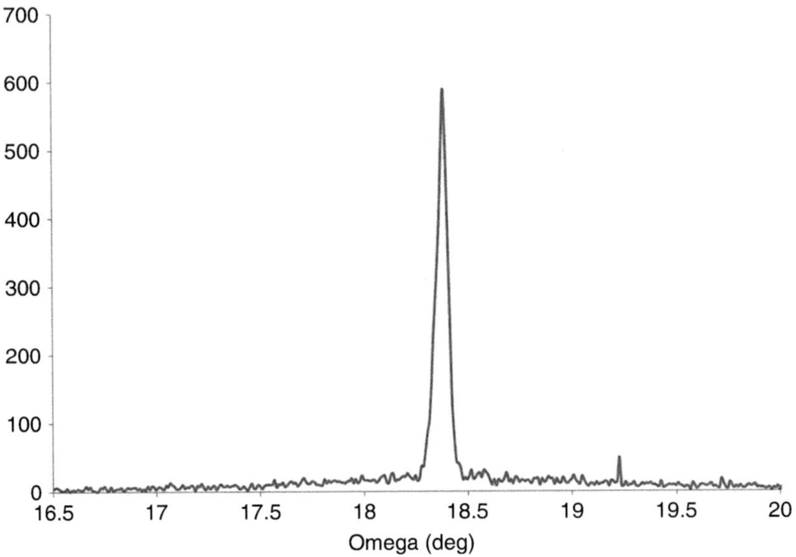

Fig. 5.7 Rocking curve scan about the 200 peak of epitaxial rutile structure NbO_2. The full width at half-maximum of the peak is 0.07° (250 arcseconds), which is identical to that of the underlying $La_{0.2}Sr_{0.8}Al_{0.6}Ta_{0.4}O_3$ (LSAT) substrate, indicating high crystalline quality of the film

Fig. 5.8 X-ray diffraction ɸ-scans of pulsed laser deposited $YMnO_3$ on GaN (0001). A 30° offset between the $11\overline{2}2$ peaks is clearly visible showing a relative in-plane rotation between the unit cells of the film and substrate. Reprinted from [16], with permission from Elsevier

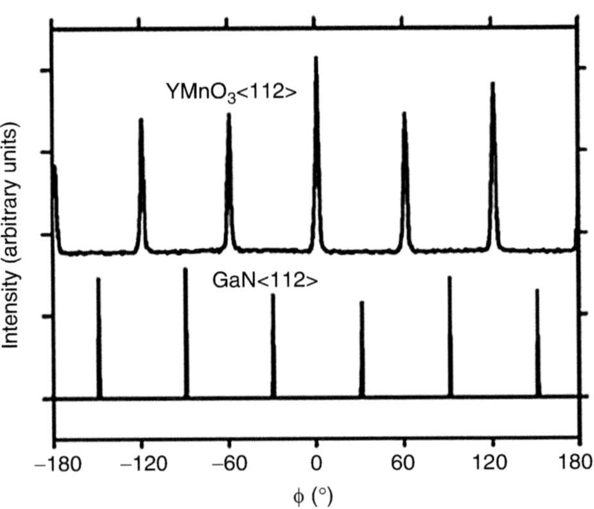

XRR is similarly a symmetric $2\theta - \theta$ scan but done with parallel beam optics and at very small grazing angles (<3° is typical). Because of the large fraction of reflected x-rays at low angles, the interference of x-rays reflected at the various interfaces in the thin film structure result in the appearance of oscillations in the reflected x-ray intensity. The period of the oscillation is directly related to the

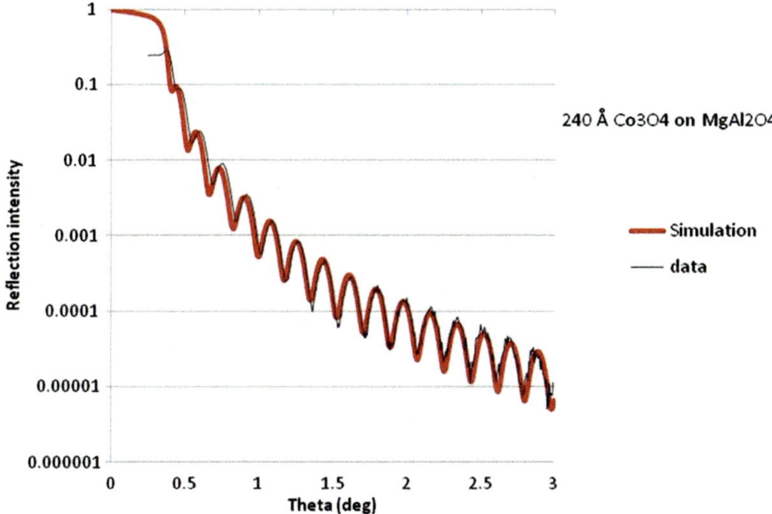

Fig. 5.9 X-ray reflectivity scan of a 240 Å epitaxial Co_3O_4 film grown on $MgAl_2O_4(110)$ substrates using molecular beam epitaxy. The *thin black curve* is the raw data while the *thick red curve* is a simulation based on ideal layers. The excellent agreement between the two curves shows the structural perfection of the surface and interface

thickness of the film (Fig. 5.9). For multilayer structures, there will be several oscillation frequencies observed and some "beating" of the reflected intensity is observed. By using a kinematic model, one can determine the thicknesses of each layer in the structure provided there is sufficient contrast in the electron densities of the different layers. The overall envelope of the reflectivity signal also provides information on the surface and interface roughness of the heterostructure. The reflectivity signal gets attenuated exponentially by the mean roughness and by modeling the reflectivity with an exponential factor, the layer roughnesses can be determined from how fast the envelope is decaying.

5.3 Electron Microscopy

There are two main types of electron microscopies: scanning electron microscopy (SEM) and tunneling electron microscopy (TEM). For a more detailed treatment, see [17] for TEM and [18] for SEM. In SEM, the image is mainly produced by inelastically scattered secondary electrons from a focused electron beam (0.1–20 keV), which is rastered or scanned across the sample. With secondary electron imaging, surface features can be imaged at up to $800,000\times$ magnification without the need for extensive sample preparation. As the electrons can also excite the atoms in the sample causing it to produce characteristic x-rays, SEM is also used for elemental composition determination using a technique known as EDX or energy dispersive x-ray spectroscopy (Figs. 5.10 and 5.11).

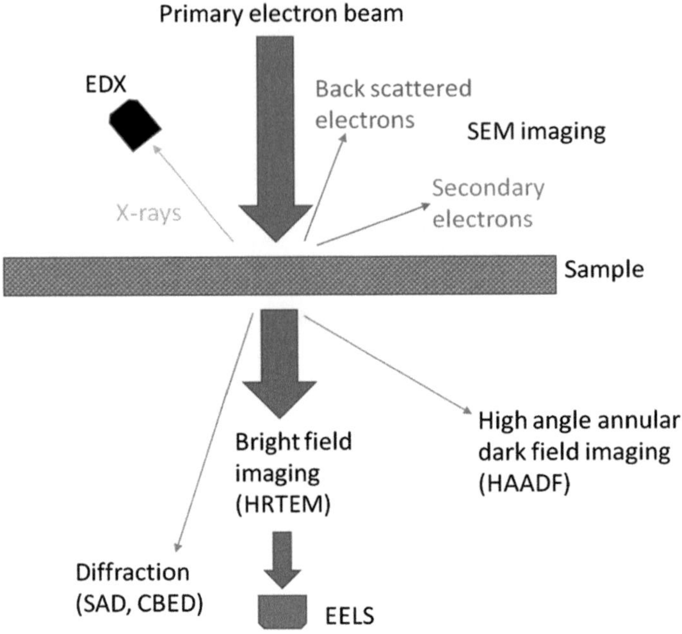

Fig. 5.10 The primary imaging and spectroscopic techniques that can be performed in electron microscopy. SEM imaging uses the secondary and backscattered electrons from the surface of the sample. EDX measures the energy spectrum of x-rays emitted as a result of excitation by the primary electrons. Elastically scattered electrons are detected in an annular detector to form dark field (Z-contrast) images. Elastically scattered electrons are also utilized to form diffraction patterns. The transmitted electrons can be imaged to form bright field images and the energy loss spectrum of the transmitted electrons can also be measured

Fig. 5.11 Scanning electron micrograph of a multilayer polycrystalline film of $SrTiO_3$ and $BaTiO_3$ grown on platinized silicon substrates. Reprinted from [19], with permission from Elsevier

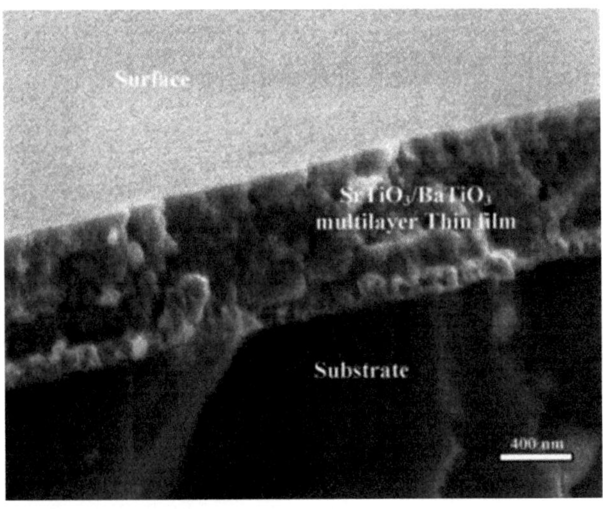

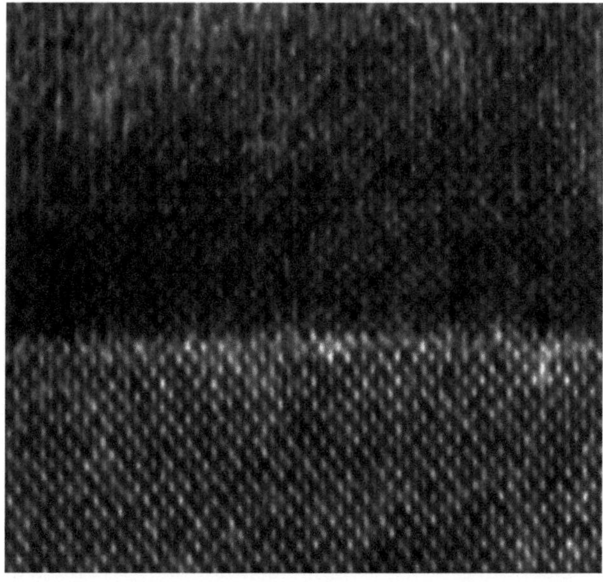

Fig. 5.12 High resolution cross-section transmission electron micrograph of epitaxial BaTiO$_3$ (*upper layer*) grown directly on Ge (*lower layer*). Image courtesy of David Smith (Arizona State University)

The second type of electron microscopy is TEM. TEM uses electron kinetic energies of 60–300 keV. These high energies result in an electron wavelength of less than 0.003 nm, making very high resolution images down to the atomic level possible. The practical resolution is much less than this theoretical limit due to various aberrations in the electron optical system. Depending on how the electron beam is generated/manipulated and what kind of signal is collected, several different sub-techniques of TEM are possible (Fig. 5.10). Bright field imaging of elastically scattered transmitted electrons can be used for low magnification imaging similar to SEM. When only diffracted electrons are imaged, a dark field image results and one obtains the so-called selected area diffraction (SAD) pattern. Transmitted electrons are also used to produce high resolution lattice images (HRTEM). Some instruments are able to use a focused electron beam that is then scanned across the sample resulting in the technique known as scanning transmission electron microscopy (STEM). STEM can be used for bright field imaging but is more commonly used for imaging of electrons scattered at higher angles. This is known as a STEM-HAADF image (high-angle annular dark field). In STEM-HAADF imaging, the brightness of the image is proportional to the square of the atomic number and is sometimes known as Z-contrast TEM [20]. If inelastically scattered electrons are analyzed, one can obtain electronic structure and chemical information on the sample using the technique known as electron energy loss spectroscopy (EELS). In combination with STEM, EELS can be used for atomic resolution mapping of the unoccupied density of states of a material [21]. EELS is often used to verify first principles calculations of electronic structure. The EDX method is also widely used in TEM measurements. For thin films, most TEM work is done in a cross-section geometry where a TEM sample is prepared from a cross-sectional slice of the thin film heterostructure (Fig. 5.12).

5.4   Electron Diffraction

Electron diffraction utilizes the wave nature of electrons and their interaction with atoms on the surface of a material to determine the crystal structure of that surface. There are two main electron diffraction techniques: LEED or low energy electron diffraction [22] and RHEED or reflection high energy electron diffraction [23]. Conventional geometries for LEED and RHEED are shown in Fig. 5.13. As a crystal structure analysis technique, it differs from x-ray diffraction in two important aspects. First, electrons strongly scatter and so the penetration depth is very typically on the order of one to five monolayers as opposed to the micron level penetration depths of x-rays. Second, electron beams can only be operated in vacuum conditions because the residual gas would also scatter the electrons resulting in diffuse images.

Both LEED and RHEED are hence used only for analyzing the crystal structure of surfaces. LEED uses electron kinetic energies of the order of ~100 eV resulting in escape depths of about 1 nm. LEED has traditionally been used for studying surface reconstructions and adsorbate geometry. A typical LEED system consists of an electron gun, which is incident normally on the sample, a set of hemispherical grids for focusing and to deflect inelastically scattered electrons, and a phosphor screen at high positive voltage (~1 kV) for imaging. Because of the surface sensitivity of LEED, it is often used qualitatively to verify surface cleanliness and to measure surface reconstructions (Fig. 5.14). LEED can also be used in the so-called LEED I–V mode where intensity as a function of electron energy is determined. With such information, it is possible to determine the atomic structure of a surface by use of calculations accounting for multiple scattering.

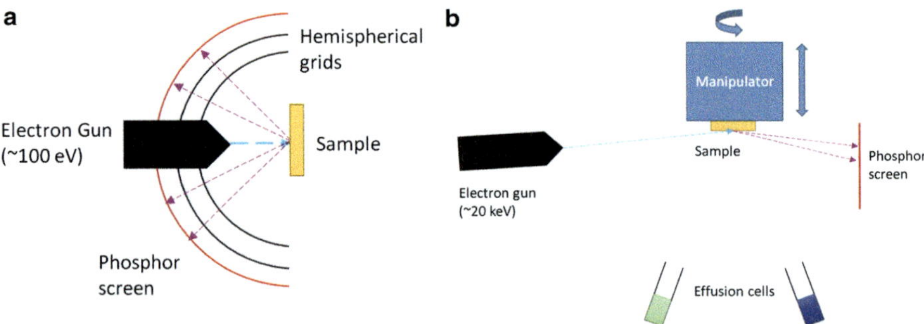

Fig. 5.13 (**a**) Basic geometry of low energy electron diffraction (LEED). The electron gun is situated behind at the center of a phosphor screen. A series of hemispherical grids to filter out inelastically scattered electrons is located in front of the screen. (**b**) Basic geometry of reflection high energy electron diffraction (RHEED). The electron gun and screen are nearly opposite each other from the sample. The nearly horizontal geometry of RHEED does not interfere with growth allowing for real time measurements during growth. RHEED requires electron beam deflection control and sample rotation/height adjustment

Fig. 5.14 Low energy
electron diffraction pattern
for MgO(111) grown on
GaN(0001) taken at an
electron energy of 96.2 eV.
The sixfold symmetry of the
film is clearly visible

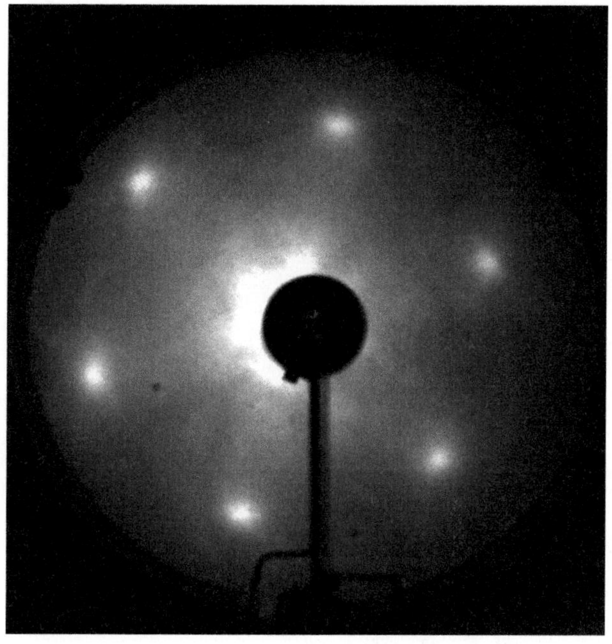

RHEED, on the other hand, is used primarily only in a qualitative way to monitor epitaxial growth in real time. Both the surface sensitivity and real time growth imaging capability of RHEED arises from its glancing incidence (typically $1°-3°$). Even though electron energies in RHEED are high (~10–30 keV), the grazing incidence results in a momentum component perpendicular to the surface of the sample that is a very small fraction of the total momentum. The near horizontal setup also allows real time monitoring during growth because it does not block the line of sight between the evaporation source and the substrate. RHEED is tradiitionally utilized together with MBE. More recently, it is being used with PLD as a result of the advent of differential pumping. RHEED can not only determine surface crystal structure, but can also be used in several other ways. One is the use of RHEED oscillations [23] where the intensity of the specular spot is monitored as an epitaxial film is grown in layer by layer fashion. The spot intensity is highest when the layer is complete and lowest when it is about half complete. By plotting the RHEED intensity oscillations, one can precisely count how many layers of material have been deposited. RHEED intensity oscillations have also been used to calibrate individual elemental fluxes for multicomponent oxides such as $SrTiO_3$ [24] and $YBa_2Cu_3O_x$ [25]. RHEED can also be qualitatively used to determine the growth mode of the film, whether flat layer by layer or island growth. Figure 5.15 shows some schematic RHEED patterns for various types of surfaces. For atomically flat surfaces, one sees a pattern of spots arranged along a circular arc (the zeroth Laue zone) (Fig. 5.16). For a nominally flat surface but with short lateral terrace widths that are uncorrelated, these Laue zone spots spread out into streaks.

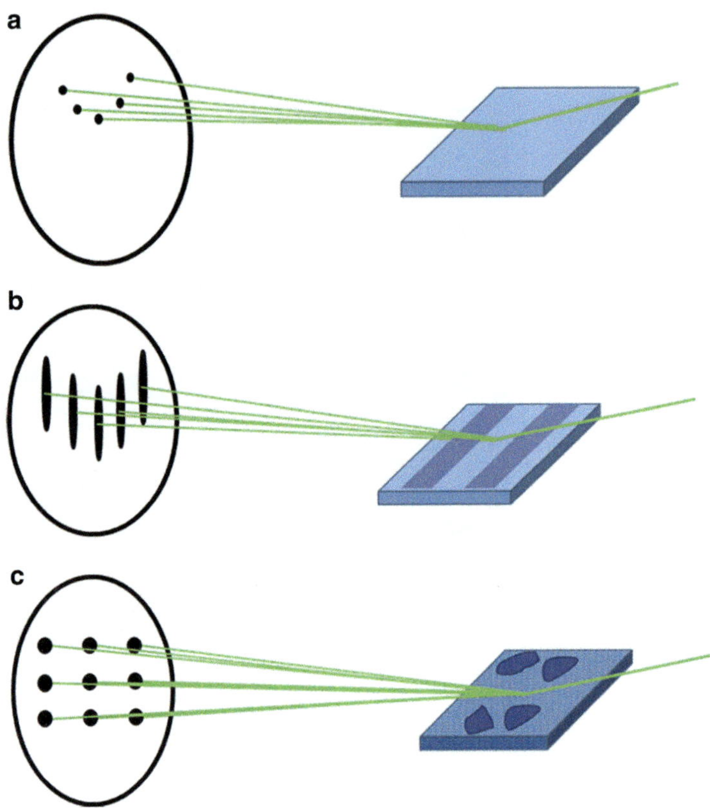

Fig. 5.15 Characteristic RHEED patterns observed from surfaces of varying degrees of flatness. (**a**) Atomically flat surface where the flat regions are wider than the coherence length of the incident electrons. (**b**) Flat surface but with uncorrelated terraces with widths smaller than the coherence length. (**c**) Rough surface with islands showing transmission diffraction pattern

Fig. 5.16 Reflection high energy electron diffraction pattern from 0.25 monolayer of Sr metal deposited on Si(100) at 550 °C. The pattern shows a combination of a $2\times$ and $3\times$ reconstruction

When the surface is rough enough that some electrons are transmitted through the islands, a two dimensional array of diffraction spots appears. A more detailed treatment of RHEED patterns can be found in the review by Chambers [26].

5.5   Optical Characterization

The two most commonly used optical characterization techniques used for functional oxide materials is photoluminescence (PL) and spectroscopic ellipsometry. Other techniques that are classified as optical include Raman spectroscopy and Fourier transform infrared spectroscopy (FTIR).

Raman spectroscopy (Fig. 5.17a) utilizes the inelastic scattering of monochromatic light when it interacts with phonons (Stokes scattering). Raman signals are intrinsically weak and necessitates the use of an intense laser. Because lasers can be focused to a small area, it is possible to perform mapping with a resolution at the micron level (micro-Raman spectroscopy). Raman spectroscopy is sensitive to anything that affects vibrational frequencies in a material including bond length, strength, and geometry. Micro-Raman can also be used to probe thin film heterostructures by varying the photon energy to change the penetration depth of the light. A good reference on Raman spectroscopy is Lewis and Edwards [30].

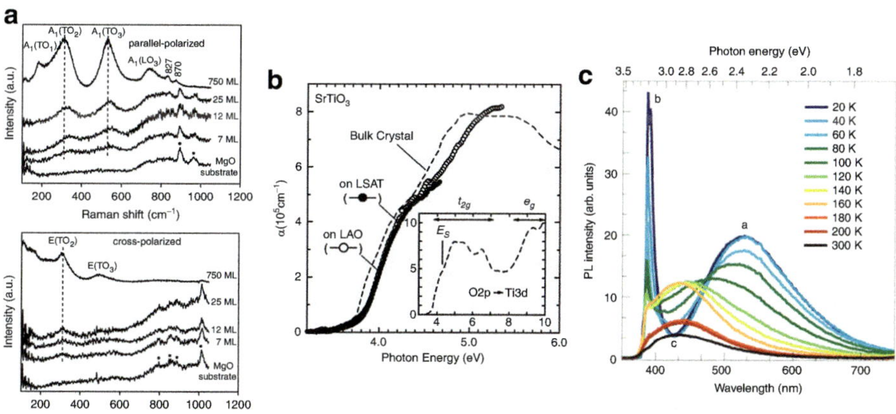

Fig. 5.17   Representative data of the common optical thin film characterization methods as applied to oxides. (**a**) Raman spectra of $BaTiO_3$ films of various thicknesses grown on MgO using parallel-polarized (*top*) and cross-polarized (*bottom*) configurations. The A_1 modes are observed to be blue shifted compared to bulk material and was attributed to stress in the film. Reprinted from [27], with permission from Elsevier. (**b**) Room-temperature optical absorption spectrum of $SrTiO_3$ films grown on $LaAlO_3$ and LSAT substrates, compared with bulk single crystal $SrTiO_3$. The films show a slightly higher absorption edge than the bulk material. Reprinted from [28], with permission from Elsevier. (**c**) Changes in the photoluminescence spectrum of Ar^+-irradiated $SrTiO_3$ as a function of temperature from 20 to 300 K. The broad feature at low temperatures is attributed to a self-trapped exciton. The sharp peak around 400 nm is due to recombination via an oxygen vacancy level. Reprinted from [29] by permission from Macmillan Publishers Ltd

Infrared spectroscopy is the study of the absorption or reflection spectrum of a material of photons in the infrared regime (0.04–0.6 eV). These energies are not enough to excite electronic levels in most materials and only affect vibrational and rotational states. Because IR light is of low intensity, the technique is often coupled with a beam splitter and interferometer and the use of Fourier transform to speed up data collection. FTIR can be used to determine film thickness, particularly dielectric thin films on semiconductors. It can also be used to identify surface adsorbate species. For more details, the reader is referred to the book by Bell [31].

Ellipsometry involves the analysis of polarized light reflected from the surface of a material (Fig. 5.17b). The reflected intensity is controlled by the frequency-dependent complex refractive index and thicknesses of the various layers in the sample. Depending on what properties of the sample are known, ellipsometry can be used to measure thickness, composition, surface roughness, electrical conductivity, particularly if the measurement is done as a function of wavelength and/or angle of incidence. Ellipsometry analysis requires the use of models whose parameters are adjusted to best fit the data. Two parameters are usually varied and the variation in the amplitude and phase of the reflected light is measured. Its geometry also allows it to be used in situ during growth as with RHEED. Several commercial ellipsometry setups for MBE systems are also available. A more detailed treatment of the use of ellipsometry in thin film analysis can be found in the review by Theeten and Aspnes [32] and in the textbook by Tompkins and Gahan [33].

Photoluminescence is the process by which a material absorbs photons of a given energy and then re-emits photons usually at lower energies. Photoluminescence analysis looks at the energy distribution of the absorbed and emitted photons as well as the time and temperature dependent characteristics of the spectra (Fig. 5.17c). This technique allows one to directly look at the electronic transition energies of the sample, especially the bandgap. PL is often used as a quick measure of general sample quality in deposited semiconductor layers because of its high sensitivity to defects and relative ease of measurement. In combination with near field scanning optical microscopy techniques, mapping of the photoluminescence properties with 100 nm resolution has been achieved. A detailed treatment of photoluminescence as applied to surfaces and interfaces of thin films can be found in the book chapter by Zobiesierski [34].

5.6   Ion Spectroscopies

Ion beams are also commonly used to study thin films and surfaces. There are two main categories of ion beam characterization depending on whether one looks at the energy and momentum of elastically scattered ions or at the species and flux of ejected secondary ions. The first of these techniques is broadly classified as ion scattering spectroscopy and is further categorized by the kinetic energy range used for the incident ions (Fig. 5.18).

For ion energies on the order of 0.5–10 keV using He ions, the technique is known as low energy ion scattering (LEIS). This technique is extremely surface

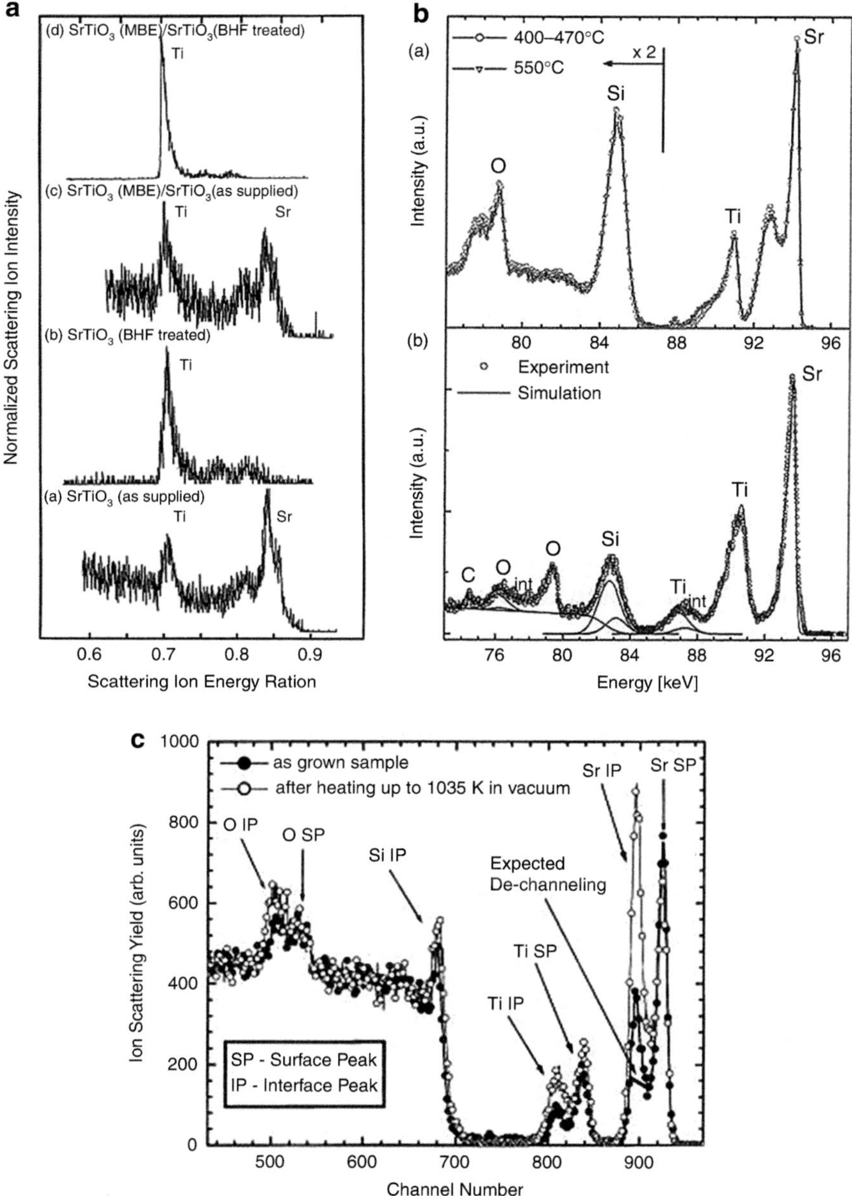

Fig. 5.18 Representative data of various ion scattering spectroscopies as applied to SrTiO$_3$. (**a**) Low energy ion scattering (LEIS or ISS) of SrTiO$_3$ substrates and homoepitaxial films to determine surface termination. The effectiveness of buffered HF treatment in producing a TiO$_2$-terminated surface can be seen. Reprinted from [35], with permission from Elsevier. (**b**) Medium energy ion scattering (MEIS) energy distribution for 20 unit cells SrTiO$_3$ on Si. The data were measured using 98–130 keV H$^+$ ions in a backscattering collection geometry. The presence of a thin Ti-Si reaction layer best fits the data. Reprinted with permission from [36]. Copyright 2006, AIP Publishing LLC. (**c**) Rutherford backscattering (RBS) measurement of 40 nm SrTiO$_3$ on Si. The sample was measured before (as-grown) and after annealing at 760 °C in ultrahigh vacuum. The disordering of the interface is clearly observable from the increase in Ti and Sr interface-related peaks. Reprinted with permission from [37]. Copyright 2002, AIP Publishing LLC

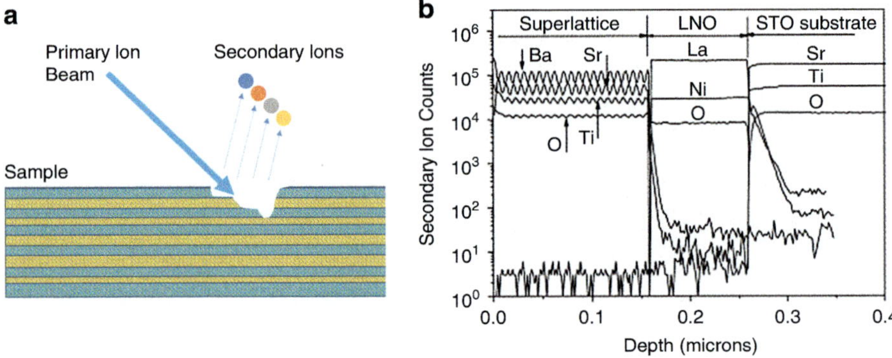

Fig. 5.19 (**a**) Schematic of the principle of secondary ion mass spectroscopy (SIMS). A primary ion beam (typically Ar^+ with kinetic energy ~20 keV) is incident on the sample and causes sputtering of the sample at a particular rate. A small fraction (~5 %) of these sputtered particles is in the form of ions. These secondary ions are collected and mass analyzed to determine the composition of the sample as a function of depth. (**b**) SIMS measurement of a 20-repeat $BaTiO_3$ (4 nm)/$SrTiO_3$ (4 nm) superlattice grown on a 100-nm conducting $LaNiO_3$ epitaxial layer on a $SrTiO_3$ single crystal substrate. Reprinted from [41], with permission by Elsevier

sensitive as ions are quickly neutralized due to their relative low energy. LEIS is commonly used to determine surface composition and surface termination of single crystals, and has also been used for surface structure determination with the advent of time of flight methods. For more detailed information see [38]. In order to study thin film heterostructures, a higher ion energy is used leading to the technique known as medium energy ion scattering (MEIS). This uses He or H ions with kinetic energies on the order of 100–300 keV and the energies and angles of the backscattered ions are measured. A full analysis of MEIS data yields information on the mass, depth, and even structural information on the atoms in the sample. MEIS can achieve depth resolutions of one atomic layer near the surface region in many materials. Structural information is obtained when spectra are taken as a function of polar and azimuthal angle. A good review of MEIS is one by Gustafsson [39]. At even higher energies (0.5–2 MeV), the technique is known as high energy ion scattering (HEIS), of which there are several methods. The most common of these is Rutherford back scattering (RBS), which is used to analyze composition and structure and thickness of thin film heterostructures. By using channeling or aligning to well-defined crystallographic directions and comparing to random angle of incidence, structural information on the sample can also be obtained. RBS is useful because of its ability to quantify in an absolute sense the stoichiometry and thickness of each layer in a heterostructure. See Feldman [40] for further information.

The second mode of incident ion characterization techniques is to collect secondary ions generated from the material as a result of the impacting primary ions. This technique is commonly known as secondary ion mass spectroscopy (SIMS) and as its name implies primarily looks at the masses and amounts of the secondary ions (Fig. 5.19). The incident ions in SIMS are usually 10–30 keV argon, oxygen or cesium ions. SIMS is widely used in the semiconductor industry to monitor impurities in semiconductor materials because it has the capability to measure in the parts per million (ppm) or even in the parts per billion (ppb) sensitivity. The primary ions are energetic enough to sputter away the material being analyzed, with a small fraction of the sputtered material being charged and so can be mass resolved in a mass spectrometer. One obvious disadvantage of SIMS is that it is destructive as the sample is sputtered away by the incident ions. A newer development is to use a very low ion current in pulsed mode and to use a time of flight mass spectrometer to analyze the secondary ions. One benefit of TOF-SIMS is compositional mapping of the sample with 100 nm lateral resolution. TOF-SIMS by itself, however, cannot do depth profiling like traditional SIMS but can be made to do so if used in combination with a second ion beam for sputtering. This method is not as sensitive as traditional dynamic SIMS though. For more information on SIMS, see the comprehensive text by Vickerman [42].

5.7   Scanning Probe Microscopy

Scanning probe microscopy (SPM) is a class of techniques where a sharp tip is rastered or scanned across the surface of a sample and a map of the interaction between surface and sample is recorded. This is a very versatile technique that allows one to map various physical properties depending on the specific probe used. The two most commonly used scanning probe techniques are scanning tunneling microscopy (STM) and atomic force microscopy (AFM) (Fig. 5.20).

In STM, the quantum mechanical tunneling of electrons through a thin energy barrier is utilized to measure the surface of a sample. A sharp tip with a small

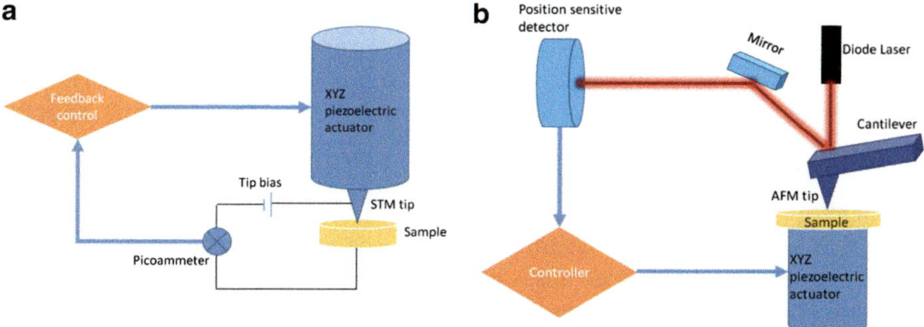

Fig. 5.20 (**a**) Basic schematic of a scanning tunneling microscope (STM). A very sharp metallic tip with a bias is approached very close to a conducting sample surface. The tunneling current across the tip-surface gap is measured and used as feedback for tip positioning. (**b**) Basic schematic of an atomic force microscope (AFM). A sharp tip is attached to a cantilever positioned above the sample surface, which is mounted on a piezoelectric actuator. The tip deflection is measured by bouncing a laser off the back of the cantilever into a position sensitive photodetector

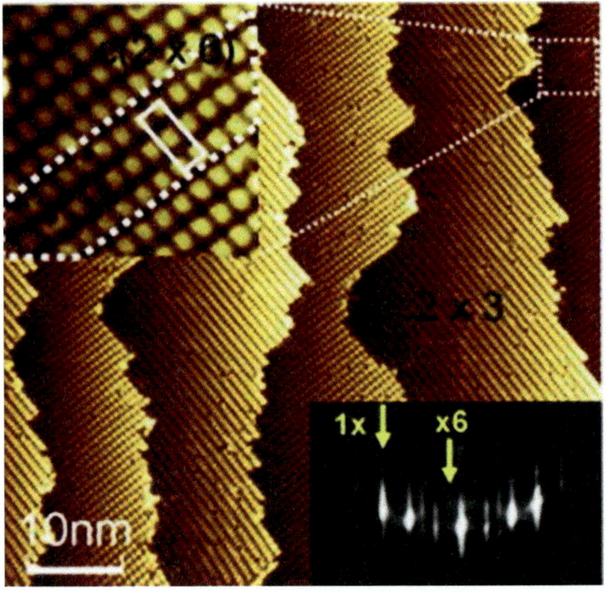

Fig. 5.21 Scanning tunneling microscope image of 1/4 monolayer Sr on Si showing a 2 × 3 reconstruction with some c(2 × 6) domains. Image taken at −2 V sample bias showing filled electronic states. *Inset* shows RHEED pattern for same surface. Reprinted with permission from [44]. Copyright 2011, AIP Publishing LLC

voltage bias is brought close to the surface of a sample until a certain tunneling current is achieved. The tip is then scanned, keeping the tunneling current constant. In this constant current mode of measurement, the topography of the sample can readily be imaged and under certain conditions, atomic resolution is possible (Fig. 5.21). By also scanning through different biases, one can in principle obtain a spectrum of both the filled and unfilled density of states near the Fermi level of the sample. When such bias scanning is done at a fixed point, the technique is called scanning tunneling spectroscopy (STS). The control of the tip position is done through the use of piezoelectric actuators that can position the tip in three dimensions precisely (Fig. 5.20a). Vibration isolation is required for atomic resolution. Because of the need to be able to tunnel between sample surface and tip, STM is only suitable for conducting samples. Also, because of the height sensitive nature of the tunneling current, rough samples cannot be imaged in STM. Tip artifacts in the images are also quite common and one should always check to make sure the observed features are real. See the book edited by Bonnell for a comprehensive treatment of STM [43].

AFM is a scanning probe technique that can be used for both conducting and insulating samples. AFM does not rely on the tunneling current between tip and sample but rather on the interatomic forces between them. A schematic of an AFM system is shown in Fig. 5.20b. The AFM probe is a cantilever whose deflection is measured by means of the position of the reflected laser spot. Deflection of the cantilever arises due to forces between the tip and sample. In this basic AFM mode, topography of the sample can be measured to a lateral resolution limited by the tip radius (~2 nm) (Fig. 5.22). There are several interatomic forces that operate between the tip and sample including electrostatic, magnetic, van der

Fig. 5.22 Atomic force microscope height image of an HF-etched and oxygen annealed $SrTiO_3$ single crystal surface. The scan shows unit cell height (0.4 nm) steps indicating an atomically flat TiO_2-terminated surface. The scan size is 5 μm

Waals, and adhesion. An excellent introduction to AFM can be found in the text by Eaton and West [45]. There are also two main imaging modes used in AFM: tapping mode and contact mode. In contact mode, the tip is slowly approached to the surface until a sudden change in deflection occurs when the tip and surface are in "contact". This is the easiest method of obtaining a topography image as it essentially measures the repulsive part of the interatomic potential which is sensitive to distance. Contact mode suffers from scan-induced artifacts including manipulation of surface adsorbates or atoms of the sample, and the inability to image soft samples such as polymers as the tip drags the material with it. To remedy this, tapping mode is utilized in which the cantilever is oscillated at its resonant frequency with relatively large amplitude. As the tip approaches the sample, the amplitude is reduced and this reduced amplitude is used as feedback to maintain constant height between tip and surface such that the tip height measures the topography of the sample. This is the method used for soft materials. Tapping mode suffers from poor resolution but a true non-contact mode of imaging can be used to circumvent this. Here, the shift in the resonance frequency of the oscillating cantilever is used to determine the force between tip and sample. Several common variants of AFM include piezoelectric force microscopy (PFM) and magnetic force microscopy (MFM). PFM utilizes an AC bias on the tip which causes a piezoelectric sample to deform with the same frequency as the tip bias. This technique is used to measure ferroelectric materials where the polarization direction will result in the piezoelectric deformation to be either in phase or out of phase with the tip bias [46]. In MFM, the tip is magnetic and the magnetic interaction will result in changes in the force between tip and sample depending on the magnetic microstructure. These force changes are measured and converted to magnetic domain images [47]. AFM has an advantage over STM in that insulating samples can be imaged. However, because interatomic forces are less distance sensitive than tunneling, AFM results in a generally lower lateral resolution and also more substantial tip shape artifacts.

5.8 Summary

In this chapter, we provided a brief overview of the more commonly used thin film characterization methods based on incident electrons, photons, ions, and scanning probe techniques. These techniques can be used to determine the crystalline structure, and electronic and chemical characteristics of epitaxial oxide systems. In most cases, a combination of one or more of these techniques is performed in order to gain a more complete picture of the sample being measured. With the widespread availability of excellent online resources for thin film characterization, it is now becoming easier for a thin film grower to become well-versed in many of these techniques. In recent years, it is almost a de facto requirement for publication in high profile journals to have a team of collaborators, each providing a different characterization technique for a given materials system. The practitioner of epitaxial oxides on semiconductors is advised to become well-versed in these methods.

References

1. H. Czichos, T. Saito, L.R. Smith, *Springer Handbook of Materials Measurement Methods* (Springer, Berlin, 2006)
2. P.M. Martin, *Handbook of Deposition Technologies for Films and Coatings: Science, Applications and Technology* (Elsevier, Amsterdam, 2010)
3. D.P. Woodruff, T.A. Delchar, *Modern Techniques of Surface Science* (Cambridge University Press, Cambridge, 1994)
4. K. Siegbahn, Science **217**, 111 (1982)
5. J.F. Moulder, J. Chastain, R.C. King, in *Handbook of X-ray Photoelectron Spectroscopy*, ed. by G.E. Muilenberg (Physical Electronics, Eden Prairie, MN, 1979)
6. M.P. Seah, W.D. Dench, Surf. Interface Anal. **1**, 2 (1979)
7. S. Tanuma, C.J. Powell, D.R. Penn, Surf. Sci. **192**, L849 (1987)
8. F.J. Himpsel, Angle-resolved measurements of the photoemission of electrons in the study of solids. Adv. Phys. **32**, 1 (1983)
9. Division of Surface Science, Institut National de la Recherche Scientifique, Online ARXPS tutorial, http://goliath.emt.inrs.ca/surfsci/arxps/
10. M. Cardona, L. Ley, in *Photoemission in Solids I*, ed. by M. Cardona, L. Ley, (Springer, New York, 1978), pp. 1–104
11. R.C. Hatch, K.D. Fredrickson, M. Choi, C. Lin, H. Seo, A. Posadas, A.A. Demkov, J. Appl. Phys. **114**, 103810 (2013)
12. VG Scienta, VUV 5000 Data Sheet version 4.1, http://www.vgscienta.com/_resources/File/VUV5000data sheet v4.1web.pdf

13. M. Kudo, in *Surface Analysis by Auger and X-ray Photoelectron Spectroscopy*, ed. by D. Briggs, J.T. Grant (IM Publications, Chichester, 2003)

14. D. Briggs, M.P. Seah, *Practical Surface Analysis: By Auger and X-ray Photoelectron Spectroscopy* (Wiley, Chichester, 1983)

15. U. Pietsch, V. Holy, T. Baumbach, *High-Resolution X-ray Scattering: From Thin Films to Lateral Nanostructures* (Springer, New York, 2004)

16. K.R. Balasubramanian, Growth and structural investigations of epitaxial hexagonal $YMnO_3$ thin films deposited on wurtzite GaN(001) substrates. Thin Solid Films **515**, 1807 (2006). doi:10.1016/j.tsf.2006.07.001

17. D.B. Williams, C.B. Carter, *Transmission Electron Microscopy: A Textbook for Materials Science* (Plenum, New York, 1996)

18. J. Goldstein, *Scanning Electron Microscopy and X-ray Microanalysis* (Kluwer, New York, 2003)

19. F.M. Pontes, E.R. Leite, E.J.H. Lee, E. Longo, J.A. Varela, Dielectric properties and microstructure of $SrTiO_3/BaTiO_3$ multilayer thin films prepared by a chemical route. Thin Solid Films **385**, 260 (2001)

20. S.J. Pennycook, M. Varela, C.J.D. Hetherington, A.I. Kirkland, MRS Bull. **31**, 36 (2006)

21. D.A. Muller et al., Science **319**, 1073 (2008)

22. V.M.A. Van Hove, W.H. Weinberg, C.-M. Chan, *Low-Energy Electron Diffraction: Experiment, Theory, and Surface Structure Determination* (Springer, Berlin, 1986)

23. A. Ichimiya, P.I. Cohen, *Reflection High-Energy Electron Diffraction* (Cambridge University Press, Cambridge, 2010)

24. J.H. Haeni, C.D. Theis, D.G. Schlom, RHEED intensity oscillations for the stoichiometric growth of $SrTiO_3$ thin films by reactive molecular beam epitaxy. J. Electroceram. **4**, 385 (2000)

25. I. Bozovic, J.N. Eckstein, Analysis of growing films of complex oxides by RHEED. MRS Bull. **20**, 32 (1995)

26. S.A. Chambers, Epitaxial growth and properties of thin film oxides. Surf. Sci. Rep. **39**, 105 (2000)

27. H. Guo et al., Structural and optical properties of $BaTiO_3$ ultrathin films. Europhys. Lett. **73**, 110 (2006)

28. A. Ohkubo et al., Combinatorial synthesis and optical characterization of alloy and superlattice films based on $SrTiO_3$ and $LaAlO_3$. Appl. Surf. Sci. **252**, 2488 (2006)

29. D. Kan et al., Blue light emission at room temperature from Ar^+-irradiated $SrTiO_3$. Nat. Mater. **4**, 816 (2005)

30. I.R. Lewis, H.G.M. Edwards, *Handbook of Raman Spectroscopy from the Research Laboratory to the Process Line* (Marcel Dekker, New York, 2001)

31. R.J. Bell, *Introductory Fourier Transform Spectroscopy* (Academic, New York, 1972)

32. J.B. Theeten, D.E. Aspnes, Ellipsometry in thin film analysis. Annu. Rev. Mater. Sci. **11**, 97 (1981)

33. H.G. Tompkins, W.A. McGahan, *Spectroscopic Ellipsometry and Reflectometry: A User's Guide* (Wiley, New York, 1999)

34. Z. Sobiesierski, Photoluminescence spectroscopy, in *Epioptics: Linear and Nonlinear Optical Spectroscopy of Surfaces and Interfaces*, ed. by J.F. McGilp, D.L. Weaire, C.H. Patterson (Springer, Berlin, 1995), pp. 133–162

35. T. Nakamura, Appl. Surf. Sci. **576**, 130–132 (1998)

36. L.V. Goncharova et al., J. Appl. Phys. **100**, 014912 (2006)

37. V. Shutthanandan et al., Appl. Phys. Lett. **80**, 1803 (2002)

38. J.W. Rabalais, *Principles and Applications of Ion Scattering Spectrometry: Surface Chemical and Structural Analysis* (Wiley-Interscience, Hoboken, NJ, 2003)

39. T. Gustafsson, Medium energy ion scattering for near surface structure and depth profiling, in *Ion Beams in Nanoscience and Technology*, ed. by R. Hellborg, H.J. Whitlow, Y. Zhang (Springer, Heidelberg, 2009), pp. 153–167

40. L.C. Feldman, Rutherford backscattering and nuclear reaction analysis, in *Ion Spectroscopies for Surface Analysis*, ed. by A.W. Czanderna, D.M. Hercules (Plenum, New York, 1991), pp. 311–362
41. H.-N. Tsai, Y.-C. Liang, H.-Y. Lee, Characteristics of sputter-deposited $BaTiO_3/SrTiO_3$ artificial superlattice films on an $LaNiO_3$-coated $SrTiO_3$ substrate. J. Cryst. Growth **284**, 65 (2005)
42. J.C. Vickerman, A. Brown, N.M. Reed, *Secondary Ion Mass Spectrometry: Principles and Applications* (Clarendon, Oxford, 1989)
43. D.A. Bonnell, *Scanning Tunneling Microscopy and Spectroscopy: Theory, Techniques, and Applications* (Wiley-VCH, New York, 1993)
44. J.H. He, G.H. Zhang, J.D. Guo, Q.L. Guo, K.H. Wu, Atomic structure of Sr-induced reconstructions on the Si(100) surface. J. Appl. Phys. **109**, 083522 (2011)
45. P. Eaton, P. West, *Atomic Force Microscopy* (Oxford University Press, Oxford, 2010)
46. A. Gruverman, S.V. Kalinin, Piezoresponse force microscopy and recent advances in nanoscale studies of ferroelectrics, in *Frontiers of Ferroelectricity*, ed. by H.L.W. Chan, S.B. Lang (Springer, New York, 2007), pp. 107–116
47. E. Meyer, H.J. Hug, MFM and related techniques, in *Scanning Probe Microscopy: The Lab on a Tip*, ed. by E. Meyer, H.J. Hug, R. Bennewitz (Springer, Berlin, 2004), pp. 97–125

Chapter 6
Growing SrTiO$_3$ on Si (001) by Molecular Beam Epitaxy

Over a decade ago, McKee and co-workers achieved a breakthrough in the epitaxial growth of single crystal perovskite SrTiO$_3$ (STO) on Si(001) by molecular beam epitaxy (MBE) using 1/2 monolayer (ML) of Sr on clean Si(001) 2 × 1 as a template [1]. At 1/2 ML coverage, Sr atoms assume positions between Si dimer rows, and inhibit formation of the amorphous SiO$_2$ layer during the subsequent STO deposition in a relatively wide range of temperatures and pressures [1–5]. The SrSi$_2$ stoichiometry of the template layer coincides with that of bulk Zintl silicide. Epitaxial growth of STO on Si(001) has enabled replacing the SiO$_2$ gate dielectric with an epitaxial oxide in a field effect transistor, and more importantly, the monolithic integration of functional perovskite oxides on Si [6–14].

Originally, crystalline STO epitaxially grown on Si was proposed as a possible gate dielectric [1, 15, 16]. Unfortunately, critical properties such as the band offset and Fermi level pinning at the Si-STO interface are unfavorable for device applications [2]. Nevertheless, since no other template layer has yet been reported that enables direct perovskite epitaxy on Si(001), and since STO is a commonly used substrate in oxide epitaxy [17], understanding its nucleation and growth directly on silicon is of great importance for the hetero-integration of functional oxide layers with Si logic. A greater understanding of the Sr template formation on Si(001) is thus crucial to controlling the growth of epitaxial oxides on silicon, and possibly extending this mechanism to other semiconductors [18–22].

The physical mechanisms behind various reconstructions of Sr at the Si surface have been explored over the last decade or so using *ab-initio* calculations [3, 23, 24]. In this chapter we will review this work and discuss the significance of 1/2 ML Sr template on Si (001). The reconstruction of Sr on Si (001) for the coverage range from 1/6 to 1 monolayer at both high and low temperatures will be discussed in Sect. 6.1. In Sect. 6.2 we discuss the electronic structure changes occurring during the Sr Zintl template formation on Si(001) as observed by the Si 2p surface core-level shifts (SCLS) as a function of Sr coverage. Next, in Sect. 6.3, we shall describe the details of the various processes enabling the growth of STO on Si (001) by MBE. In Sect. 6.4, the residual strain control in STO films by post-deposition annealing in oxygen will be discussed. Specifically, we consider the

A.A. Demkov and A.B. Posadas, *Integration of Functional Oxides with Semiconductors*, DOI 10.1007/978-1-4614-9320-4_6, © The Author(s) 2014

effect of oxygen partial pressure and annealing duration on the thickness of the amorphous interfacial SiO$_2$ layer, and how that, in turn, affects the lattice constant of STO films grown on Si (001). In Sect. 6.5 the atomic structure of the Si/STO (001) interface will be considered with emphasis on density functional theory (DFT) modeling. Finally, in Sect. 6.6 we will describe the latest advances in developing manufacturing processes to grow STO on Si (001).

6.1 The Zintl Template: Reconstruction of Sr on Si (001)

There are several growth mechanisms resulting in direct STO epitaxy on Si (001), but all depend on the initial 1/2 ML Sr template layer grown by MBE. The atomic structure of such a template is largely determined by the phase diagram of a sub-monolayer of the alkaline-earth metal on the Si (001) surface [25]. For example, in the process developed at Motorola [26, 27] Sr is present at the surface from the very early stage since it is used to clean and prepare large area wafers for the subsequent oxide growth using the so-called Sr-deoxidation process [28]. Experimentally, it is found that for a coverage between 1/6 ML and 1/3 ML of Sr on Si (001), a 3 × 2 reconstruction is dominant. In the range of 1/3–1/2 ML of Sr, a 2 × 1 reconstruction is most stable, and a 5 × 1 reconstruction occurs at approximately 0.7 ML coverage. For ~0.8 ML coverage a 7 × 1 reconstruction is stabilized, and a 3 × 1 reconstruction is observed at a higher coverage at least at very low temperatures [29, 30]. There are several theoretical and experimental studies of alkaline earth metals on the Si (001) surface [3, 23, 24, 31–40]. One of the earliest important studies is that of Wang et al., who identified the favorable bonding sites for Ba adsorption on Si(001) [23]. They report that the lowest energy bonding site of a Ba ad-atom is the fourfold site in the trough between two Si dimer rows. The result also holds for Sr at low coverage on Si(001) [3, 24].

6.1.1 Low Temperature Sr-Induced Surface Reconstruction

In order to better understand what happens when an electropositive alkaline earth atom is placed on Si (001), it is instructive to discuss a qualitative picture of the surface reconstruction on clean Si (001). Each atom on the ideal bulk-terminated Si (001) surface has two partially occupied dangling bonds leading to a rather unfavorable high energy (radical) configuration. The basic energy reduction mechanism is therefore the formation of surface dimers [41]. This dimer formation results in saturating one dangling bond per surface atom, and thus significantly lowers the energy. The cell doubles becoming 2 × 1, and the surface is comprised of dimer rows separated by troughs, all running along a <110> direction with respect to the bulk lattice. On a surface with single steps, dimer rows on adjacent terraces are oriented 90° with respect to each other owing to the diamond lattice of the crystal.

Note, that upon the "dimerization" the second nearest neighbors form first neighbor bonds, thus causing significant strain in the surface layer. The energy of a symmetric dimer can be further reduced by introducing a slight tilt [42]. This tilt leads to a charge transfer from the 'down' Si atom to the 'up' Si atom. The overall mechanism is driven by the lifting of degeneracy due to symmetry lowering (similar to Jahn-Teller or Peierls effects). This will be important to our understanding of the charge transfer between the Sr ad-atom and the Si (001) surface. If the alkaline earth atom, which is more electropositive than Si, donates some of its charge to the Si (001) surface, it is expected that a tilted dimer configuration is no longer favorable since this extra charge will occupy a high energy anti-bonding orbital. A symmetric rather than a tilted dimer configuration is now preferred since the π^*orbital is lower in energy than the p_z-like state localized at the 'down' Si atom of a dimer. In the spirit of this qualitative analysis we note that one dimer can accommodate two electrons, and if the anti-bonding state were filled the tendency not only to tilt but even to dimerize or "buckle" would be reduced. Thus we conjecture that one Sr atom at the surface will "un-tilt" one Si dimer and at a coverage exceeding one Sr per dimer the dimer may "unbuckle". This charge transfer (CT) conjecture can be easily tested by doing a model "charged system" simulation using a vacuum slab configuration for a 2×2 cell with two dimers on both surfaces and no Sr atoms. The relaxed structure shows that two dimers are tilted in opposite directions. To simulate the effect of the charge-transfer we introduce two, four, six, and eight extra electrons to the system and observe that one, two, three and four dimers, respectively, un-tilt and partially unbuckle, which confirms our qualitative analysis. However, when ten electrons are introduced, bonds in the bulk region of the silicon slab start breaking. This indicates that all excited surface states are filled and the additional electron now occupies the anti-bonding bulk states. This simulation, though qualitative in nature, illustrates a very important effect and suggests an interesting correlation. Note that one Sr atom has two valence electrons in the 5s orbital and may un-tilt and partially un-buckle one Si dimer leading to a *local* $SrSi_2$ stoichiometry, which incidentally, is the same as that of bulk Zintl silicide $SrSi_2$ [43, 44]. One would thus expect that a half-monolayer coverage that has a *global* $SrSi_2$ stoichiometry would be special. When translated into the "growth language", this means that a 1/2 ML Sr coverage is the turning point for the adsorption of Sr on Si(001), and indeed it is. However, as we shall show below, the electronic properties of the $SrSi_2$ template layer are very different from those of its intermetallic bulk cousin.

In a pioneering paper, Wang et al. [23] used density functional calculations and the energy-mapping technique to find the potential energy surface (and possible bonding sites) for a single Ba ad-atom on the Si(001) surface. They found that the most favorable bonding location was a fourfold site between two dimer rows. The energy of this site is 0.8 eV lower than that of any of the other potential wells identified on the surface. Later simulations by Demkov and Zhang [24] for Sr on Si (001) confirmed this bonding site as the most stable. For an excellent discussion of the alkaline earth metal interactions with Si(001) and its surface diffusion the reader is referred to [23]. Figure 6.1 shows the relaxed atomic geometries for Sr coverages of 1/6 ML, 1/3 ML, 1/2 ML, and 1 ML reported in [24]. In that work, a 3×4

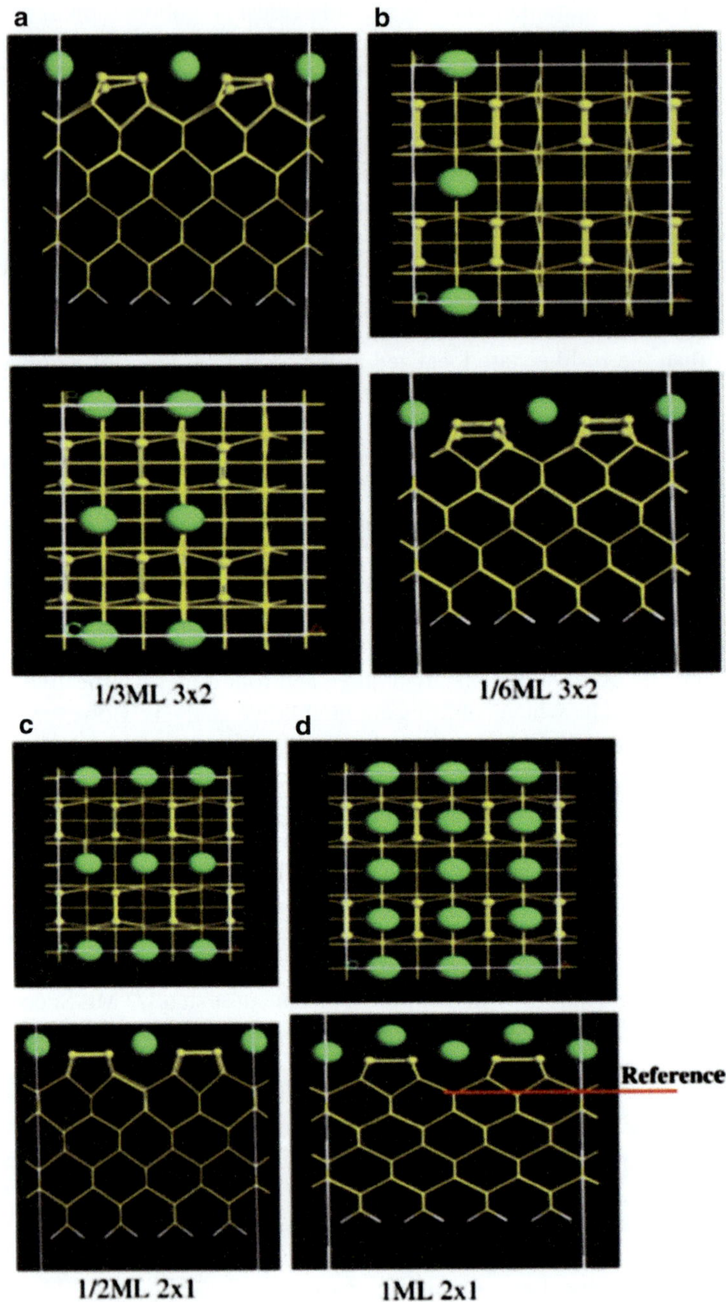

Fig. 6.1 The relaxed Sr/Si(001) surface models for the Sr coverage of 1/6 ML, 1/3 ML, ½ ML, and 1 ML. For each model we show the top and the side view. For the 1 ML structure (**d**) *red line* indicates a reference plane with respect to which the atomic positions are given (see text). Reprinted with permission from [24]. Copyright 2008, AIP Publishing LLC

surface simulation cell was used, so adding 2, 4, 6 and 12 Sr atoms to the surface corresponds to 1/6 ML, 1/3 ML, 1/2 ML and 1 ML coverage, respectively. For a Sr coverage below 1/2 ML, one can simply fill the best bonding sites on the surface with Sr atoms. When more than a 1/2 ML of Sr is placed on top of the Si(001) surface, an extra Sr atom finds a place in the dimer row between two adjacent Si dimers. This site was also identified by Wang et al. [23].

6.1.2 Coverage Up to ½ Monolayer

For the 1/3 ML coverage the relaxed atomic geometry found in [24] is shown in Fig. 6.1a. There are four Sr atoms and six Si dimers in the 3×4 surface cell. The simple CT conjecture discussed above, suggests that four dimers should assume symmetric configurations. Indeed, precisely four dimers are symmetric and the remaining two are still tilted, with the resulting symmetry being 3×2. This reconstruction, obtained as a result of the self-consistent conjugate gradient minimization [24], seems to follow our simple conjecture. However, for a coverage of 1/6 ML a somewhat different and rather unique surface reconstruction is found. The CT rule predicts two symmetric dimers out of six. However, as shown in Fig. 6.1b, all six Si dimers become un-tilted! Four straight dimers are positioned slightly higher than the other two resulting again in an overall symmetry of 3×2. This reconstruction roughly agrees with the qualitative rule—one Sr atom affects one Si dimer. However, the system prefers a more symmetric reconstruction. More importantly, the lowest energy atomic configurations for the Sr coverage of 1/6 ML and 1/3 ML both result in a 3×2 reconstruction.

The 1/2 ML coverage is achieved by adding six Sr atoms to the 3×4 cell [24]. All six dimers become symmetric after the relaxation in agreement with the CT rule, and we find a 2×1 reconstruction. In Fig. 6.1c the 1/2 ML results are shown, and again six Sr atoms cause six dimers to un-tilt. All the optimum bonding sites (the trough position) are now filled, which in terms of the electronic structure corresponds to the surface π^* band being fully occupied. Any additional charge would have to be accommodated in states of a different nature, and one would expect a "phase transition" in terms of the surface phase diagram [26]. Therefore 1/2 ML is a limiting coverage, and thus is a rather special case. It is worth noting, that stable Sr silicide has the stoichiometry of $SrSi_2$, the crystal structure of this intermetallic compound is shown in Fig. 6.2. Unlike the case of $BaSi_2$, Si atoms are arranged in a three-coordinated net and not in pyramids. This Si net structure is also very different from the four-coordinated diamond net of bulk Si, and is made possible entirely by the Zintl charge transfer from Sr to Si. Bulk $SrSi_2$ has a cubic structure with an experimental lattice constant of 6.54 Å, and belongs to space group $P4_332$ (group number 212). There are eight Si atoms and four Sr atoms in the primitive unit cell. All silicon atoms are threefold coordinated and the Si-Si bond length is about 2.31 Å (compare to 2.35 Å in bulk Si). Demkov and Zhang have

Fig. 6.2 The crystal structure of Zintl intermetallic silicide SrSi$_2$ (G.E. Pringle, Acta Cryst. B **28**, 2326 (1972)). Sr atoms are shown as *green balls*, while the Si framework is shown with a wire diagram. Reprinted with permission from [24]. Copyright 2008, AIP Publishing LLC

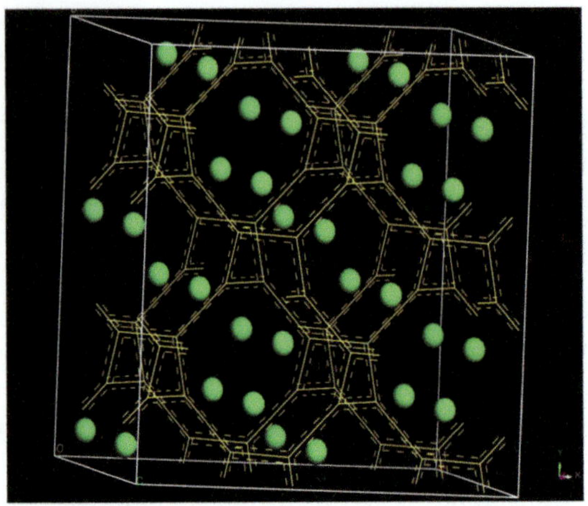

calculated the electronic structure of SrSi$_2$ using the local density approximation (LDA). They found it to be metallic [24], however, electrical measurements suggest it to be semiconducting with a very small gap of 0.035 eV [45]. A recent DFT calculation in the generalized gradient approximation (GGA) also gives a small gap (0.06 eV) [46]. At 1/2 ML coverage the stoichiometry of the surface layer SrSi$_2$ is equivalent to that of the bulk silicide, and it clearly is a semiconductor.

6.1.3 Coverage from ½ to 1 Monolayer

To simulate higher coverage, more Sr atoms should be placed at the surface. As all low energy sites have been already occupied at a coverage of 1/2 ML, these extra Sr atoms need to occupy a meta-stable site atop the Sr/Si template, namely the site over the dimer row between two Si dimers. When extra Sr atoms partially occupy these sites at random, the Low-Energy Electron Diffraction/Reflection High-Energy Electron Diffraction (LEED/RHEED) signal becomes weaker and a diffuse background emerges [47]. In Fig. 6.1c, d we show two 2×1-ordered structures with 1/2 ML and 1 ML coverage, respectively. As has been shown by Demkov and Zhang [24], the ordered 2×1 structure with 1 ML of Sr is unstable with respect to a different class of reconstructions accompanied by the total unbuckling of Si dimers, and driven by what we call a size constraint.

Looking at Sr on the Si(001) surface in the case of 1 ML coverage shown in Fig. 6.1d, we notice that Sr atoms form an ordered 2×1 structure, and the underlying Si(001) surface also keeps its 2×1 reconstruction. As we have discussed in the previous section, only half of a ML is needed to fill the π^* surface states, and at a higher coverage one might expect the dimers to unbuckle. It is then natural to ask why Sr atoms do not form a 1×1 pattern instead of a 2×1 pattern in this case. Clearly, the 2×1 structure is at least a local minimum. To overcome

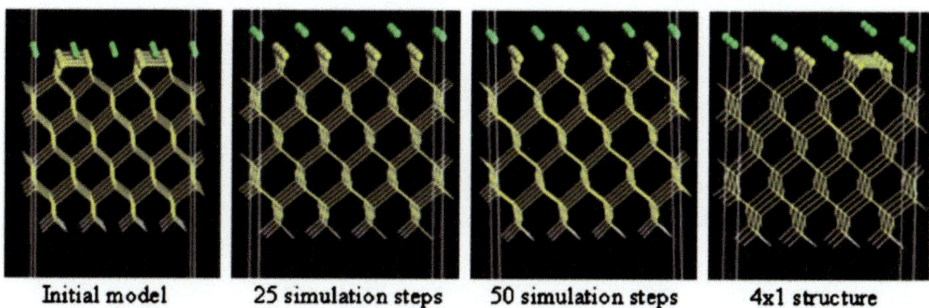

Fig. 6.3 The structural progression of the 2×1 to 4×1 Sr/Si surface reconstruction during a conjugate gradient energy minimization. The final 4×1 structure can be viewed as $1\times$ patches separated by a dimer row. Reprinted with permission from [24]. Copyright 2008, AIP Publishing LLC

what appears to be a barrier around a local minimum (the structure shown in Fig. 6.1d) one can force all Sr atoms to have the same height closer to the Si surface (gently push them down), and then perform a conjugate gradient minimization in the hope of finding the global minimum of this system. In Fig. 6.3 we illustrate the minimization process, showing the initial, final, and two intermediate structures from [24]. Even though this is not a molecular dynamics minimization, and therefore the interpretation of the trajectory is not straightforward, it is nevertheless a rather instructive exercise. It is clear that Sr atoms at this coverage first indeed break Si dimer bonds (Fig. 6.3b) due to the extra electrons occupying anti-bonding states. We also observe that Sr atoms "almost" form a 1×1 pattern at one stage (Fig. 6.3c). Finally, one row of Sr atoms "pops up" allowing Si atoms underneath to re-dimerize! In this simulation, the initial 2×1 reconstruction transforms into a 4×1 reconstruction pattern. The total energy is reduced by 2.0 eV per cell or 0.17 eV per Sr atom. *This is a qualitatively different reconstruction!* Now the major driving force is the size mismatch between the Si surface and Sr atoms. The covalent and atomic radii of Sr are 1.91 Å, and 2.15 Å, respectively, and the unreconstructed Si (001) surface unit cell lattice vector is only 3.84 Å. Qualitatively speaking, since the bonding is more "charge transfer" in nature than truly covalent, the surface area of Si is insufficient to hold 1 ML of Sr atoms. Putting a full Sr ML in one-to-one registry with unreconstructed Si (001) would result in a large compressive strain in the metallic layer. The system lowers its energy by vertically displacing one row of metal atoms, allowing more space for the remaining layer. The charge transfer is a sensitive function of the inter-atomic distance, and is reduced between the displaced Sr row and the surface, making the dimer energetically preferred in this location. The Si-Si dimer bond in the case of a 2×1 reconstruction is 2.61 Å (compare to 2.35 Å of the bulk Si-Si bond, or 2.40 Å of a clean surface dimer). The in-trough Sr atom is 3.84 Å above the reference Si plane, and the over-the-ridge Sr is 5.23 Å above that plane (see Fig. 6.1d). Note that in bulk Si the vertical separation between the adjacent atomic layers is 1.36 Å. In the case of a 4×1 reconstruction, the surface dimer bond is 2.51 Å, the "low" Sr is 4.01 Å above the reference plane, and the "high" over-the-ridge Sr is 5.17 Å above it.

Fig. 6.4 Three different "components" used to calculate the thermodynamic potential of the Sr/Si surface phases. The crystal structure of SrSi$_2$, the fourth element of our analysis is shown in Fig. 6.2. Reprinted with permission from [24]. Copyright 2008, AIP Publishing LLC

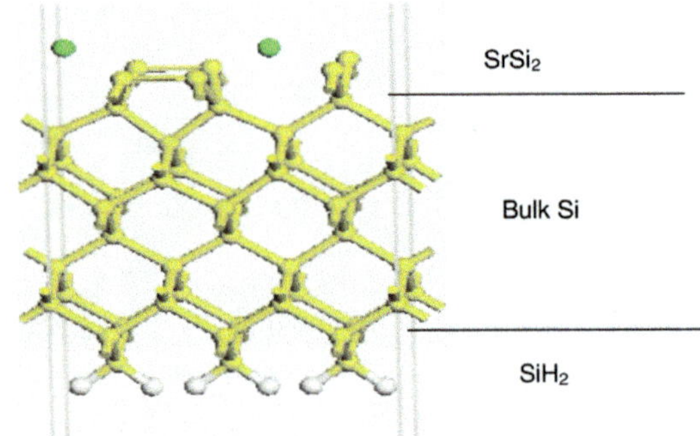

A 4×1 phase described above suggests a different class of possible reconstruction patterns that is characterized by 1×1-reconstructed areas separated by a dimer row. In this particular simulation [24] the separation between two dimer rows is three unit cells, and is due to the size of our initial cell (3×4). In general, we can make the distance between two Si dimer rows two, four, five or six unit cells and construct Sr/Si surface models with 3×1, 5×1, 6×1 and 7×1 reconstructions. Although the initial simulation used a Sr coverage of 1 ML, it was found that if the row of Sr atoms directly above the row of Si dimers was removed, the remaining structure was stable and showed a similar reconstruction pattern. Based on this model, one would expect that the onsets of the 3×1, 5×1 and 7×1 reconstructions are 2/3 ML, 4/5 ML and 6/7 ML. The 3×1, 5× and 7× phases have indeed been reported [29, 30].

6.1.4   Thermodynamics

The stability of the Sr/Si surface reconstructions under equilibrium conditions can be determined by thermodynamic considerations. The thermodynamic analysis of a mixed species system requires the introduction of chemical potentials [48]. The thermodynamic potential of the surface is equated to the Gibbs free energy change for a chemical reaction resulting in the final surface structure. In addition to the reaction, the appropriate reservoirs need to be chosen, and possible constraints specified. The system can be divided into three parts as shown in Fig. 6.4: SiH$_2$, bulk Si and SrSi$_2$ regions. The thermodynamic potential is then calculated by the following formula:

$$F^{suf} = E^{\mathrm{mod}} - \left[N_{SiH_2} E_{SiH_2} + N_{Si} E_{Si} + N'_{Si}(E_{Si} + \mu_{Si}) + N_{Sr}(E_{Sr} + \mu_{Sr}) \right], \quad (6.1)$$

where E^{mod} is the calculated total energy of the entire surface model, and N_{SiH_2}, N_{Si}, N'_{Si} and N_{Sr} are the numbers of SiH$_2$, "bulk" Si, Si in Sr/Si layer, and Sr species, respectively. E_{Si} and E_{Sr} are the energy per atom in bulk crystal silicon and metallic Sr, respectively. Since we fix the number of hydrogen atoms and silicon atoms that

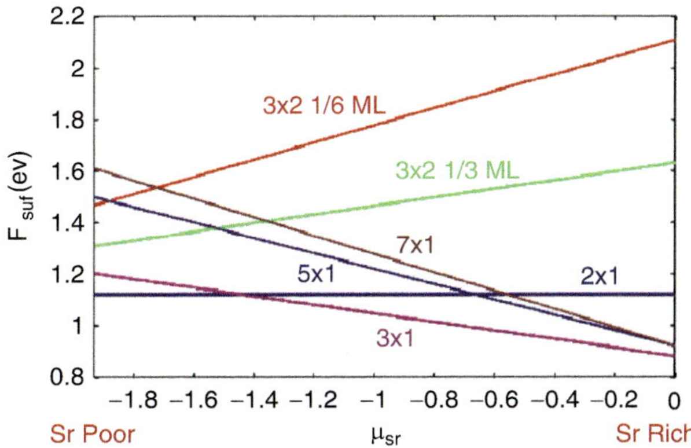

Fig. 6.5 The thermodynamic potential for the considered surface phases of the Sr/Si system as function of the Sr chemical potential. Reprinted with permission from [24]. Copyright 2008, AIP Publishing LLC

are attached to at the bottom surface of the slab, and we fix their positions to those of the bulk during the simulation, we assume that the energy of this pseudo SiH_2 compound does not change between systems. We calculate the energy E_{SiH_2} of the SiH_2 unit in a separate calculation of a Si slab with both surfaces hydrogen terminated. The chemical potentials of Sr and Si atoms μ_{Sr} and μ_{Si} are referenced to crystal silicon and Sr metal, and define the two-dimensional phase space of the problem. The reaction of bulk silicide formation gives a natural constraint represented as a phase boundary along which μ_{Sr} and μ_{Si} are not independent but are subject to the following constraint relation:

$$2\mu_{Si} + \mu_{Sr} = -E_f^{SrSi_2} \tag{6.2}$$

where $-E_f^{SrSi_2}$ is the heat of formation of crystalline $SrSi_2$ (we calculated $-E_f^{SrSi_2}$ to be -1.93 eV). Assuming the surface is in equilibrium with the silicide, and using (6.2) one of the chemical potentials can be eliminated in (6.1), so only one independent variable is left. We choose μ_{Sr} to describe the relative stability of various phases of the Sr/Si surface system. The range of the Sr chemical potential is limited to:

$$-E_f^{SrSi_2} \leq \mu_{Sr} \leq 0 \tag{6.3}$$

Zero of the chemical potential corresponds to equilibrium with the metal source (Sr-rich conditions). In Fig. 6.5 we show the grand thermodynamic potential of various surface phases in the Sr/Si system as a function of the chemical environment described by the potential μ_{Sr}. In this phase diagram the 3×1, 5×1 and 7×1 phases are calculated using surface models with the Sr coverage of 2/3 ML, 4/5 ML and 6/7 ML. We see that the 3×1 phase is preferred under Sr rich conditions. However, the 2×1 phase is favorable under Sr poor conditions. It is worth mentioning that the 3×2 phase becomes more stable at extremely Sr poor conditions. This picture agrees with available experiment [29, 30].

A different way of looking into the thermodynamics of Sr deposition is to consider a chemical reaction between the reconstructed surface and a reservoir of Sr metal. The energetics of Sr adsorption on Si(001) can be described defining the reaction energy as follows:

$$E_{re} = E_{SrSi-surface} - NE_{Sr} - E_{Si-surface} \qquad (6.4)$$

In (6.4), $E_{SrSi-surface}$ is the total energy of the relaxed Sr/Si(001) system, E_{Sr} is the energy per atom in Sr metal calculated using the same pseudo-potential and basis set as used for the combined Sr/Si system, and $E_{Si-surface}$ is the total energy of the slab modeling the reconstructed Si(001) surface (the bottom layer is passivated with hydrogen). Analysis of the reaction energy as function of the number of Sr atoms suggests that below 1/2 ML adding a Sr atom to the surface from the metal source results in a significant energy gain, reflecting a strong Sr-surface interaction. However, above 1/2 ML coverage, the reaction energy saturates and does not change significantly with the addition of extra Sr. This implies that the surface has reached equilibrium with the metal reservoir. One can also consider the specific reaction energy E_{re}/N describing the reaction energy as a function of coverage. It turns out that between 1/6 ML and 1/2 ML, E_{re}/N is almost flat indicating an ordered phase formation for a coverage over 1/6 ML, which is a 3×2 Sr/Si surface phase. The most important result is that it takes half of a monolayer of Sr on the Si surface to complete the surface metallization from the thermodynamic point of view (note, that the surface is semiconducting).

6.1.5 Reconstruction at High Temperature

Interestingly, at high temperature the behavior of Sr on Si (001) surface is totally different. Reiner et al. have explored this regime depositing Sr on 4° miscut silicon [49]. The miscut eliminates one of the two terrace orientations [49]. Using in-situ RHEED they report 2×3 and 1×3 structures for 1/6 and 1/3 ML coverage, respectively. They also used DFT calculations to determine the corresponding atomic structures. The results from their calculations could describe the 1×2 structure observed at 650 °C, but only if there is a transition of the symmetry of the Si surface from 2×1 to 1×2 during Sr deposition. A 2×1 → 1×2 transition of the Si surface occurs when the top layer of Si is removed, since this rotates the dimer bonds by 90°. The RHEED data indicate that this movement of the top layer of Si is occurring during Sr deposition, but only at high temperature. Obviously, this requires a massive rearrangement of the Si surface. Removal of the top layer of Si had also been proposed to explain scanning tunneling microscopy (STM) studies of sub-monolayer Ba structures on Si (001) [50], and reported to occur during As deposition [51]. Theoretical calculations suggest the existence of a class of surface structures created by removing two adjacent dimers in a row (i.e., removing two dimers along the original 1× direction), reconstructing the exposed silicon atoms

Fig. 6.6 Top view and side view of selected dimer vacancy structures. (**a**) Top view of $c(2 \times 6)$ 1/6 ML structure. (**b**) Top view of 2×3 1/6 ML structure. (**c**) Side view of 2×3 or $c(2 \times 6)$ 1/6 ML structure. The Sr atom is *large* and *yellow*, the original silicon dimer (1/3 ML silicon) is in *light blue*, the second-layer silicon atoms are in *green*, and lower level silicon are in *dark blue*. Reprinted with permission from [34]. Copyright 2009 by the American Physical Society

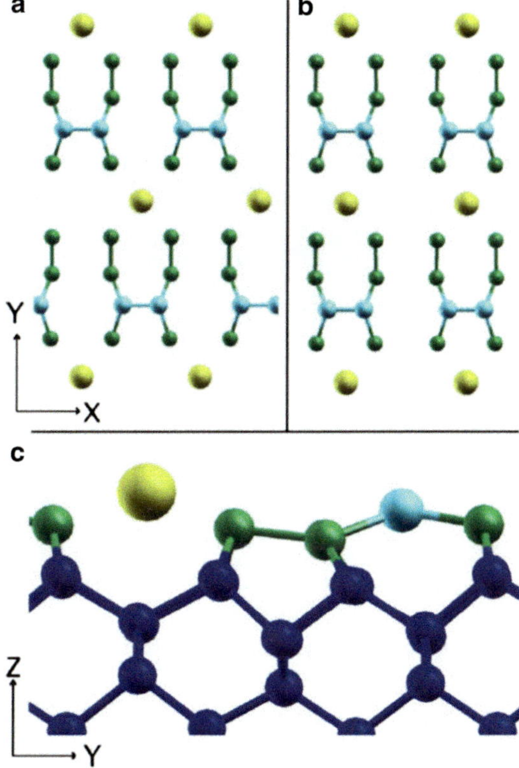

into new dimers perpendicular to the original dimers, and placing a Sr atom in the created hole as shown in Fig. 6.6. The details of this work can be found in [34]. The overall conclusion is that 1/2 ML Sr deposited at 650 °C on single termination miscut Si (001) surfaces replaces the top monolayer of Si, causing a $2 \times 1 \rightarrow 2 \times 3 \rightarrow 1 \times 2$ transition of the surface symmetry. This movement of Si is driven by the formation of a 2×3 structure at 1/6 ML Sr, which removes 2/3 of the surface Si atoms.

This phenomenon is most likely related to step bunching, often observed in the presence of metals on vicinal Si (001) at elevated temperature [52]. For example, As on a Si(001) vicinal surface causes the reversible formation of small facets, or (at lower step densities) quadruple-layer height steps, which has been observed in both RHEED and LEED, when cooling through the As desorption temperature [53, 54]. On the other hand, annealing an In-covered vicinal Si(100) surface between 250 and 650 °C causes step bunching at low coverages and faceting at high coverages [55]. Depositing gold on Si (001) at 500–700° results in formation of straight 2n-height steps [56]. While deposition of silver at 700 K on 4° vicinal Si(001) results in formation of multiple-layer steps as seen by SPA-LEED and STM [57]. Fölsch et al. argued that the reconstruction is driven by the minimization of the surface free-energy, which is achieved by balancing two competing contributions:

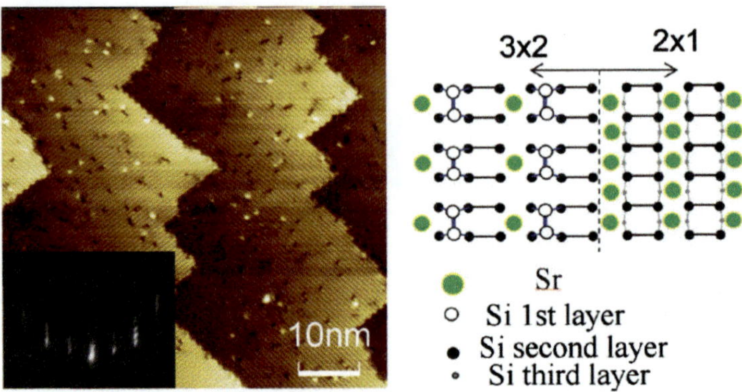

Fig. 6.7 (*Left*) At the saturation coverage of 0.5 ML, the surface is fully covered by the 1/2-Sr phase. The corresponding RHEED patterns (*inset*) show only ×2 streaks. (*Right*) The schematic drawing showing the structural model of the 1 × 2-Sr phase and the relationship between 2 × 3-Sr and 1 × 2-Sr. Reprinted with permission from [59]. Copyright 2011, AIP Publishing LLC

the strain relaxation of the (332) Ag reconstructed surface due to the creation of steps and the step energy determined by bond breaking and strain along the steps [58]. Interestingly, these multi-steps are distributed homogeneously over the entire substrate, thus preserving the 4° inclination of the macroscopic surface locally.

These Sr-induced reconstructions on Si(100) at elevated temperature have been recently studied by He and co-workers using a combination of STM and RHEED [59]. They also found that for the Sr coverage increasing from zero to 1/2 ML, the surface exhibits phase transitions from 2×1-Sr to 2×3-Sr and then back to 1×2-Sr. The STM image of the 1×2 structure is shown in Fig. 6.7. The bias-dependent, high resolution STM images unambiguously support a dimer-vacancy structural model for the 2×3-Sr phase, and the coverage-dependent evolution of the surface from 2×3-Sr to 1×2-Sr can also be nicely explained by extending the 2×3-Sr model to the 1×2-Sr.

6.1.6 Electronic Structure

A few words should be said about the electronic structure of Sr/Si phases. In our qualitative analysis we have proposed the CT rule, according to which, one Sr ad-atom on the surface un-tilts or flattens one Si dimer. This implies that two electrons from the Sr atom will occupy the π^* orbitals of the dimer. Thus the system is metallic for coverages below 1/2 ML and as we are filling up the surface π^* band, at 1/2 ML a semiconductor state is recovered. When the coverage goes over 1/2 ML additional electrons start occupying the bulk conduction band of Si, assuming just filling the bands but keeping them unchanged (a rigid band approximation).

Indeed, Demkov and Zhang [24] found that the 2×1 model at 1/2 ML coverage is semi-conducting in agreement with this qualitative analysis, and the 3×2 model at 1/3 ML coverage is metallic. However, they reported changes in the overall band edge structure due to the electrostatic effects. We will discuss the electronic structure changes caused by Sr deposition in more detail in the next section.

6.1.7 Conclusions

At low temperature, depending on coverage, there are two distinct mechanisms for the Sr-surface interaction, one quantum mechanical and one geometric. For Sr coverage below 1/2 ML the electronic structure effects allow the system to reduce its energy through the charge transfer between Sr and Si atoms, resulting in flattening of Si dimers. For a higher coverage we identify several 3×, 5×, and 7× models based on the size mismatch between Sr atoms and the Si substrate. The electronic structure analysis and thermodynamic considerations suggest that at this high Sr coverage we reach the complete metallization of the surface. Therefore it can be viewed as a close-packed arrangement of spheres, conventionally adopted for simple metals, fitted to a box of a certain size. This is a principally different reconstruction mechanism. The analysis of the relative stability of various phases suggests a 3×2 to 2×1 to 3×1 progression of the surface phases with increasing Sr coverage. On the other hand, experiments performed on vicinal Si at high temperature, suggest the 2×1 → 2×3 → 1×2 progression for coverages from zero to 1/2 ML. High resolution STM [59] agrees with a dimer-vacancy model for the 2×3 reconstruction suggested by Reiner and co-workers [49].

6.2 Looking for the Zintl Template

X-ray photoemission has been extensively used to study Si(001), and ample literature exists describing the bulk and surface core-level states both experimentally and theoretically [60–68]. As we have discussed in the previous section, at room temperature in vacuum, a clean Si(001) surface exhibits 2×1 reconstruction with buckled asymmetric dimers [69], giving rise to two surface states one of which is filled (the so-called "up atom" of a dimer) while the other is empty (the "down atom"). Experimentally, the Si $2p$ core level spectrum of 2×1 Si(001) can be described (fitted) with seven spin-orbit-split pairs of surface core-level components. Specifically, as shown in Fig. 6.8, the spectrum is decomposed into one bulk component (B), and six surface components (S_u, C, S_d, S′, D and L) including the up (S_u) and down (S_d) surface states [67]. The C component is positioned in energy between the bulk and S_u components and originates from one half of the third and fourth layer atoms [62]. The S′ component has been associated with either the second layer of Si atoms [64] or with one half of the third layer plus the fourth layer

Fig. 6.8 Decompositions of the Si 2*p* spectra taken at a photon energy (*hv*) of 145 eV and an emission angle (θ*e*) of 0° with different SCLS's for *Sd* of (**a**) *Ed* = 150 meV, (**b**) 80 meV, and (**c**) 10 meV. *Open circles* represent the raw data, and *solid lines* are the fitting results, fitted components, and integral backgrounds. The data in the *insets* of (**c**) and (**d**) are magnified vertically 9.5 times. Reprinted with permission from [67]. Copyright 2003 by the American Physical Society

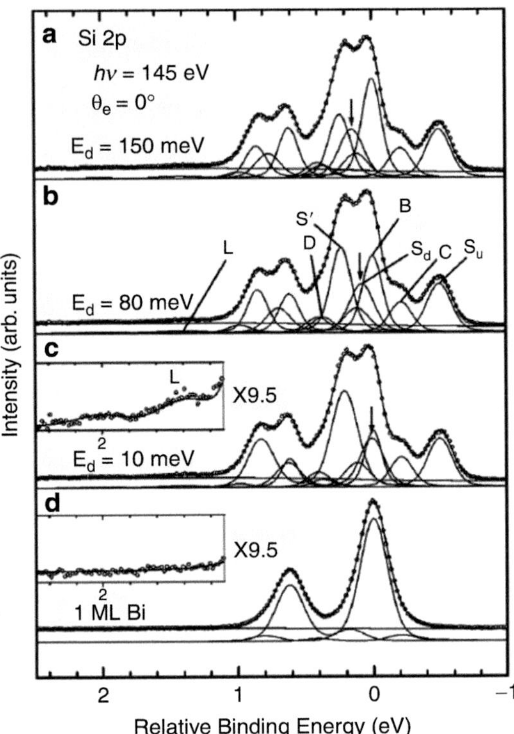

of Si atoms [62]. The D and L components are observed on the higher binding energy side of the spectrum with their origin debated in the literature. However, the D component is likely due to surface defects [62, 70], while the L component is related to a surface loss process *via* inter-band transitions in surface bands [67].

The detailed changes in the Si *2p* core level spectrum when ad-atoms are placed on the surface depend on the specific adsorbing atom and show a unique trend since core level shifts are related to the electronic structure of the system [71–73]. For example, when Mg and Ca are adsorbed on Si(001) to form MgSi$_2$ and CaSi$_2$, the Si *2p* core-level shifts as a function of Ca and Mg coverage indicating charge transfer from alkaline-earth metals to Si(001) [73]. As we discussed in the previous section, at sub-monolayer coverage, Sr on Si (001) results in un-tilting of the dimers due to Zintl charge transfer from the electropositive metal to Si [24]. This structural change is an essential factor in creating the template for the subsequent STO growth [40]. The electronic structure of Sr on Si(001) in general has been studied using x-ray standing wave [74] and x-ray photoemission techniques [74, 75] More recently, Choi et al. and Seo et al. focused on understanding the relationship between the surface reconstruction and the electronic structure changes induced by Sr deposition [35, 36].

6.2.1 Experimental Details

For the study in [35], B-doped ($\sim10^{16}$ cm^{-3}) prime Si wafers were cut into 20 mm$\times$20 mm pieces and ultrasonically cleaned in acetone, deionized water and isopropanol for 5 min each, then exposed to ultraviolet (UV)-ozone lamp to remove carbon impurities at the surface. The Si substrates were then introduced into a customized DCA 600 MBE system with a base pressure of 3×10^{-10} Torr. To remove the native SiO_2 layer, the substrates were heated up to 875 °C and annealed for 3 h under ultrahigh vacuum. After annealing, a sharp 2×1 reconstruction pattern is observed by in situ RHEED. The samples were then transferred in situ to the x-ray and ultraviolet photoelectron spectroscopy (XPS/UPS) analysis chamber (VG Scienta R3000). The C $1s$ and O $1s$ core level spectra were measured to verify the cleanness of the Si(001) surface. After confirming that there was no detectable SiO_2 layer at the surface, the sample was moved to the MBE chamber for subsequent Sr deposition. The Sr flux was calibrated using a quartz crystal monitor and fine-tuned using RHEED oscillations to yield a rate of 1 ML per minute, where we define 1 ML as the atomic surface density of an ideal unreconstructed Si(001) surface (1 ML $= 6.78\times10^{14}$ atoms/cm^2).

Measurements of the surface core level shifts and work function as a function of Sr coverage at room temperature were performed with in situ XPS using monochromatic Al Kα radiation ($h\upsilon = 1{,}486.6$ eV) and in situ UPS using a bright monochromatic He plasma light source (He I radiation $h\upsilon = 21.22$ eV). The analyzer was calibrated using a two-point measurement of the Ag $3d_{5/2}$ core level at 368.28 eV and the Fermi edge of Ag at 0.00 eV. The resolution of the XPS spectra is limited by the x-ray source line width, which is approximately 300 meV, while that of UPS is analyzer-limited and is <30 meV. To assist with the interpretation of the XPS spectra, DFT calculations of SCLS were performed within the LDA [76].

Figure 6.9 shows the RHEED patterns for clean 2×1 Si(001) and Si with various coverage of Sr. As the Sr coverage increases, the RHEED pattern evolves from a 2×3 reconstruction between 1/6 ML and 1/4 ML of Sr into a 2×1 structure between 1/3 ML and 1/2 ML of Sr coverage. At 1/2 ML Sr coverage, the RHEED pattern is qualitatively identical to that of clean 2×1 Si(001). After Sr deposition and RHEED imaging, the samples were transferred in situ to the photoemission analysis chamber.

Si $2p$ core-level photoemission spectra of 2×1 reconstructed Si(001) and 1/2 ML Sr-induced 2×1 reconstruction on Si(001) were taken at normal emission at room temperature and analyzed as shown in Fig. 6.10. The Si $2p$ core-level spectra were modeled using six components labeled S_u, S_d, C, SS, L, and B (for bulk). For buckled asymmetric Si dimers, the S_u and S_d components represent up and down dimer atoms, respectively. The intensity and full width at half maximum (FWHM) of the two dimer components were constrained to be equal to each other, in accordance with the known structure of a clean Si(001) surface. The C component, which can be readily resolved in Si $2p$ core-level synchrotron measurements, is reported to be from the third layer beneath the dimer rows (S3 in Fig. 6.10a)

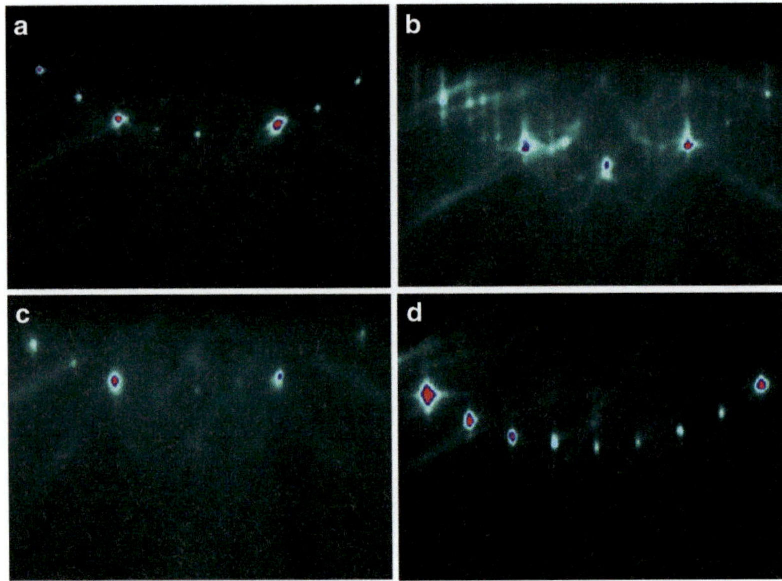

Fig. 6.9 RHEED patterns of the Si(001) surface as a function of Sr coverage for (**a**) 0, (**b**) 1/6, (**c**) 1/3, and (**d**) 1/2 ML Sr coverage on 2 × 1 Si(001) deposited at 600 °C. All patterns are viewed along the Si<110> azimuth. Reprinted with permission from [35]. Copyright 2013, AIP Publishing LLC

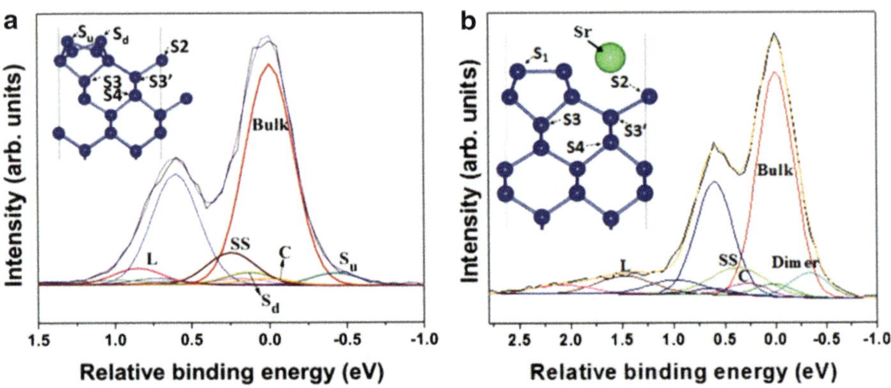

Fig. 6.10 Si $2p$ core-level spectra of (**a**) clean 2 × 1 Si(001), and (**b**) 1/2 ML Sr deposition on clean 2 × 1 Si(001) by in situ XPS at room temperature. *Insets* are theoretical structures of clean 2 × 1 Si(001) and 1/2 ML Sr on Si(001) obtained from the DFT calculations. Reprinted with permission from [35]. Copyright 2013, AIP Publishing LLC

and the fourth layer atoms below S3 [62]. For the present case, the C components in Fig. 6.10 were somewhat difficult to assign precisely due to the resolution limit. The SS component was assigned to sub-surface components of the third and fourth layers (S3′ and S4 in Fig. 6.10a). The L component was needed to fit the tail of the spectrum on the higher binding energy side [67]. All components included a spin-orbit-split pair located 0.605 eV higher in binding energy with a branching ratio fixed at the theoretical value of 0.5. The energy positions of all components were expressed in terms of the binding energy relative to binding energy of the well-resolved bulk component. The peaks were fit with the Voigt function consisting of 90 % Gaussian and 10 % Lorentzian using Casa XPS software [77]. The value of the FWHM from the bulk component was constrained to be the same as that of S_u, C, and S_d components. For SS and L, the FWHM was constrained to be 1.5 times wider than that of bulk because the SS component is a combination of two very closely spaced features (third and fourth layers) that cannot be effectively resolved, while the L component has a broad tail in the higher binding region.

6.2.2   Surface Core-Level Shifts of the Zintl Template

For clean 2×1 reconstructed Si(001), the surface core-level shifts (SCLS) of the S_u, S_d, and SS components were -0.43 eV, 0.12 eV, and 0.22 eV, respectively and are shown in Fig. 6.10a. These values are in good agreement with recent synchrotron-based measurements [62, 66]. After 1/2 ML Sr deposition on clean 2×1 Si(001), the asymmetric tilt of the dimer was eliminated as a result of Sr atoms donating their two electrons to the Si(001) surface, giving rise to a single merged peak in place of previously separate S_u and S_d components. The SCLS of the merged dimer peak is -0.33 eV as shown in Fig. 6.10b.

An unexpected result was a 0.49 eV shift to higher binding energy of the bulk component of the Si $2p$ core-level spectra when 1/2 ML of Sr was deposited on Si(001). Because the Sr donates two electrons to the Si(001) surface, from a purely electrostatic point of view, the bulk core-level is expected to move to lower binding energy because the extra charge raises the electrostatic potential. Additionally, the shape of Si $2p$ spectra broadens in the higher binding region after 1/2 ML Sr deposition. To verify the charge transfer, the location of the valence band edge was measured for both clean Si(001) and 1/2 ML Sr on Si(001) as shown in Fig. 6.11. The valence band edge positions were determined using the linear extrapolation method [78]. The energy shift of the valence band for 1/2 ML Sr on Si is approximately 0.42 eV toward higher binding energy when compared to that of clean Si(001). This value of the valence band shift confirmed that the entire spectrum shifted, supporting the presence of charge transfer from the Sr atoms.

To identify the possible origin of this core level shift to higher binding energy which is induced by the Sr deposition, Seo et al. calculated the Si $2p$ SCLS's for the Si(001) $p(2\times 2)$ and 1/2 ML Sr-adsorbed 2×1 Si(001) surfaces.

In the initial state approximation, the $2p$ core level binding energy was calculated from the difference between the $2p$ energy level ε_{2p} and the Fermi level ε_F. However, when a core hole is created at a Si atom, the system is excited and electrons tend to screen the core hole positive charge [79]. This relaxation energy gain was included on top of the initial state effect in the final state calculations.

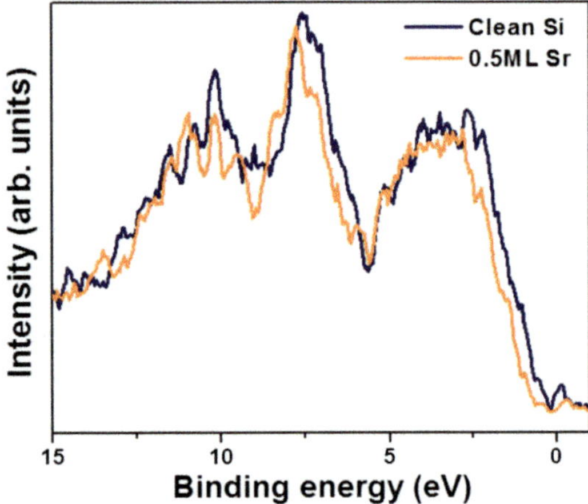

Fig. 6.11 Valence band edge of the clean 2 × 1 Si (001) and 1/2 ML Sr on Si (001) measured using XPS. The zero of energy is set at the Fermi level. Reprinted with permission from [35]. Copyright 2013, AIP Publishing LLC

Therefore, the difference between the initial and final state calculations could be used as a measure of the screening ability of the system [62, 68]. The larger relaxation effect tends to push the 2*p* peak towards lower binding energy.

The 2*p* core level binding energy in the final state theory was calculated as

$$E_B^{final} = E'(n_c - 1) - E(n_c)$$

where $E(n_c)$ is the ground state energy and $E'(n_c - 1)$ is the system's energy with a screened core hole. In order to ensure the overall charge neutrality of the system, one electron was added to the system (complete screening picture). First, the SCLS of the Si (001) $p(2 \times 2)$ surface was calculated and the positions of the S$_u$ and S$_d$ components were found to be −0.53 eV and −0.12 eV, respectively, in good agreement with the experimental results in Fig. 6.10a, as well as existing theoretical work [61, 62, 68]. For Sr on Si(001), it is found that the tilted dimer rows of the Si (001) $p(2 \times 2)$ surface are flattened due to the charge transfer [24]. As a result, there is only one dimer peak remaining at −0.35 eV which is in good agreement with the −0.33 eV found in the experiment as shown in Fig. 6.10b.

Considering bulk SrSi$_2$ [80], the 2*p* chemical shift induced by the charge transfer from Sr to Si was calculated to be −0.35 eV. Strikingly, however, the experiment on 1/2 ML Sr on Si showed that the bulk 2*p* peak and all other surface 2*p* peaks shift in an opposite way compared to the bulk SrSi$_2$ case. Using the initial state theory, the bulk 2*p* binding energy shift was found to be almost zero. This can be explained by the bulk 2*p* level rising up due to the surface dipole layer induced by the charge transfer, simultaneously with the Fermi level rising upward due to the change of the surface electronic structure. On the other hand, using the final state theory, the bulk 2*p* binding energy was calculated to increase by 0.42 eV in excellent agreement with experiment. Since there is a negligible shift in the initial state calculation, it should be noted that the shift of 0.42 eV originates from the reduced relaxation energy gain (screening effect) when 1/2 ML Sr is deposited.

Fig. 6.12 Work function variation as a function of Sr coverage on 2 × 1 Si(001) from experiment (*filled squares*) and theory (*open circles*). For the work function calculation for the Sr coverage of 1/6 ML, 1/4 ML, and 1/3 ML, we use the structural model proposed in [51] and [2], respectively. Reprinted with permission from [35]. Copyright 2013, AIP Publishing LLC

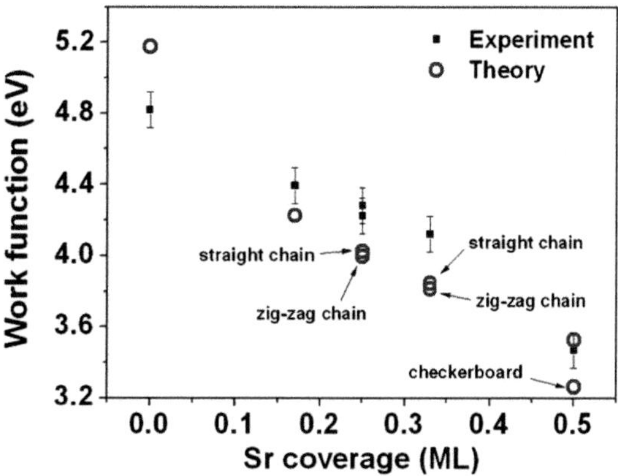

6.2.3   Effect of Sr on Work Function

Figure 6.12 shows the work function of Si(001) as a function of Sr coverage from experiment and theory [35]. Choi et al., measured the work function of clean 2×1 Si (001) to be 4.82 eV by using UPS [81, 82]. The measured work function decreases with increasing Sr coverage. The deposition of 1/2 ML of Sr causes a change of electronic structure of Si(001), which leads to a decrease in the work function by 1.35 eV and, considering experimental uncertainty, is in excellent agreement with theory. Similar work function changes caused by the interface electronic properties were also observed when organic molecules were deposited on metals or oxides [83, 84].

6.2.4   Conclusions

In conclusion, the key features of the Sr-based Zintl template on Si(001) were established by investigating the change in the electronic structure of 1/2 ML Sr-induced 2×1 reconstruction on Si(001) from that of clean 2×1 Si(001) using XPS, UPS and DFT. As a strong evidence of the charge transfer from 1/2 ML of Sr into the 2×1 Si(001) substrate, the bulk component of the Si $2p$ spectrum shifts toward higher binding energy by 0.49 eV and the previously separate S_u and S_d dimer components merge into a single surface component at −0.33 eV with respect to the bulk This indicates that, though intact, Si dimers of the Zintl template are un-tilted, in agreement with DFT predictions. Theoretical calculations using the final state theory are quantitatively consistent with the experimental results for the shift of the Si $2p$ spectrum and the decrease of the work function of the system upon Sr deposition.

6.3 Growing SrTiO$_3$ on Si

Epitaxial growth of STO on Si was first reported by Tambo et al. in early 1998 [85]. Their method involves the use of a relatively thick SrO buffer layer that is grown by MBE on the Si without the use of a Zintl template layer. After desorbing a chemically grown SiO$_2$ on Si, SrO is grown directly on Si from a Sr effusion cell at a substrate temperature of 300 °C and 5×10^{-8} Torr O$_2$ partial pressure. Because SrO is more thermodynamically stable than SiO$_2$, and because SrO crystallizes very easily owing to its ionic nature, SrO can be formed on Si, but has an SiO$_2$ interfacial layer. By growing a sufficiently thick SrO layer (100 Å), the interface with Si is sufficiently far away for it to not affect subsequent growth of STO. Tambo et al. grow STO by co-deposition using a substrate temperature of 500 °C and an oxygen partial pressure of 8×10^{-8} Torr. While successful at putting STO on a Si substrate, this method is not considered direct epitaxy of STO on Si as the buffer layer is essentially bulk-like SrO.

As mentioned in Sect. 6.1, the first true epitaxy of STO on Si was reported by McKee et al. at Oak Ridge National Laboratory in late 1998 [1]. They used the concept of layer by layer interfacial energy minimization to achieve direct epitaxy of STO on Si, ensuring that each new atomic layer is thermodynamically stable in contact with the previous layer while maintaining atomic registry [1]. In their ground breaking work, they report the use of a sub-monolayer Sr (they claim the use of ¼ ML of Sr) deposited at 600 °C on clean 2×1 Si(100) resulting in a c(4×2) reconstruction as observed by RHEED. This forms the Zintl silicide layer described in the previous section. Subsequent to the submonolayer silicide formation, the substrate temperature is lowered to 200 °C. An additional 1/2 ML of Sr metal is deposited, then oxygen is introduced with Sr deposition continuing, forming a SrO monolayer without perturbing the underlying silicide. Then, a TiO$_2$ layer is deposited followed by another SrO layer to form one unit cell of STO. Although, not explicitly mentioned in the original paper, this first unit cell of STO is presumably annealed at higher temperature as depositing a TiO$_2$ layer at 200 °C is known to result in an amorphous structure. The oxygen partial pressures used were also not specified in the original work. Additional STO was deposited by repeating the TiO$_2$ and SrO deposition and anneal to achieve the desired STO thickness. Additional details of the growth process were later published by Jeon et al. [86].

Using the Oak Ridge method as a basis, researchers at Pennsylvania State University led by Darrell Schlom further studied and optimized the STO on Si growth process. Details of the Penn State process were first reported in 2002 [87]. In their process, they first deposit 1/2 ML of Sr on clean Si(100) at 700 °C to form the Zintl template with 2×1 reconstruction. The substrate temperature is then reduced to about 120 °C where additional Sr metal (3/8 ML) is deposited prior to oxygen introduction. The Sr metal deposition results in a 3×1 surface reconstruction. When oxygen is introduced, the 3×1 pattern changes to a 1×1 pattern as the Sr metal is oxidized. When the oxygen partial pressure reaches 5×10^{-9} Torr, the Sr shutter is re-opened and oxygen pressure continues ramping to about 2×10^{-8} Torr. During this time, a total of 2 ML of SrO has been deposited, including the initial 3/8 ML. A single TiO$_2$ layer is then deposited at 120 °C and 2×10^{-8} Torr O$_2$ then all metal and oxygen sources are shut off and the sample is annealed in vacuum for 30 min at 500 °C. By the end of this process a 1.5 unit cells of crystalline, epitaxial STO (SrO-TiO$_2$-SrO) is formed directly on Si. Additional STO can then be grown at 500 °C and 1×10^{-8} Torr O$_2$ using a shuttered, alternating layer deposition of TiO$_2$ and SrO layers with TiO$_2$ first. Co-deposition of Sr and Ti also works but requires a slightly higher substrate temperature to obtain the same crystalline quality. In subsequent work, the growth process was tweaked slightly so that three SrO layers at 4×10^{-8} Torr O$_2$ and two TiO$_2$ layers at 2×10^{-7} Torr O$_2$ were deposited at ~100 °C. This layer of 2.5 unit cells of STO was crystallized at 550 °C in vacuum. To achieve thicker STO with no interfacial SiO$_2$, this process of low temperature deposition and crystallization was repeated several times in chunks of five unit cells [88, 89]. A very similar process was utilized by the research group of Charles Ahn at Yale University to obtain epitaxial STO on Si [90].

IBM Zurich also developed an optimized STO on Si growth process based on the original Oak Ridge method, which they first reported in 2004. In their work, the 1/2 ML Sr Zintl template was deposited at 650 °C. The substrate was then cooled down to ~100 °C and additional Sr metal is deposited (0.5–0.8 ML) to form a 3×1 reconstruction. Oxygen is then introduced and (Ba,Sr)O, with composition such that it is lattice matched to Si, was deposited rather than SrO. The low temperature (Ba,Sr)O deposition was done under an oxygen partial pressure of 2×10^{-8} Torr and a total of 3 ML was deposited, forming a crystalline, epitaxial layer. Amorphous STO (up to ten unit cells) is then deposited at ~100 °C on top of the (Ba,Sr)O under an oxygen partial pressure of 7×10^{-8} Torr. The STO is crystallized by a 10-min 500 °C vacuum anneal [91–93].

Motorola first reported being able to grow epitaxial STO on Si in 1999, although no details of the process were described in their early reports [15, 16]. One key difference was the development of a Sr-assisted SiO$_2$ removal process where Sr is deposited to saturation coverage on Si containing a UV oxide at temperatures above 700 °C and then heated to about 850 °C where the Sr catalyzes the formation of SiO and facilitates the complete removal of SiO$_2$ from the Si surface [28]. Upon cooling back down to below 600 °C, a small amount (~1/4 ML) of Sr remains on the Si surface. Details of the Motorola process were finally revealed in 2003. After SiO$_2$ removal using the Sr-deoxidation procedure, additional Sr is deposited at 600 °C to complete the 1/2 ML Sr template, which can be determined by carefully observing the RHEED pattern until one sees a clear 2×1 reconstruction. The substrate temperature was further lowered to 200–300 °C for subsequent STO deposition without an initial Sr metal or SrO intermediate layer. Oxygen was introduced to a partial pressure of 5×10^{-8} Torr and then Sr and Ti were co-deposited while the oxygen pressure continues to increase to the low 10^{-7} Torr range [26]. After a thickness of one to three unit cells was deposited, the metal and oxygen sources are shut off and the sample is vacuum annealed at 600 °C for a few minutes.

This process is repeated several times to achieve up to ten unit cells of STO. Beyond ten unit cells, STO can be grown in crystalline form at 700 °C and 1×10^{-7} Torr O$_2$. A variant of the above process was reported by Liang et al. [94] where the first five unit cells of STO were deposited while simultaneously ramping oxygen pressure (from 1×10^{-8} to 2×10^{-7} Torr) and temperature (from 300 to 500 °C) linearly and then subsequent STO was deposited at 500 °C and 2×10^{-7} Torr O$_2$. Liang et al. also studied the difference between co-deposition and alternating layer growth and found that co-deposition produces better films. The Motorola process was also used by the group of Bruce Wessels at Northwestern University to fabricate their STO on Si samples [95, 96].

In 2006, the research group of Saint-Girons at Ecole Centrale de Lyon developed their own variant of the Motorola version of the STO on Si growth process [97]. As with all such processes, the first step is to deposit 1/2 ML Sr on clean Si(100). The Lyon group used the Motorola-developed Sr deoxidation process to remove SiO$_2$ and to add additional Sr at 600 °C to complete the 1/2 ML Sr Zintl template. Three unit cells of amorphous STO are then deposited at 250 °C with an oxygen pressure of 5×10^{-8} Torr. Without removing oxygen, the amorphous STO is crystallized at 550 °C for 10 min. Subsequent STO is then deposited at 4×10^{-7} Torr O$_2$ at 550 °C. In 2009, the Lyon group further optimized their process to prevent formation of interfacial SiO$_2$ [98]. In their new process, after the 1/2 ML Sr Zintl template formation, the substrate is cooled to 360 °C and then the Zintl template is oxidized at 6×10^{-8} Torr O$_2$ to form a Sr$_{0.5}$O layer. On top of this, two to three unit cells of STO is deposited at 6×10^{-8} Torr O$_2$ and then the oxygen partial pressure is increased quickly to 1×10^{-6} Torr while keeping the temperature at 360 °C. Subsequent STO is then deposited at these growth conditions.

The Demkov group at the University of Texas at Austin developed their own modification of the Motorola STO on Si process in 2010, the details of which were first reported in 2012 [99]. They also utilize Sr-assisted deoxidation to remove the SiO$_2$ layer from Si at 800 °C and then heat the Sr-covered deoxidized wafer to 850 °C for 5 min to remove more Sr to achieve a Sr coverage of ¼ ML. The substrate is then cooled down to 575 °C where the sticking coefficient of Sr metal below 1 ML coverage is unity [75]. An additional ¼ ML of Sr is then deposited at 575 °C. By using this method, the formation of the Sr Zintl template is highly reproducible. For STO deposition, the substrate is first cooled to 200 °C and oxygen is introduced to a pressure of 6×10^{-8} Torr. Sr and Ti are then co-deposited while the oxygen pressure is ramped to 3×10^{-7} Torr. By carefully tuning the Sr/Ti fluxes, STO growth even at 200 °C is highly disordered but still crystalline as observed by RHEED. Semi-crystalline STO with thicknesses of three to ten unit cells is typically grown and then fully crystallized by a vacuum anneal at 550 °C for 5 min. Subsequent STO growth can proceed at 500 °C under an oxygen pressure of 4×10^{-7} Torr, which was found to form a ~10 Å SiO$_x$ interfacial layer, or by repeating the semi-crystalline growth at 200 °C and vacuum anneal at 500 °C several times to form a SiO$_2$-free interface.

6.4 Strain Management

As described in the previous section, STO can be grown epitaxially on Si without forming an interfacial SiO_2 layer by first depositing half a monolayer of Sr metal on Si, which partially protects the Si surface from rapid oxidation while essentially maintaining the underlying surface structure of Si [26]. If a thin amorphous STO layer is then grown on the Sr-passivated Si surface near room temperature and subsequently crystallized in vacuum, an epitaxial layer of STO can be formed directly on Si. Thin STO grown in this manner has a compressive strain of 1.7 % at room temperature, with the STO unit cell rotated by 45° relative to the Si conventional unit cell. Once the initial STO layer is crystallized, subsequent STO growth can be done in two ways. If one continues the STO deposition near room temperature and annealing in vacuum after each deposition step, a thicker STO layer with no interfacial SiO_2 can be obtained after repeating the process a sufficient number of times. Below the critical thickness, STO grows coherently strained to Si [100, 101]. Strained STO on Si is polar and has also been reported to be ferroelectric [102, 103].

Another way of obtaining thicker STO is to simply treat the initial STO layer as if it were a regular STO substrate and deposit more STO under the usual high oxygen partial pressure at high substrate temperature. Using this method, however, results in the formation of a thin amorphous SiO_2 layer at the interface of STO and Si, since the excess oxygen is able to diffuse through the STO into the underlying silicon. STO films grown in this manner are found to have an in-plane lattice constant that is larger than that of bulk STO [96, 103]. This in-plane expansion is attributed to the thermal expansion coefficient mismatch (the thermal expansion coefficient of STO is 8.8×10^{-6} K^{-1} which is four times larger than that of Si) [97]. The presence of the SiO_2 interlayer itself has beneficial effects from the MOSFET point of view: sufficient conduction band offset and improved channel mobility, but at the price of a larger equivalent oxide thickness. Furthermore, as has been demonstrated by Choi et al. [99], post-deposition oxygen annealing of the $STO/SiO_2/Si$ stack can provide a way of controlling the strain relaxation of the STO layer by controlling the thickness of the SiO_2 interlayer.

Choi et al. studied the strain relaxation behavior of STO films grown on Si through post-deposition annealing as a function of oxygen pressure and time [99]. Using epi-grade prime Si(001) wafers and a variation of the Sr-assisted deoxidation process developed by Motorola, they grew STO using the Zintl template. All films in this study were grown with the same recipe and had approximately 62 Å of total STO thickness. The particular growth process used in the study resulted in the formation of a ~20 Å SiO_2 interlayer prior to any post-deposition annealing under a controlled oxygen partial pressure. For the post-deposition anneal, the oxygen partial pressure was varied from 2×10^{-7} Torr to 1×10^{-5} Torr using an annealing temperature of 650 °C. The annealing time was varied from 10 to 90 min. The lattice parameters of the STO films were measured at room temperature by RHEED

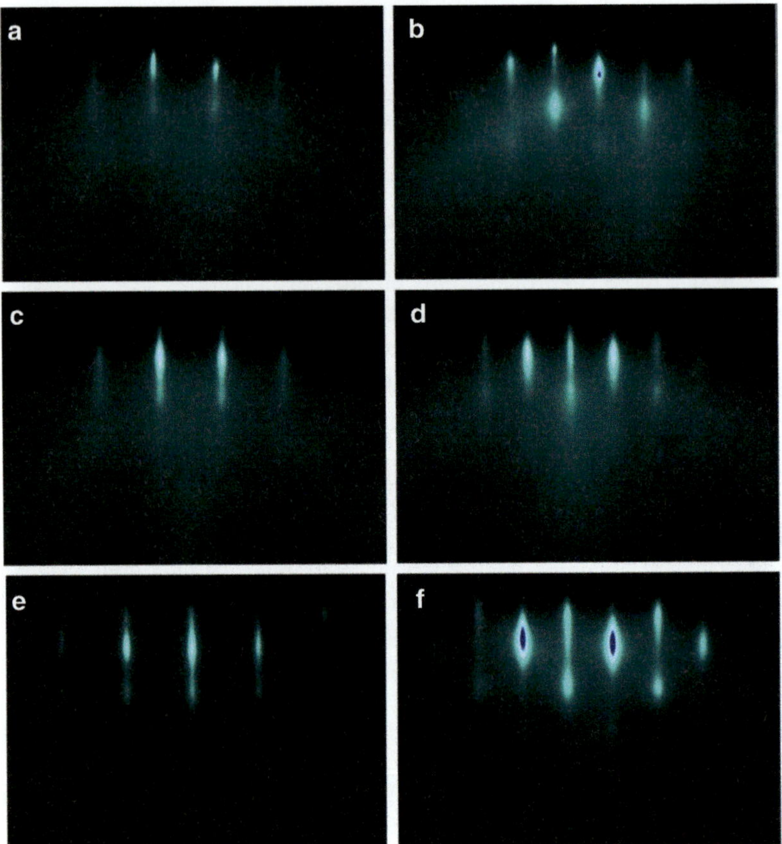

Fig. 6.13 RHEED patterns of the STO films on Si after each growth step along <010> (b, d, f, h) and <110> (c, d, g, i) directions. (**a**) and (**b**) are RHEED patterns after the initial three ML of amorphous STO are crystallized. (**c**) and (**d**) are RHEED patterns after the main 13 ML STO growth at 565 °C. (**e**) and (**f**) are RHEED patterns after a post-deposition oxygen anneal at 650 °C. Reprinted with permission from [99]. Copyright 2012, AIP Publishing LLC

and by XRD. The layer thicknesses were measured by a combination of x-ray reflectivity (XRR) and transmission electron microscopy (TEM). For selected films, in-plane and out-of-plane high resolution XRD measurements were carried out at the National Synchrotron Light Source beamline X20A ($\lambda = 1.5407$ Å).

Figure 6.13a–d shows the RHEED patterns along the <010> and <110> directions of STO after each growth step. All of the STO films show qualitatively the same RHEED patterns. The RHEED pattern shown in Fig. 6.13e, f is for a sample annealed at 650 °C for 30 min in a 1×10^{-6} Torr oxygen environment. After post-deposition annealing, the streaks became sharper, indicating improvement in the crystallinity and flatness of the STO film.

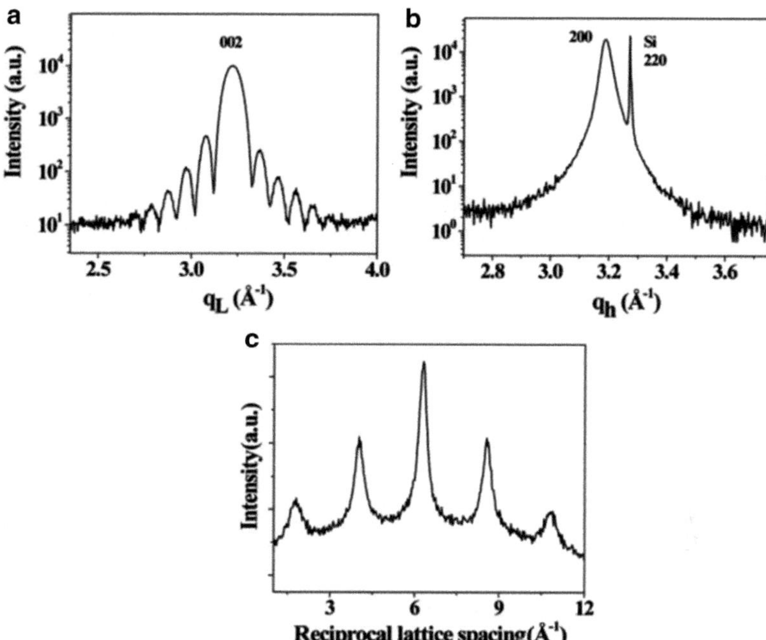

Fig. 6.14 Typical x-ray diffraction and RHEED data used to determine STO lattice constants. (**a**) X-ray diffraction L scan about STO 002 peak; (**b**) H scan about STO 200 peak; (**c**) RHEED profile of STO film on Si. The data shown are for an un-annealed film. Reprinted with permission from [99]. Copyright 2012, AIP Publishing LLC

An X-ray diffraction L scan around the 002 diffraction peak of an un-annealed STO film on Si is shown in Fig. 6.14a. The 002 peak position is used to obtain the out-of-plane lattice constant. The in-plane lattice constant is then calculated using the Poisson's ratio of bulk STO. The in-plane lattice parameter can be calculated following the equation, $a = a_{STO,\ bulk} + (c - a_{STO,\ bulk}) \cdot ((\nu - 1)/2\nu)$, where $a_{STO,\ bulk}$ is the bulk lattice parameter of STO; c is the experimental value obtained from the X-ray diffraction 002 peak of the STO films; and ν is the Poisson's ratio of bulk STO (0.232) [104]. The un-annealed film has lattice parameters $a = 3.935$ Å and $c = 3.902$ Å, indicating an in-plane expansion of STO on Si relative to bulk due to the thermal expansion coefficient mismatch during the cool down process [97].

The in-plane lattice constant of an un-annealed sample was also measured using grazing incidence x-ray diffraction at the National Synchrotron Light Source, with a value of 3.936 Å being obtained, confirming the in-plane expansion of the un-annealed STO film (Fig. 6.14b). As a further check to confirm the observed trends in XRD lattice constants, in-plane lattice constants of the deposited films were also obtained from RHEED patterns at room temperature, with a typical measurement shown in Fig. 6.14c. RHEED confirms that the un-annealed film has an expanded in-plane lattice constant and that the lattice constant trend

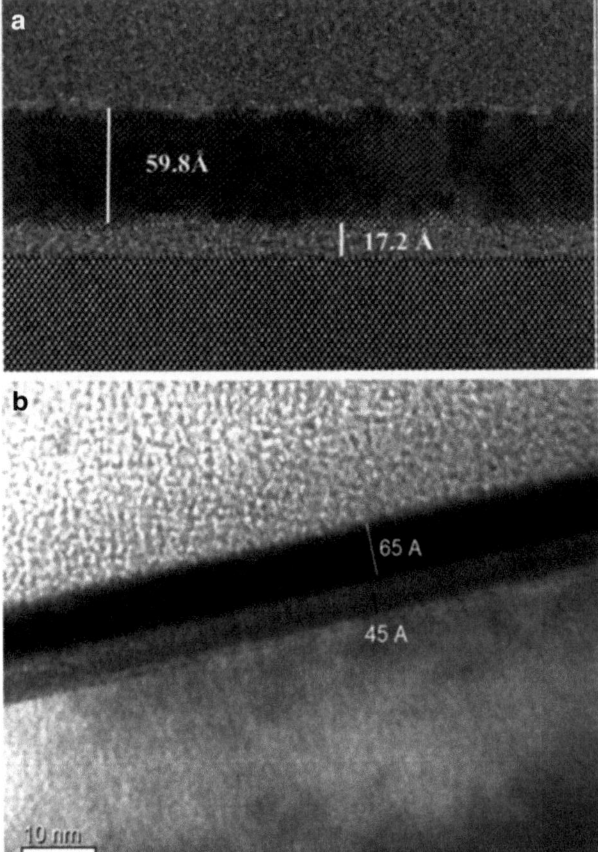

Fig. 6.15 Cross sectional TEM images of STO films grown on Si. (**a**) High-resolution lattice image of an un-annealed film; (**b**) typical low-resolution Z-contrast image of an STO film annealed in oxygen. The specific image is for a sample annealed under 5×10^{-7} Torr of oxygen at 650 °C for 10 min. Reprinted with permission from [99]. Copyright 2012, AIP Publishing LLC

observed in RHEED is the same as that of the values calculated using the Poisson's ratio. During interlayer formation, the STO lattice assumes a size appropriate to the growth temperature because the exothermic oxidation process disrupts the epitaxy. The larger thermal expansion coefficient of STO compared to Si (four times larger) results in an in-plane expansion during cool down from the growth temperature because STO is clamped to the SiO$_2$ interlayer.

To obtain the SiO$_2$ thickness, x-ray reflectivity (XRR) was used and the data was analyzed with the simulation program SimulReflec [105]. Due to being a multilayer system, various conditions such as density, roughness, and thickness had to be considered. The fitting process was repeated with varying initial conditions to minimize error. Cross-section TEM was also performed on the samples to measure the thickness of the amorphous interlayer. Figure 6.15a shows a high resolution image of an un-annealed sample showing highly crystalline STO layers with an initial ~20 Å SiO$_2$ interlayer. Figure 6.15b shows a typical low resolution cross-sectional TEM image of an annealed STO film on Si(001). This particular film was subjected to post-annealing in 5×10^{-7} Torr oxygen pressure at 650 °C for 10 min.

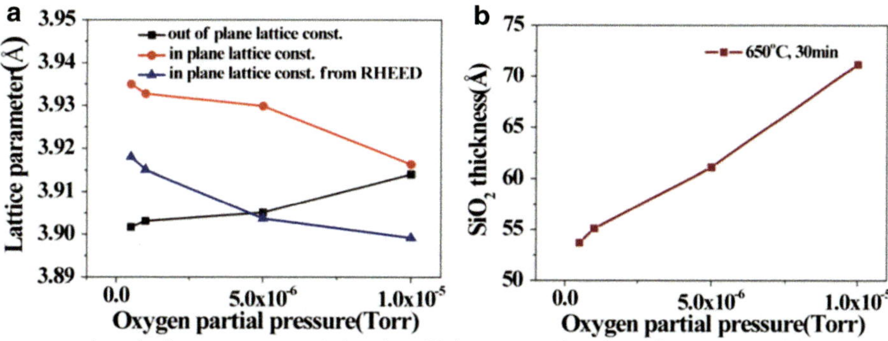

Fig. 6.16 (**a**) In-plane lattice constants and (**b**) SiO$_2$ thickness as a function of oxygen partial pressure. All films were annealed at 650 °C for 30 min in different oxygen environments. Reprinted with permission from [99]. Copyright 2012, AIP Publishing LLC

The annealing resulted in an SiO$_2$ thickness of 45 Å as measured by TEM. The SiO$_2$ thickness obtained from the TEM images agrees well (to within 10 %) with that from the x-ray reflectivity simulation program for all the samples measured.

A summary of the effect of oxygen partial pressure on the SiO$_2$ thickness and the STO lattice constants is shown in Fig. 6.16. All films in Fig. 6.16 were annealed at 650 °C for 30 min under different oxygen partial pressures. In Fig. 6.16a, both the in-plane and out-of-plane lattice constants show a systematic variation as the oxygen partial pressure is increased. The in-plane lattice constant of the STO film is initially larger than that of bulk STO, due to the difference in thermal expansion between Si and STO film. However, as oxygen partial pressure is increased, the in-plane lattice constant of the STO films decreases while the out-of-plane lattice constant increases. The STO films thus become more cubic, indicating that the STO films experience relaxation toward the bulk, stress-free lattice parameters. At the same time, the SiO$_2$ thickness increases with increasing oxygen partial pressure, as shown in Fig. 6.16b. This suggests that the relaxation of STO towards its bulk lattice constant is concurrent with the growth of the SiO$_2$ interlayer during the annealing process, with the STO becoming increasingly decoupled from Si as oxygen partial pressure is increased.

Figure 6.17 summarizes the effect of annealing time on SiO$_2$ thickness and lattice constants of STO. All films in Fig. 6.17 were annealed at 650 °C for different lengths of time in an oxygen environment of 5×10^{-7} Torr. From 10 to 60 min of annealing, the SiO$_2$ thickness and lattice constants were unchanged to within the limits of experimental error. However, annealing for 90 min results in an unexpected compression of the in-plane, and expansion of the out-of-plane lattice constants. This unusual behavior was found to be reproducible. It is not yet clear why the in-plane lattice constants decreased as SiO$_2$ thickness was increased for the longer annealing duration at this oxygen pressure. It should also be noted that there is a marked difference in the evolution of the RHEED patterns for 90 min annealing duration and 10 min annealing duration. Figure 6.18a shows the RHEED pattern of the film which was annealed at 650 °C at 5×10^{-7} Torr for 90 min. This film was

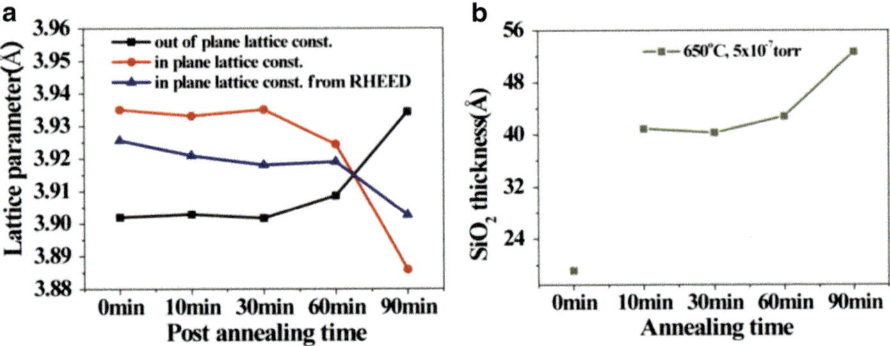

Fig. 6.17 (a) In-plane lattice constants and (b) SiO$_2$ thickness as a function of annealing duration. All films were annealed in oxygen environment of 5×10^{-7} Torr at 650 °C. Reprinted with permission from [99]. Copyright 2012, AIP Publishing LLC

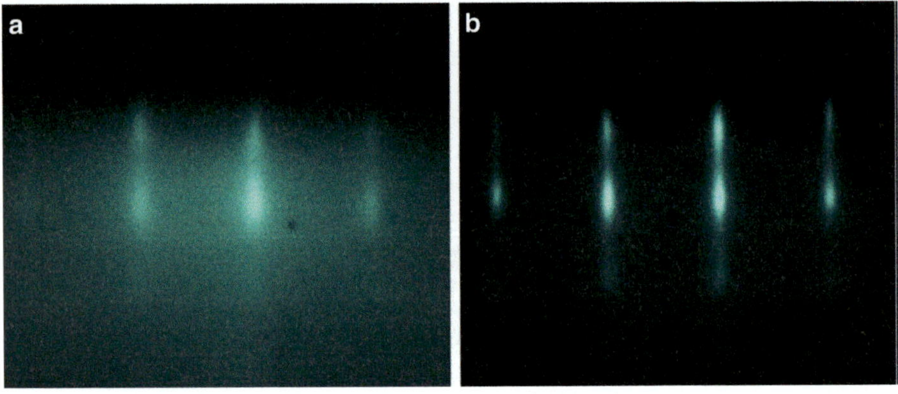

Fig. 6.18 RHEED patterns along <110> direction of a film which was (a) post-annealed at 650 °C in an oxygen environment of 5×10^{-7} Torr for 90 min, and of film which was (b) post-annealed at 650 °C in an oxygen environment of 1×10^{-5} Torr for 10 min. Reprinted with permission from [99]. Copyright 2012, AIP Publishing LLC

53 Å SiO$_2$ thick. During this longer annealing, the streaks became weaker and the background of the RHEED pattern became brighter, indicating degradation of the STO crystallinity.

However, for the film annealed at 650 °C in an oxygen environment of 1×10^{-5} Torr for 10 min (Fig. 6.17b), the streaks remained very sharp, indicating that the STO crystallinity is still good even though the SiO$_2$ thickness has increased to 71 Å. One possible explanation is that the longer annealing duration could allow for Sr and/or Ti to diffuse down to the SiO$_2$ layer (or Si to diffuse upward) and react with Si to form strontium silicate or titanium silicide [106–109]. This indicates that STO on Si may be a thermodynamically unstable but kinetically-limited state, which should be taken into consideration when using STO on Si as a virtual substrate.

All STO films grown at high temperature and high oxygen partial pressure develop a thin SiO_2 interlayer during growth. During post-deposition annealing, the thickness of this SiO_2 interlayer increases, affecting the strain relaxation behavior of the STO layer. Prior to annealing, the STO layers are initially expanded in-plane as a result of thermal expansion mismatch. As oxygen partial pressure is increased, the STO lattice constants relax towards their bulk, cubic values concurrent with an increase in the SiO_2 interlayer thickness. The use of post-deposition annealing can be used to tune the strain in STO films to within a half a percent. However, for prolonged annealing times (over 90 min), it is found that STO films show evidence of decomposition as manifested by an unexpected decrease of the in-plane lattice constant as SiO_2 thickness is increased.

6.5   Physical and Electronic Structure of the STO-Si Interface

6.5.1   Thermodynamic Considerations

We begin with a discussion of wetting that is critical for growth of high quality STO on Si. Until recently [109], an interfacial layer (SrO in [1] and SiO_2 in [16]) almost always separated Si from the perovskite. The growth mode necessary to produce a high quality film is a so-called two-dimensional (2D) or Frank-Van der Merwe growth [110]. In this mode the film grows layer-by-layer, and if the lattice mismatch between the film and the substrate is sufficiently small (typically less than 1 %) the film grows epitaxially. In the case of a larger lattice mismatch strain in the epitaxial layer builds up with the layer thickness, and is eventually relieved either through island formation (the Stransky-Krastanov mode) or plastically. Strained films below the critical thickness may however be grown; for the layer-by-layer growth to occur the film should *wet* the substrate. The cubic perovskite cell of STO contains Sr atoms at the cube's corners, one Ti atom at the body center, and oxygen atoms in the centers of all faces. The matching of the perovskite cell to silicon is often described as a 45° rotation of the perovskite with respect to the conventional cubic cell of Si. Indeed, a (1×1) unit cell of the unreconstructed Si (001) surface has a lattice vector of 3.84 Å $(a/\sqrt{2}, a = 5.43$ Å$)$, which is only 1.7 % smaller than the 3.905 Å cell of STO. Though many groups have reported the growth of thin STO films on Si using MBE, achieving the 2D epitaxial growth of dissimilar materials (hetero-epitaxy) is not a simple matter. The thermodynamic condition necessary for wetting is as follows: the surface energy of the substrate should exceed the sum of the surface energy of the epitaxial layer and the energy of the interface. Therefore, to examine the possibility of Si wetting by STO we need to know the surface energies of Si and STO, and estimate the energy of the interface between the two.

The surface energy of Si is easy to compute using first principles methods [111–114]. For consistency, Zhang et al. performed a slab calculation for a (2×2) Si cell and found 1,710 erg/cm^2 in good agreement with experiment [2]. They calculated the surface energy of STO following the method of Padilla and Vanderbilt [115]. The surface energy of STO was found to vary from 860 to 2,400 erg/cm^2, depending on the termination and the growth conditions as captured by the chemical potential of TiO$_2$. The details of the calculation can be found in [116, 117]. The lowest possible surface energy of 800 erg/cm^2 is achieved under SrO-rich conditions for the SrO-terminated surface. This is consistent with the fact that in most cases the MBE growth conditions are Sr rich. These numbers limit the set of interfacial structures that could result in wetting. Any interface with an energy cost higher than 900 erg/cm^2 means 3D growth.

We now discuss two models proposed by Zhang et al. for the Si-STO interface [2], and based on the MBE growth sequence used in experiment [108]. They and other workers (see below) considered many other structures but these two are important for the following discussion of the band alignment. The MBE growth starts with the deposition of a 1/2 ML Sr template [26] on the (2×1) reconstructed clean Si (001) surface (Sr atoms occupy trough sites). The template is then oxidized, and Sr and Ti are co-deposited in the presence of oxygen, resulting in STO film growth. Our first structure (later referred to as structure I or 1/2 ML interface) discussed here is built by connecting a TiO$_2$ layer of STO with the oxidized template (the oxygen to silicon ratio is 1:1). The rest of STO is then built up starting from this layer. Zhang et al. used a 2×2 supercell geometry; the lateral lattice constant was set to the theoretical value for Si. Internal coordinates and the vertical cell size were optimized using CASTEP and checked with VASP. The relaxed interface structure has 2×1 symmetry, and is shown in Fig. 6.19a. The Si dimer structure is preserved, with a slight increase of the dimer bond length to 0.25 nm (0.24 nm for the 2×1 surface). The Si-O, and Ti-O bond lengths are 0.164 nm and 0.212 nm, respectively (compare with 0.161 in SiO$_2$, and 0.198 nm in TiO$_2$ rutile). The Sr-O bond length is 0.264 nm (0.253 nm for the SrO rock salt structure). The Sr plane is shifted upward from the oxygen plane by 0.035 nm.

The second structure (further referred to as structure II or 1 ML interface) was built by connecting the SrO-terminated STO slab with the unreconstructed Si (001) surface. With respect to the model described above this interface has a stoichiometric SrO layer at the interface (full monolayer of Sr). The relaxation leads to the formation of slightly stretched Si dimers. The structure again has 2×1 symmetry (Fig. 6.19b), and is very similar to the 1/2 ML structure. However, Sr atoms located above Si dimers are displaced upward by 0.024 nm with respect to those above the troughs. This reflects the size mismatch between Sr and the Si (001) surface. The covalent and atomic radii of Sr are 1.91 Å, and 2.15 Å, respectively, and the unreconstructed Si (001) surface unit cell lattice vector is only 3.84 Å. Qualitatively speaking, the surface area of the Si surface is too small to hold 1 ML of Sr atoms.

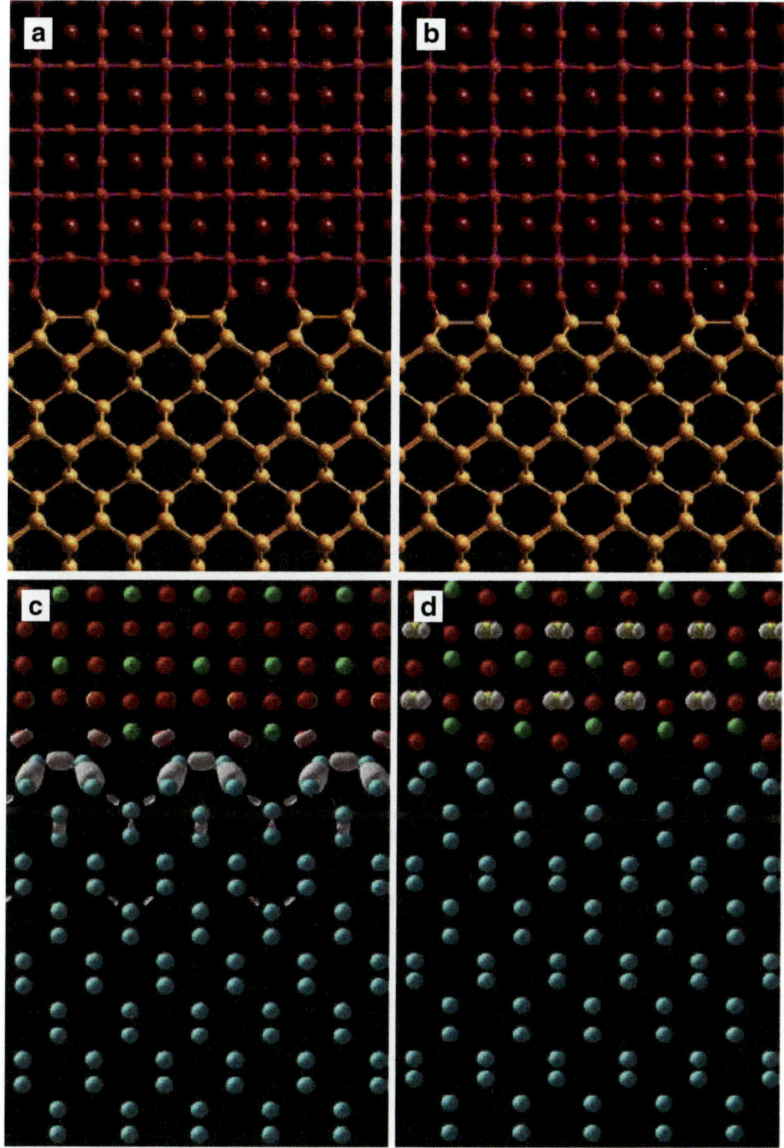

Fig. 6.19 (**a**) A (2 × 1) structure of the Si-STO interface with a ½ ML of Sr at the interface. (**b**) A (2 × 1) structure of the Si-STO interface with 1 ML of Sr at the interface. The electron density obtained by integrating over the states within a 1 eV window below the Fermi level; (**c**) the (2 × 1) structure with ½ ML of Sr at the interface. The states localized in the plane of the interface are clearly seen. (**d**) The (2 × 1) structure with 1 ML of Sr at the interface. No localized interface states are observed in the gap region (see text). From [2]

The energy of the stoichiometric 1 ML interface for the system containing different atomic species can be computed in a fashion similar to a surface calculation [115–117]:

$$E = \frac{1}{2}\left\{ E_{slab} - N_{Si}\mu_{Si} - N_{TiO_2}\mu_{TiO_2} - N_{SrO}\mu_{SrO} \right\}, \qquad (6.5)$$

here the energy is given per surface unit cell, and the factor of ½ is due to having two interfaces in the super-cell. The chemical potential of Si is set to the bulk Si energy. The chemical potentials of SrO and TiO$_2$ are related by the equilibrium condition $\mu_{SrO} + \mu_{TiO_2} = \mu_{STO}$, with the STO chemical potential is set to its bulk value. Thus the grand thermodynamic potential of the interface is a function of just one variable, which we choose to be μ_{TiO_2}. For the non-stoichiometric interface (1/2 ML) (6.5) needs to be modified and additional reaction channels need to be considered [117]. It turns out that for the 1/2 ML structure the grand thermodynamic potential can be written as a function of the oxygen chemical potential only. Under the typical MBE environment, the interfacial energy of the 1/2 ML interface is about 900 erg/cm^2 or borderline with respect to wetting. It can be lowered by reducing temperature and increasing oxygen pressure (the latter condition is not easily achievable for the oxidation of Si needs to be prevented [26]). Zhang et al. have found the interface energy for the (2×1) 1 ML interface to be 574 erg/cm^2 under Sr-rich conditions. It is important that the lowest interface energy of the 1 ML (2×1) structure is realized for the same growth conditions as the lowest STO surface energy (SrO-termination, Sr rich-growth). Therefore the sum of the interface energy and the surface energy of the STO film is now only 1,433 erg/cm^2. Comparing 1,710 erg/cm^2 with 1,433 erg/cm^2 we conclude that STO should wet Si under appropriate conditions. The wetting "window" exists only within a 40 % range of the allowed chemical potential values on the Sr-rich side. It is important to realize that contrary to our intuition the thermodynamic "window" for wetting is not all that large.

6.5.2 Conduction Band Offsets and Interface Structure

As the original proposal of McKee et al. was to use STO as a gate dielectric in a field effect transistor (FET), perhaps, the most important contribution of [2] was a clear demonstration that the conduction band offset between Si and STO is too small. The gate action of a FET is based on the existence of an energy discontinuity between the electronic bands of Si and the dielectric (it is 3.2 eV in the conduction band in the case of the Si/SiO$_2$ capacitor). This discontinuity, or band offset, provides an energy barrier for both electron injection from the gate electrode into the active area of the device and for parasitic electronic leakage current in the MOS capacitor at low voltage. To obtain a first order estimate of the band discontinuity at the Si-STO interface one can try the commonly used model [118, 119] due to

Tejedor-Flores-Tersoff (TFT) [120, 121]. In this model the conduction band offset is given by:

$$\phi_n = (\chi_a - \Phi_a) - (\chi_b - \Phi_b) + S(\Phi_a - \Phi_b) \qquad (6.6)$$

Here χ is the electron affinity, Φ is the charge neutrality level measured from the vacuum level, S is an empirical pinning parameter describing the screening by the interfacial states, and subscripts a and b refer to Si and STO, respectively. If $S = 0$ we get the strong pinning or the Bardeen limit, and if $S = 1$ we have no pinning or the Schottky limit. The electron affinities of Si and STO are 4.0 eV and 3.9 eV, respectively. The estimated charge neutrality level is 4.9 eV for Si and between 5.8 and 6.4 eV for STO (both are given with respect to the vacuum level) [2]. Thus within this simple theory one expects a 1.0–1.6 eV conduction band offset in the Bardeen limit, and a small 0.1 eV offset in the Schottky limit.

Note that the STO charge neutrality level estimate of Zhang et al. is different from that reported by Robertson and Chen [118, 119], whose result was obtained *via* searching for the zero of the Green's function [121]. However, the energy integration range was restricted to roughly six conduction bands corresponding to a minimum-basis tight-binding calculation, as the integral defined in [118, 119] had a logarithmic divergence. To get a better feel for the position of the branch point, Demkov et al. computed the complex band structure of STO and found the branch point 0.73 eV above the valence band top [122]. Given the uncertainty of the LDA band gap, the rescaled (using experimental band gap) charge neutrality level is 6.4 eV below vacuum. The length of the imaginary wave vector at the branch point along the (001) direction is 0.5 Å^{-1}. It is interesting to note that despite a smaller band gap the evanescent states die off much faster in STO (within merely 2 Å) than in monoclinic HfO$_2$ (3.3 Å). This anomalously rapid decay suggests a large value of the pinning parameter typical for the Schottky type alignment.

The following is a more realistic approach to the band offset estimation [2]. First Zhang et al. used a direct density of states analysis technique [123] to compute the valence band offset, and infer the conduction band offset using the experimental band gap values (1.17 eV and 3.2 eV for Si and STO respectively). They calculated the total valence band density of states for a 4.5 nm thick 2×1 Si-STO slab in vacuum. Site-projected densities were then computed separately for Si and STO atoms in the slab. The valence band discontinuity is then readily obtained (see Fig. 6.20a, b top panels). With this technique the conduction band offset was 0.87 eV, and 0.23 eV for models I and II, respectively (see Fig. 6.19a, b). Then, following Van de Walle and Martin [124] they used the electrostatic potential across the slab as a reference. The results are shown in Fig. 6.20a, b (lower panels). The agreement between the two methods is fair, considering the fast oscillation of the reference potential. In the spirit of the simple TFT theory one would conclude that structure I is closer to the Bardeen limit with the S value ranging between 0.1 and 0.47 (an empirical estimate gives 0.28 [118, 119]), while structure II corresponds to the Schottky limit. The fact that in structure II STO has a bulk SrO-layer

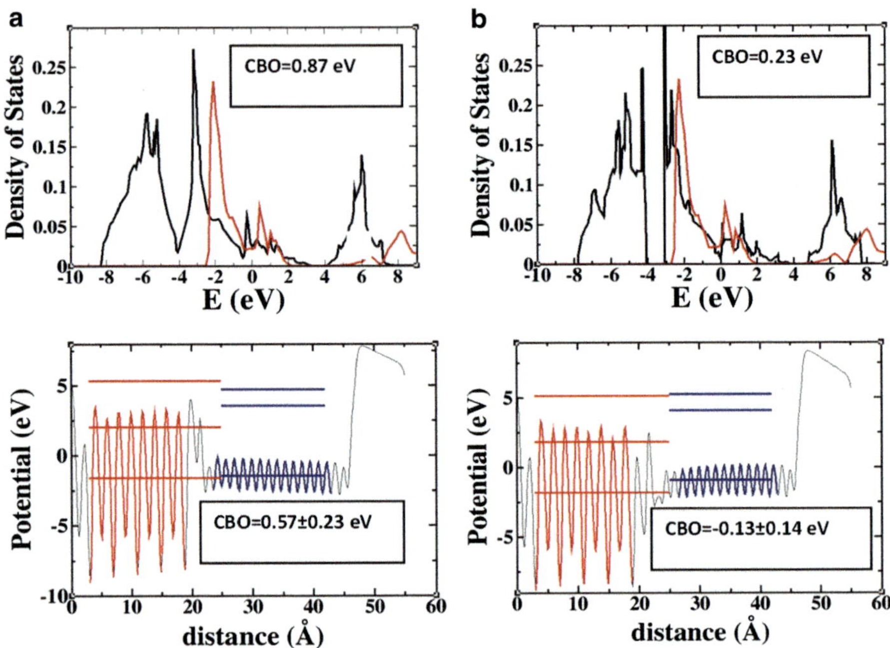

Fig. 6.20 The band discontinuity at the Si-STO interface; (**a**) For the (2×1) structure with ½ ML of Sr at the interface we find a sizable conduction band offset in agreement with the strong pinning or the Bardeen limit. The *top panel* shows the projected density of states analysis (the *red curve* shows the STO contribution, and the *black curve* the Si contribution). The conduction band offset of 0.87 eV is estimated using experimental values for the band gaps. The *bottom panel* shows the reference potential calculation. STO is on the left side of the simulation cell, and Si is on the right side. The average value of the potential on each side and both bands are indicated with *horizontal lines*. The valence band of STO is at 2.02 eV, the valence band of Si is at 3.58 eV resulting in the valence band offset of 1.56 eV. Using experimental values for the band gaps the conduction band offset is 0.57 eV. (**b**) For the (2×1) structure with a full Sr monolayer at the interface we find a very small conduction band offset in agreement with the unpinned Schottky model. The *bottom panel* shows the reference potential calculation. STO is on the left side of the simulation cell, and Si is on the right side. The valence band offset of 2.26 eV is found, and the conduction band offset is estimated to be -0.13 eV. From [2]

termination for which there are no surface states in the gap [115, 117] lends additional credence to such an assessment. This picture is indeed correct. In Fig. 6.19c, d we show the electron density obtained by integrating over the states within a 1.0 eV window below the Fermi level. In the case of structure I, states localized on Si dimers are clearly seen, while no localized charge is observed at the interface for structure II. The localized states of structure I fall into the STO gap. The origin of these states can be explained as follows. Note that the interface layer has the SrSi₂ stoichiometry corresponding to 1/2 ML of Sr deposited on the Si (001) 2×1 reconstructed surface at the template stage. The top of the valence band for such a template is precisely the dimer localized surface state. The relatively large

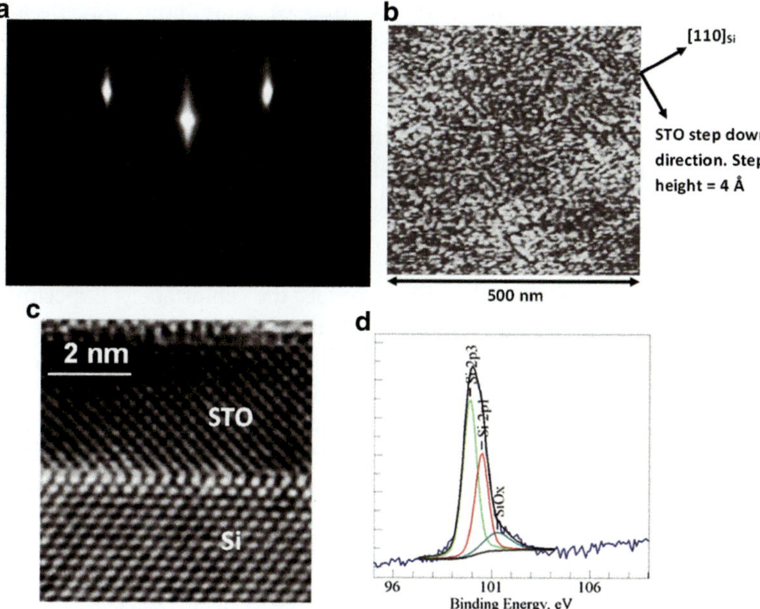

Fig. 6.21 (**a**) The RHEED pattern of a 2 ML STO film along the STO [100] direction. (**b**) An STM image of 5 ML STO film on Si, image was taken under 3 V sample bias (empty state). (**c**) The high resolution TEM image of a 10 ML STO film the epitaxial registry and high crystallinity are evident. (**d**) The XPS spectrum of the Si 2p peak for a 2 ML STO film on Si, no oxidation state higher than 2+ is present. From [2]

conduction band offset found for structure I does not solve the issue for gate dielectric applications, however, because of the Fermi level pinning by the interface charge.

The foregoing theoretical findings are consistent with the experimental data. First, experiment shows that STO can wet the Si surface and 2D nucleation is possible, but the process window is small. Details of the crystal growth and interface characterization can be found elsewhere [26, 108]. The RHEED of the STO surface taken during the growth, and the in-situ STM of a 5 ML thick film are shown in Fig. 6.21a, b. The RHEED image of the 2 ML STO presents the "smiley face" configuration, indicating a very flat surface. In addition, the disappearance of the Si half order reflections indicates that the surface is covered with STO instead of Si dimers. A very flat STO surface is also seen in the STM image, with clear one-unit cell high steps (4 Å). The full coverage of the substrate with such a thin layer of STO and the fact that STO itself exhibits a very flat surface strongly suggest that in agreement with calculations the STO layer is wetting the Si surface and the growth is in a 2D fashion, provided that the two materials are in direct contact. The high-resolution cross-section TEM image of the 10 ML film (Fig. 6.21c) illustrates local epitaxy.

Second, the STO/Si interface can be free of amorphous transition layers typically observed for oxides grown on Si [15, 125]. Figure 6.3c already suggests that an amorphous interfacial layer is minimal, if present at all, and confirmation of the absence of such a layer is established by a combination of XPS and EELS analysis [26, 108]. The XPS spectrum in the Si $2p$ region shown in Fig. 6.21d indicates that Si atoms in the oxidation state no higher than 2+ are present at the interface (the escape length of a photoelectron is only 2 nm). This confirms the absence of SiO$_2$ at the interface.

Third, the conduction band offset between STO and Si is small for the structure without chemically induced localization at the interface. Experimentally, Chambers and co-workers examined the band discontinuity at the Si-STO interface [126, 127]. They reported a small conduction band offset for n-type, and a negligible one for p-type Si substrates, in qualitative agreement with the results of Zhang et al. for the thermodynamically more stable interface structure II. Amy et al. measured valence band offsets using XPS and reported values of 2.38 and 2.65 eV depending on the oxygen content [128]. These values suggest a negative conduction band offset. Importantly, from the theoretical point of view the band offset at the Si-STO interface is very sensitive to the interface stoichiometry.

This last point has been recently reiterated by Kolpak and Ismail-Beigi who used DFT to examine a large number of possible STO/Si interfaces [129, 130]. Importantly, through comprehensive analysis they found that though the electronic properties are sensitive to the composition and structure of the interface, some universal features could be identified. The first key observation is that there is always charge transfer from the more electropositive Si to the oxygen atoms of the first oxide layer. In other words, the interface dipole is set by the ionicity difference. In general, the magnitude of the interface dipole increases with the concentration of Sr in the interfacial layer and decreases with the concentration of oxygen. The net dipole is always oriented towards the silicon substrate, and ranges from 0.04 to 0.69 eÅ/(unit cell). The second observation is that there is a large displacement between oxygen and metal (polarization) in the first oxide layer. In each case, the cations move away from the interface and anions move closer to it. While the sign of polarization is always the same (away from the substrate), its size varies depending on the interface structure and composition. In addition, the authors discuss the possibility of controlling the strain-induced polarization in the STO film *via* the interface composition. It is found that while the magnitude of polarization can indeed be altered, the direction is fixed by the chemical, mechanical and electrostatic boundary conditions. This suggests that despite tetragonal strain caused by epitaxy STO is polar but not ferroelectric.

Another recent study of the STO/Si interface has been performed by Hellberg et al. who used a combination of DFT and x-ray diffraction measurements to understand the formation as well as the structure of the interface [131]. The key physical idea of their approach is the realization that the interface cannot be optimized alone; the surface must be adjusted as well due to its electrostatic coupling with the interface. Hellberg and co-authors varied the stoichiometry of

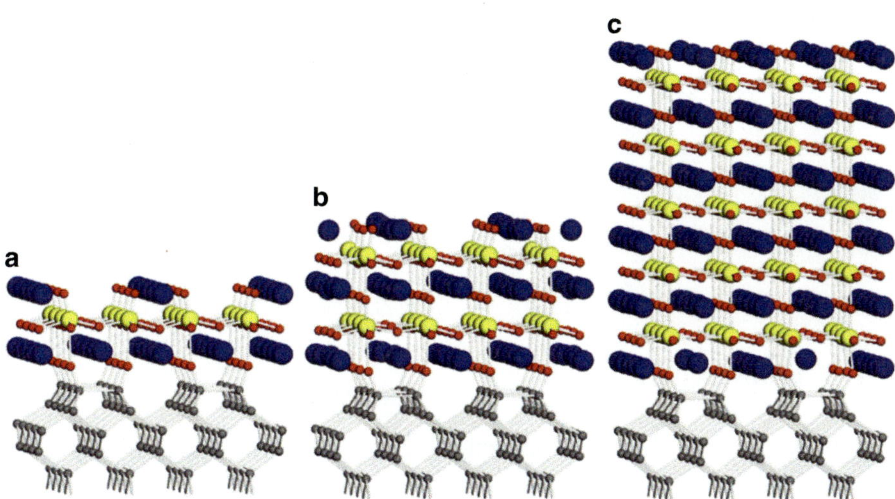

Fig. 6.22 The lowest energy structures of STO films on Si(001) at thicknesses of 1 (**a**), 2 (**b**), and 5 (**c**) unit cells. Sr atoms are *blue*, Ti are *yellow*, O are *red*, and Si are *grey*. The film thicknesses are defined by the number of TiO_2 layers. The SrO interface in (**a**) is energetically preferred; it is positively charged, and the surface assumes a compensating negative charge by forming Sr vacancies, resulting in a $Sr_{1/2}O$ surface. As the films grow, the charge of the interface is reduced by the formation of Sr vacancies, and the Sr content at the surface increases. At a thickness of two unit cells, the optimal interface and surface combination is $Sr_{15/16}O$ (interface) and $Sr_{9/16}O$ (surface), while at five unit cells it is $Sr_{11/16}O$ (interface) and $Sr_{13/16}O$ (surface). Note the polarization of the film as indicated by the upward displacement of the Ti atoms. The polarization decreases with increasing film thickness while the rotation of the O octahedra increases. Reprinted with permission from [131]. Copyright 2012 by the American Physical Society

both the interface and surface to determine the overall structure minimizing the free energy at each film thickness. The lowest energy interface they found is a full SrO layer, shown in Fig. 6.22a. As described below, the interface is positively charged. The surface assumes a compensating negative charge by changing its stoichiometry, creating a strong electric field which polarizes the film [100, 132]. As growth proceeds, the energy cost of the electric field grows with the film thickness. To reduce the field, Sr vacancies form at the interface, reducing its charge. The Sr vacancy density at the interface increases continuously with increasing film thickness, shown in Fig. 6.22b, c. At a thickness of five unit cells, the predicted structure agrees well with x-ray diffraction measurements of a five unit cell film grown coherently on Si(001) [26, 100, 103].

The interface with a full SrO layer shown in Fig. 6.22a, is positively charged: Each Sr has nominal charge +2, but each O-Si has charge -1 due to the ionization of the Si by the O [2]. This results in a planar charge density of +1 per perovskite cell at the interface. The surface compensates the interfacial charge by forming Sr vacancies. The $Sr_{1/2}O$ surface has a net charge density of -1 per surface cell, and overall the system is insulating. As the thickness of the film grows (as shown in Fig. 6.22b, c) the energy cost of the charged interface becomes too large, and Sr vacancies form. Writing the interface stoichiometry as Sr$_x$O, DFT calculations indicate that x decreases with increasing thickness. The two unit cell film has a Sr$_{15/16}$O interface, and the five unit cell film has a Sr$_{11/16}$O interface. The optimal surface, with stoichiometry Sr$_y$O, always compensates the interfacial charge. The energy is lowest when the sum of the charges of the interface and surface is zero, or $x + y = 3/2$ [131].

6.6   Commercial Process

Gu, Lubyshev, Batzel et al. at IQE, Inc. recently confirmed that STO/Si pseudo-substrate technology can be transitioned to high volume manufacturing [133]. They reported growth of STO on Si conducted on a manufacturing (not R&D) MBE platform that is capable of growth on substrates up to 8 in. in diameter. Growth was performed on an Oxford Instruments V-100 MBE system using Si wafers with diameters up to 8 in. The Si substrates were prepared in a UV ozone cleaning system to remove the residual organic contamination on the surface. After loading substrates into the MBE chamber, a Sr-assisted thermal desorption procedure was used to remove the native SiO$_2$ on the surface prior to the epitaxial growth [134]. RHEED exhibited bright and clear (2×2) patterns from the Si (100) substrate, indicating that the surface oxide was completely desorbed and the surface was clear of any native amorphous SiO$_2$ layers. The perovskite growth was initiated by depositing a thin (2.5 to 5 ML) layer of STO followed by recrystallization at 550 °C. Bulk STO growth was carried out at high temperature (between 700 and 800 °C) with a growth rate around 4 Å/min. STO epilayers with thicknesses of 100, 190, 280, 390, 510, and 1,200 Å were reported. Rutherford back-scattering confirmed excellent stoichiometric control. The XRD data for the films indicated that the STO epilayer was single crystalline, single phase, and with excellent crystalline quality as evidenced by small FWHM value of STO (200) x-ray diffraction rocking curve of 0.06° for a 1,200 Å film. AFM images showed that smooth, defect-free STO surface can be attained under optimized conditions, with a rms roughness value of ~0.6 Å. The dependence of strain on STO film thickness has been studied, and the critical thickness was found to be around 280 Å. Cross-sectional TEM micrographs revealed an abrupt interface between STO and Si, with a 2.6 nm thick SiO$_2$ interfacial layer. Thickness uniformity across an 8 in. wafer was ~1 %. This work is clearly a demonstration of the manufacturability of this material system.

Although MBE has been the primary means of growing epitaxial oxides on silicon, largely due to its control of atomic layer-by-layer deposition, there is often difficulty maintaining a clean interface with the relatively high temperature of MBE growth. For many applications, the presence of an amorphous SiO$_x$ layer at the

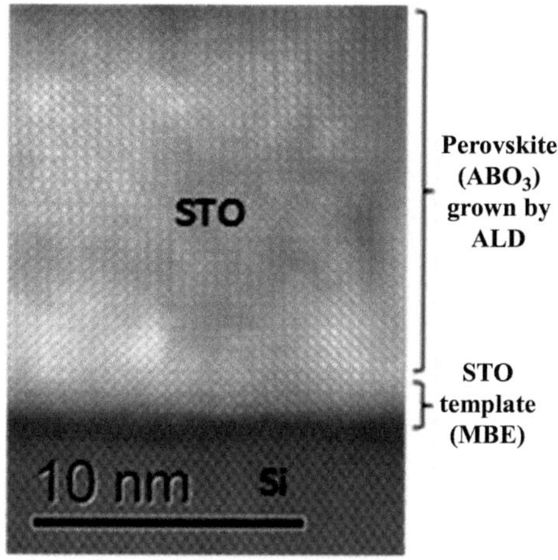

Fig. 6.23 Cross-sectional TEM of a 15-nm thick STO film on Si (001) using the combined MBE-ALD technique. From [135]

STO-Si interface is undesirable. Atomic layer deposition (ALD) has been recently proposed as a potential method to grow epitaxial metal oxides on Si without this amorphous layer [135–137]. Because of the low deposition temperatures used in ALD, chemical reactions at the metal oxide/silicon interface (e.g., STO/Si) are minimized [135]. McDaniel et al. used MBE to grow a four-unit-cell (~1.5 nm) STO (001) layer directly on Si (001), which serves as a surface template for subsequent ALD growth. STO films were grown at 250 °C using strontium bis (triisopropylcyclopentadienyl) $[Sr(iPr_3Cp)_2]$ (HyperSr), titanium tetraisopropoxide $[Ti(O\text{-}iPr)_4]$ (TTIP), and water as co-reactants. HyperSr and TTIP were vaporized at 130 °C and 40 °C, respectively, and water was held at room temperature. The water dosing was regulated using an in-line needle valve. In situ XPS revealed that the ALD process did not induce additional Si-O bonding at the STO-Si interface. However, post-deposition annealing at higher temperatures (>275 °C) gave rise to a small increase in Si-O bonding, as indicated by XPS. Figure 6.23 shows a cross-section transmission electron microscopy (TEM) image of 15 nm of ALD-grown single crystalline STO on the MBE-grown STO template on Si. The TEM confirms that there is negligible SiO_2 formation as a result of the ALD process. Therefore, careful consideration of growth temperature and annealing conditions may allow epitaxial oxide films to be grown by ALD on Si(001) substrates without an amorphous SiO_x layer. Using an STO-buffer layer as a surface template should be extendable to other crystalline perovskite films to be monolithically integrated with silicon by ALD. Successful demonstration of this combined MBE-ALD technique has enabled monolithic integration of several crystalline oxide films with Si(001), including photocatalytic anatase (TiO_2) [136], STO [135], and high-k $LaAlO_3$ [137].

In summary, over the past 15 years, the robust STO/Si pseudo-substrate technology has been developed. Multiple methods of deposition have been demonstrated, including wafer scale manufacturing MBE and chemical deposition such as ALD. Understanding of microscopic physics and chemistry of the perovskite/Si interface enabled precise control of the growth process and physical properties of the STO film. In addition to large area high quality single crystal STO, the technology offers a straightforward way of integrating many perovskite oxides and other materials on Si (001). Efforts at integrating functional oxides on Si using STO/Si pseudo-substrates is the topic of the next chapter.

References

1. R.A. McKee, F.J. Walker, M.F. Chisholm, Phys. Rev. Lett. **81**, 3014 (1998)
2. X. Zhang, A.A. Demkov, H. Li, X. Hu, Y. Wei, J. Kulik, Phys. Rev. B **68**, 125323 (2003)
3. C.R. Ashman, C.J. Först, K. Schwarz, P.E. Blöchl, Phys. Rev. B **69**, 075309 (2004)
4. Z. Yu, Y. Liang, C. Overgaard, X. Hu, J. Curless, H. Li, Y. Wei, B. Craigo, D. Jordan, R. Droopad, J. Finder, K. Eisenbeiser, D. Marshall, K. Moore, J. Kulik, P. Fejes, Thin Solid Films **462**, 51 (2004)
5. Y. Liang, S. Gan, Y. Wei, R. Gregory, Phys. Status Solidi B **243**, 2098 (2006)
6. C. Rossel, B. Mereu, C. Marchiori, D. Caimi, M. Sousa, A. Guiller, H. Siegwart, R. Germann, J.-P. Locquet, J. Fompeyrine, D.J. Webb, C. Dieker, J.W. Seo, Appl. Phys. Lett. **89**, 053506 (2006)
7. J.W. Reiner, A. Posadas, M. Wang, M. Sidorov, Z. Krivokapic, F.J. Walker, T.P. Ma, C.H. Ahn, J. Appl. Phys. **105**, 124501 (2009)
8. V. Vaithnayathan, J. Lettieri, W. Tian, A. Sharan, A. Vasudevarao, Y.L. Li, A. Kochhar, H. Ma, J. Levy, P. Zschack, J.C. Woicik, L.Q. Chen, V. Gopalan, D.G. Schlom, J. Appl. Phys. **100**, 024108 (2006)
9. G. Niu, S. Yin, G. Saint-Girons, B. Gautier, P. Lecoeur, V. Pillard, G. Hollinger, B. Vilquin, Microelectron. Eng. **88**, 1232 (2011)
10. A.K. Pradhan, J.B. Dadson, D. Hunter, K. Zhang, S. Mohanty, E.M. Jackson, B. Lasley-Hunter, K. Lord, T.M. Williams, R.R. Rakhimov, J. Zhang, D.J. Sellmyer, K. Inaba, T. Hasegawa, S. Mathews, B. Joseph, B.R. Sekhar, U.N. Roy, Y. Cui, A. Burger, J. Appl. Phys. **100**, 033903 (2006)
11. J. Wang, H. Zheng, Z. Ma, S. Prasertchoung, M. Wuttig, R. Droopad, J. Yu, K. Eisenbeiser, R. Ramesh, Appl. Phys. Lett. **85**, 2574 (2004)
12. A. Posadas, M. Berg, H. Seo, A. de Lozanne, A.A. Demkov, D.J. Smith, A.P. Kirk, D. Zhernokletov, R.M. Wallace, Appl. Phys. Lett. **98**, 053104 (2011)
13. H. Seo, A.B. Posadas, C. Mitra, A.V. Kvit, J. Ramdani, A.A. Demkov, Phys. Rev. B **86**, 075301 (2012)
14. M.D. McDaniel, A. Posadas, T. Wang, A.A. Demkov, J.G. Ekerdt, Thin Solid Films **520**, 6525 (2012)
15. Z. Yu, R. Droopad, J. Ramdani, J. Curless, C. Overgaard, J. Finder, K. Eisenbeiser, J. Wang, J. Hallmark, W. Ooms, Mater. Res. Soc. Symp. Proc. **567**, 427 (1999)
16. K. Eisenbeiser, J. Finder, Z. Yu, J. Ramdani, J. Curless, J. Hallmark, R. Droopad, W. Ooms, L. Salem, S. Bradshaw, C.D. Overgaard, Appl. Phys. Lett. **76**, 1324 (2000)
17. M.K. Singh, S. Ryu, H.M. Jang, Phys. Rev. B **72**, 132101 (2005)
18. K. Eisenbeiser, R. Emrick, R. Droopad, Z. Yu, J. Finder, S. Rockwell, J. Holmes, C. Overgaard, W. Ooms, IEEE Electron Device Lett. **23**, 300 (2002)
19. L. Largeau, J. Cheng, P. Regreny, G. Patriarche, A. Benamrouche, Y. Robach, M. Gendry, G. Hollinger, G. Saint-Girons, Appl. Phys. Lett. **95**, 011907 (2009)

20. J. Cheng, A. Chettaoui, J. Penuelas, B. Gobaut, P. Regreny, A. Benamrouche, Y. Robach, G. Hollinger, G. Saint-Girons, J. Appl. Phys. **107**, 094902 (2010)
21. A.A. Demkov, H. Seo, X. Zhang, J. Ramdani, Appl. Phys. Lett. **100**, 071602 (2012)
22. A. Slepko, A.A. Demkov, Phys. Rev. B **85**, 195462 (2012)
23. J. Wang, J.A. Hallmark, D.S. Marshall, W.J. Ooms, P. Ordejon, J. Junquera, D.L. Sanchez-Portal, E. Artacho, J.M. Soler, Phys. Rev. B **60**, 4968 (1999)
24. A.A. Demkov, X. Zhang, J. Appl. Phys. **103**, 103710 (2008)
25. R. McKee, F.J. Walker, M. Buongiorno Nardelli, W.A. Shelton, G.M. Stocks, Science **300**, 1726 (2003)
26. H. Li, X. Hu, Y. Wei, Z. Yu, X. Zhang, A.A. Demkov, J. Edwards Jr., K. Moore, W. Ooms, J. Appl. Phys. **93**, 4521 (2003)
27. R. Droopad, Z. Yu, H. Li, Y. Liang, C. Overgaard, A.A. Demkov, X. Zhang, K. Moore, K. Eisenbeiser, M. Hu, J. Curless, J. Finder, J. Cryst. Growth **251**, 638 (2003)
28. Y. Wei, X. Hu, Y. Liang, D. Jordan, B. Craigo, R. Droopad, Z. Yu, A.A. Demkov, J. Edwards Jr., W. Ooms, J. Vac. Sci. Technol. B **20**, 1402 (2002)
29. W.C. Fan, N.J. Wu, A. Ignatiev, Phys. Rev. B **42**, 1254 (1990)
30. Y. Liang, S. Gan, M. Engelhard, Appl. Phys. Lett. **79**, 3591 (2001)
31. C.R. Ashman, C.J. Först, K. Schwarz, P.E. Blöchl, Phys. Rev. B **70**, 155330 (2004)
32. A.J. Ciani, P. Sen, I.P. Batra, Phys. Rev. B **69**, 245308 (2004)
33. M.P.J. Punkkinen, M. Kuzmin, P. Laukkanen, R.E. Perälä, M. Ahola-Tuomi, J. Lång, M. Ropo, M. Pessa, I.J. Väyrynen, K. Kokko, B. Johansson, L. Vitos, Phys. Rev. B **80**, 235307 (2009)
34. K. Garrity, S. Ismail-Beigi, Phys. Rev. B **80**, 085306 (2009)
35. M. Choi, A.B. Posadas, H. Seo, R.C. Hatch, A.A. Demkov, Appl. Phys. Lett. **102**, 031604 (2013)
36. H. Seo, M. Choi, A.B. Posadas, R.C. Hatch, A.A. Demkov, J. Vac. Sci. Technol. B **31**, 04D107 (2013)
37. K.F. Garrity, M. Padmore, Y. Segal, J.W. Reiner, F.J. Walker, C.H. Ahn, S. Ismail-Beigi, Surf. Sci. **604**, 857 (2010)
38. X. Yao, X. Hu, D. Sarid, Z. Yu, J. Wang, D.S. Marshall, R. Droopad, J.K. Abrokwah, J.A. Hallmark, W.J. Ooms, Phys. Rev. B **59**, 5115 (1999)
39. X. Hu, C.A. Peterson, D. Sarid, Z. Yu, J. Wang, D.S. Marshall, R. Droopad, J.A. Hallmark, W.J. Ooms, Surf. Sci. **426**, 69 (1999)
40. D.M. Goodner, D.L. Marasco, A.A. Escuadro, L. Cao, M.J. Bedzyk, Phys. Rev. B **71**, 165426 (2005)
41. R.E. Schlier, H.E. Farnsworth, J. Chem. Phys. **30**, 918 (1959)
42. D.J. Chadi, J. Vac. Sci. Technol. **16**, 1290 (1979)
43. E. Zintl, W. Dullenkoopf, Z. Phys. Chem. B **16**, 183 (1932)
44. E. Zintl, Angew. Chem. **52**, 1 (1939)
45. M. Imai, T. Naka, T. Furubayashi, H. Abe, T. Nakama, K. Yagasaki, Appl. Phys. Lett. **86**, 032102 (2005)
46. Z. Chen, M. Yu, T. Chen, J. Appl. Phys. **113**, 043515 (2013)
47. M. Hu, *private communication*
48. G.-X. Qian, R.M. Martin, D.J. Chadi, Phys. Rev. B **38**, 7649 (1988)
49. J.W. Reiner, K.F. Garrity, F.J. Walker, S. Ismail-Beigi, C.H. Ahn, Phys. Rev. Lett. **101**, 105503 (2008)
50. X. Hu, X. Yao, C.A. Peterson, D. Sarid, Z. Yu, J. Wang, D.S. Marshall, R. Droopad, J.A. Hallmark, W.J. Ooms, Surf. Sci. **445**, 256 (2000)
51. O.L. Alerhand, J. Wang, J.D. Joannopoulos, E. Kaxiras, R.S. Becker, Phys. Rev. B **44**, 6534 (1991)
52. H.-C. Jeong, E.D. Williams, Surf. Sci. Rep. **34**, 171 (1999)
53. P.R. Pukite, P.I. Cohen, Appl. Phys. Lett. **50**, 1739 (1987)
54. T.R. Ohno, E.D. Williams, Jpn. J. Appl. Phys. **28**, L2061 (1989)

55. L. Yi, Y. Wei, I.S.T. Tsong, Surf. Sci. **304**, 1 (1994)
56. T. Shimakura, H. Minoda, K. Yagi, Surf. Sci. **407**, L657 (1998)
57. S. Fölsch, G. Meyer, M. Henzler, Surf. Sci. **394**, 60 (1997)
58. S. Fölsch, D. Winau, G. Meyer, K.H. Rieder, Appl. Phys. Lett. **67**, 2185 (1995)
59. J. He, G. Zhang, J. Guo, Q. Guo, K. Wu, J. Appl. Phys. **109**, 083522 (2011)
60. O.V. Yazyev, A. Pasquarello, Phys. Rev. Lett. **96**, 157601 (2006)
61. M.P.J. Punkkinen, K. Kokko, L. Vitos, P. Laukkanen, E. Airiskallio, M. Ropo, M. Ahola-Tuomi, M. Kuzmin, I.J. Väyrynen, B. Johansson, Phys. Rev. B **77**, 245302 (2008)
62. P.E.J. Eriksson, R.I.G. Uhrberg, Phys. Rev. B **81**, 125443 (2010)
63. E. Landemark, C.J. Karlsson, Y.C. Chao, R.I.G. Udhrberg, Phys. Rev. Lett. **69**, 1588 (1992)
64. T.-W. Pi, C.-P. Cheng, I.-H. Hong, Surf. Sci. **418**, 113 (1998)
65. P. De Padova, R. Larciprete, C. Quaresima, C. Ottaviani, B. Ressel, P. Perfetti, Phys. Rev. Lett. **81**, 2320 (1998)
66. G. Le Lay, A. Cricenti, C. Ottaviani, P. Perfetti, T. Tanikawa, I. Matsuda, S. Hasegawa, Phys. Rev. B **66**, 153317 (2002)
67. H. Koh, J.W. Kim, W.H. Choi, H.W. Yeom, Phys. Rev. B **67**, 073306 (2003)
68. E. Pehlke, M. Scheffler, Phys. Rev. Lett. **71**, 2338 (1993)
69. J.A. Kubby, J.E. Griffith, R.S. Becker, J.S. Vickers, Phys. Rev. B **36**, 6079 (1987)
70. R.I.G. Uhrberg, J. Phys. Condens. Matter **13**, 11181 (2001)
71. C.P. Cheng, I.H. Hong, T.W. Pi, Phys. Rev. B **58**, 4066 (1998)
72. M. Kuzmin, R.E. Perälä, P. Laukkanen, I.J. Väyrynen, Phys. Rev. B **72**, 085343 (2005)
73. T.-W. Pi, C.-P. Cheng, G.K. Wertheim, J. Appl. Phys. **109**, 043701 (2011)
74. A. Mesarwi, W.C. Fan, A. Ignatiev, J. Appl. Phys. **68**, 3609 (1990)
75. A. Herrera-Gomez, F.S. Aguirre-Tostado, Y. Sun, P. Pianetta, Z. Yu, D. Marshall, R. Droopad, W.E. Spicer, J. Appl. Phys. **90**, 6070 (2001)
76. G. Kresse, J. Furthmüller, Phys. Rev. B **54**, 11169 (1996)
77. http://www.casaxps.com
78. A.D. Katnani, G. Margaritondo, Phys. Rev. B **28**, 1944 (1983)
79. L. Köhler, G. Kresse, Phys. Rev. B **70**, 165405 (2004)
80. M. Imai, T. Kikegawa, Chem. Mater. **15**, 2543 (2003)
81. K. Shudo, T. Munakata, Phys. Rev. B **63**, 125324 (2001)
82. Y. Enta, T. Kinoshita, S. Suzuki, S. Kono, Phys. Rev. B **36**, 9801 (1987)
83. A. Crispin, X. Crispin, M. Fahlman, M. Berggren, W.R. Salaneck, Appl. Phys. Lett. **89**, 213503 (2006)
84. P.C. Rusu, G. Giovannetti, C. Weijtens, R. Coehoorn, G. Brock, J. Phys. Chem. C **113**, 9974 (2009)
85. T. Tambo, T. Nakamura, K. Maeda, H. Ueba, C. Tatsuyama, Jpn. J. Appl. Phys. **37**, 4454 (1998)
86. S. Jeon, F.J. Walker, C.A. Billman, R.A. McKee, H. Hwang, IEEE Electron Device Lett. **24**, 218 (2003)
87. J. Lettieri, J.H. Haeni, D.G. Schlom, J. Vac. Sci. Technol. A **20**, 1332 (2002)
88. D.M. Schaadt, E.T. Yu, V. Vaithyanathan, D.G. Schlom, J. Vac. Sci. Technol. B **22**, 2030 (2004)
89. J.Q. He, C.L. Jia, V. Vaithyanathan, D.G. Schlom, J. Schubert, A. Gerber, H.H. Kohlstedt, R.H. Wang, J. Appl. Phys. **97**, 104921 (2005)
90. J.W. Reiner, A. Posadas, M. Wang, T.P. Ma, C.H. Ahn, Microelectron. Eng. **85**, 36–38 (2008)
91. G.J. Norga, A. Guiller, C. Marchiori, J. Locquet, H. Siegwart, D. Halley, C. Rossel, D. Caimi, J. Seo, J. Fompeyrine, MRS Proc. **786**, E7.3 (2003)
92. G.J. Norga, C. Marchiori, A. Guiller, J. Locquet, H. Siegwart, C. Rossel, D. Caimi, T. Conard, J. Fompeyrine, Appl. Phys. Lett. **87**, 262905 (2005)
93. G.J. Norga, C. Marchiori, C. Rossel, A. Guiller, J.P. Locquet, H. Siegwart, D. Caimi, J. Fompeyrine, J.W. Seo, C. Dieker, J. Appl. Phys. **99**, 084102 (2006)

94. Y. Liang, Y. Wei, X.M. Hu, Z. Yu, R. Droopad, H. Li, K. Moore, J. Appl. Phys. **96**, 3413 (2004)
95. F. Niu, A. Meier, B.W. Wessels, J. Electroceram. **13**, 149 (2004)
96. F. Niu, B.W. Wessels, J. Cryst. Growth **300**, 509 (2007)
97. G. Delhaye, C. Merckling, M. El-Kazzi, G. Saint-Girons, M. Gendry, Y. Robach, G. Hollinger, L. Largeau, G. Patriarche, J. Appl. Phys. **100**, 124109 (2006)
98. G. Niu, G. Saint-Girons, B. Vilquin, G. Delhaye, J.-L. Maurice, C. Botella, Y. Robach, G. Hollinger, Appl. Phys. Lett. **95**, 062902 (2009)
99. M. Choi, A. Posadas, R. Dargis, C.-K. Shih, A.A. Demkov, D.H. Triyoso, N. David Theodore, C. Dubourdieu, J. Bruley, J. Jordan-Sweet, J. Appl. Phys. **111**, 064112 (2012)
100. J.C. Woicik, H. Li, P. Zschack, E. Karapetrova, P. Ryan, C.R. Ashman, C.S. Hellberg, Phys. Rev. B **73**, 024112 (2006)
101. M.P. Warusawithana, C. Cen, C.R. Sleasman, J.C. Woicik, Y. Li, L. Fitting Kourkoutis, J.A. Klug, H. Li, P. Ryan, L.-P. Wang, M. Bedzyk, D.A. Muller, L.-Q. Chen, J. Levy, D.G. Schlom, Science **324**, 367 (2009)
102. D.P. Kumah, J.W. Reiner, Y. Segal, A.M. Kolpak, Z. Zhang, D. Su, Y. Zhu, M.S. Sawicki, C.C. Broadbridge, C.H. Ahn, F.J. Walker, Appl. Phys. Lett. **97**, 251902 (2010)
103. F.S. Aguirre-Tostado, A. Herrera-Gómez, J.C. Woicik, R. Droopad, Z. Yu, D.G. Schlom, P. Zschack, E. Karapetrova, P. Pianetta, C.S. Hellberg, Phys. Rev. B **70**, 201403 (2004)
104. H. Ledbetter, M. Lei, S. Kim, Phase Trans. **23**, 61 (1990)
105. http://www-llb.cea.fr/prism/programs/simulreflec/simulreflec.html
106. G.J. Yong, R.M. Kolagani, S. Adhikari, W. Vanderlinde, Y. Liang, K. Muramatsu, S. Friedrich, J. Appl. Phys. **108**, 033502 (2010)
107. A. Posadas, R. Dargis, M.R. Choi, A. Slepko, A.A. Demkov, J.J. Kim, D.J. Smith, J. Vac. Sci. Technol. B **29**, 03C131 (2011)
108. X. Hu, H. Li, Y. Liang, Y. Wei, Z. Yu, D. Marshall, J. Edwards Jr., R. Droopad, X. Zhang, A.A. Demkov, K. Moore, Appl. Phys. Lett. **82**, 203 (2003)
109. V. Shutthananda, S. Thevuthasan, Y. Liang, E.M. Adams, Z. Yu, R. Droopad, Appl. Phys. Lett. **80**, 1803 (2002)
110. J.Y. Tsao, *Materials Fundamentals of Molecular Beam Epitaxy* (Academic, San Diego, CA, 1993)
111. A.A. Stekolnikov, J. Furthmuller, F. Bechstedt, Phys. Rev. B **65**, 115318 (2002)
112. M.J. Beck, A. van de Walle, M. Asta, Phys. Rev. B **70**, 205337 (2004)
113. Y. Umeno, A. Kushima, T. Kitamura, P. Gumbsch, J. Li, Phys. Rev. B **72**, 165431 (2005)
114. S.D. Solares, S. Dasgupta, P.A. Schultz, Y.-H. Kim, C.B. Musgrave, W.A. Goddard III, Langmuir **21**, 12404 (2005)
115. J. Padilla, D. Vanderbilt, Surf. Sci. **418**, 64 (1998)
116. A.A. Demkov, Phys. Status Solidi B **226**, 57–67 (2001)
117. X. Zhang, A.A. Demkov, J. Vac. Sci. Technol. B **20**, 1664 (2002)
118. J. Robertson, C.W. Chen, Appl. Phys. Lett. **74**, 1168 (1999)
119. J. Robertson, J. Vac. Sci. Technol. B **18**, 1785 (2000)
120. C. Tejedor, F. Flores, E. Louis, J. Phys. C **10**, 2163 (1977)
121. J. Tersoff, Phys. Rev. B **30**, 4874 (1984)
122. A.A. Demkov, L. Fonseca, E. Verret, J. Tomfohr, O.F. Sankey, Phys. Rev. B **71**, 195306 (2005)
123. A.A. Demkov, R. Liu, X. Zhang, H. Loechelt, J. Vac. Sci. Technol. B **18**, 2388 (2000)
124. C.G. Van de Walle, R.M. Martin, Phys. Rev. B **39**, 1871 (1989)
125. G.D. Wilk, R.M. Wallace, J.M. Anthony, J. Appl. Phys. **89**, 5243 (2001)
126. S.A. Chambers, Y. Liang, Z. Yu, R. Droopad, J. Ramdani, K. Eisenbeiser, Appl. Phys. Lett. **77**, 1662 (2000)
127. S.A. Chambers, T. Droubay, T.C. Kaspar, M. Gutowski, J. Vac. Sci. Technol. B **22**, 2205 (2004)
128. F. Amy, A.S. Wan, A. Kah, F.J. Walker, R.A. McKee, J. Appl. Phys. **96**, 1635 (2004)

129. A.M. Kolpak, S. Ismail-Beigi, Phys. Rev. B **85**, 195318 (2012)
130. A.M. Kolpak, F.J. Walker, J.W. Reiner, Y. Segal, D. Su, M.S. Sawicki, C.C. Broadbridge, Z. Zhang, Y. Zhu, C.H. Ahn, S. Ismail-Beigi, Phys. Rev. Lett. **105**, 217601 (2010)
131. C.S. Hellberg, K.E. Andersen, H. Li, P.J. Ryan, J.C. Woicik, Phys. Rev. Lett. **108**, 166101 (2012)
132. A. Antons, J.B. Neaton, K.M. Rabe, D. Vanderbilt, Phys. Rev. B **71**, 024102 (2005)
133. X. Gu, D. Lubyshev, J. Batzel, J.M. Fastenau, W.K. Liu, R. Pelzel, J.F. Magana, Q. Ma, L.P. Wang, P. Zhang, V.R. Rao, J. Vac. Sci. Technol. B **27**, 1195 (2009)
134. B.K. Moon, H. Ishiwara, Jpn. J. Appl. Phys. **33**, L472 (1994)
135. M.D. McDaniel, A. Posadas, T.Q. Ngo, A. Dhamdhere, D. Smith, A.A. Demkov, J.G. Ekerdt, J. Vac. Sci. Technol. A **31**, 01A136 (2013)
136. M.D. McDaniel, A. Posadas, T.Q. Ngo, A. Dhamdhere, D. Smith, A.A. Demkov, J.G. Ekerdt, J. Vac. Sci. Technol. B **30**, 04E111 (2012)
137. T.Q. Ngo, A. Posadas, M.D. McDaniel, D.A. Ferrer, J. Bruley, C. Breslin, A.A. Demkov, J.G. Ekerdt, J. Cryst. Growth **363**, 150 (2013)

Chapter 7
Integration of Functional Oxides on SrTiO₃/Si Pseudo-Substrates

As mentioned previously, SrTiO₃ (STO) is a widely used substrate for epitaxial growth of many oxide materials. Therefore, the STO/Si pseudo-substrate described in the previous chapter offers a convenient way of integrating functional oxides on Si. Typically, a very thin STO layer of only four unit cells (16 Å) is sufficient to ensure high quality crystal growth. In this chapter we will describe recent efforts to integrate photocatalytic, ferromagnetic and ferroelectric materials on Si using this strategy. We shall focus on molecular beam epitaxial (MBE) growth and will very briefly mention some recent work involving atomic layer deposition (ALD).

7.1 Integration of Anatase TiO₂ on STO/Si

Anatase TiO_2 is the subject of extensive research efforts owing to its energy and environmental applications [1–4]. The conduction band edge of TiO_2 is well-matched to the redox potential of water, making it an excellent candidate for hydrogen production via photocatalytic water splitting [5]. This material system is promising due to its relatively high efficiency, chemical stability in water, non-toxicity, and low production cost [1]. Among the three polymorphs of TiO_2 (rutile, anatase, and brookite), the most abundant phase in nature is rutile (space group $P4_2/mnm$). However, it has been found that anatase (space group $I4_1/amd$) shows significantly higher photocatalytic activity than rutile [5]. This has been attributed to the higher reactivity of the anatase (001) surface [6, 7], higher mobility of the charge carriers [8, 9], and longer electron-hole pair lifetime [10].

The main challenge for photocatalytic applications is the rather large band gap of anatase TiO_2 (approximately 3.2 eV making it ultraviolet-active) and high recombination rate of the photo-excited electron-hole pairs [11]. A variety of methods for band gap engineering have been proposed to utilize solar-abundant visible light instead of ultraviolet, including nitrogen-doping [12], co-doping [13, 14], and surface hydrogenation [15]. In order to overcome these challenges, oxide interface

A.A. Demkov and A.B. Posadas, *Integration of Functional Oxides with Semiconductors*, DOI 10.1007/978-1-4614-9320-4_7, © The Author(s) 2014

engineering has attracted considerable attention [16–19]. In addition to band gap engineering at the interface, a longer lifetime of photo-excited electron-hole pairs could be achieved, for example, by spatially separating the carriers across the interface using a staggered band alignment. However, a clear understanding of the interface effects on the photocatalytic activity of mixed oxide catalysts is lacking, in particular due to a limited number of model systems where such a connection can be traced [20].

Since the thermodynamically stable bulk phase of TiO_2 at room temperature and ambient pressure is rutile, single crystal anatase is typically synthesized only in the form of nanoparticles [2, 6]. However, recent advances in oxide heteroepitaxy, have made it possible to grow high-quality, single crystal anatase films on perovskite substrates such as STO or $LaAlO_3$ (LAO) [21, 22]. This provides an excellent model system for controlled studies of the photocatalytic behavior of anatase under various conditions [23–25]. Burbure et al. have shown that for anatase-TiO_2/ $BaTiO_3$ structures, dipole fields from the underlying ferroelectric domains separate holes and electrons, leading to spatially selective photochemical reactions at the anatase surface [23]. Kazazis and co-workers have reported that the photocatalytic activity at the surface of anatase deposited on Si (111) can be controlled by changing the Fermi level position of the Si substrate [24]. Moreover, a number of interesting physical phenomena and potential applications of anatase/perovskite oxide heterostructures have been reported, including thermoelectric and spintronic applications [26–30].

Despite a considerable amount of experimental work in the literature, a detailed theoretical understanding of the electronic structure of the anatase-TiO_2/perovskite interface has been developed only recently [31, 32]. Most notably, Chambers et al., using x-ray photoelectron spectroscopy (XPS), have reported that there is no measurable valence band offset between anatase TiO_2 (001) and STO (001) [32]. In apparent contradiction with the XPS result, their density functional theory (DFT) calculations suggested a valence band offset of 0.5 eV. The authors interpreted this discrepancy as the inability of DFT to fully account for the interface properties. However, detailed analysis performed by Seo and coworkers suggests that, most likely, the atomic structure used in the calculation differs from the experimental one [31]. For MBE-grown epitaxial c-axis oriented anatase on an STO/Si (001) pseudo-substrate [33], they were able to elucidate the real physical structure of the interface using Z-contrast scanning transmission electron micros-copy (STEM). The interface geometry inferred from STEM was used in the DFT calculations of the band offset. The theoretical structure was validated by compar-ing the theoretical density of states with the measured O K edge electron energy loss spectroscopy (EELS) structure.

In [31], anatase was deposited using MBE on an STO layer that had been epitaxially grown on Si (001) using methods described in Chapter 6. For the anatase TiO_2 deposition, the pseudo-substrate temperature was increased to 650 °C, while oxygen pressure was slowly ramped to 1×10^{-6} Torr. Samples with total thick-nesses of 10–40 monolayers (MLs) of anatase (one anatase unit cell equals four MLs) were deposited. The growth of all the layers was monitored in-situ using

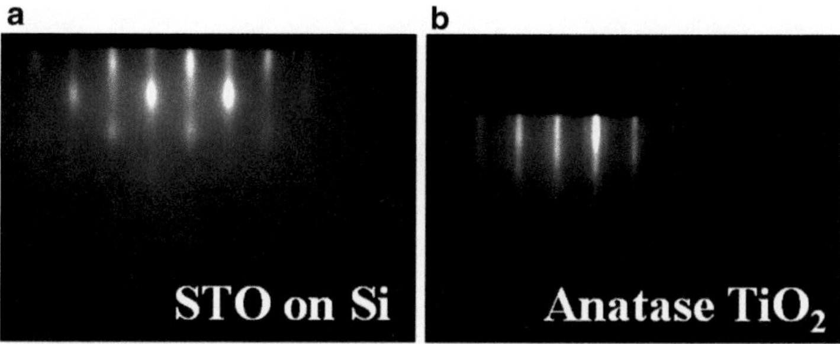

Fig. 7.1 RHEED patterns for (**a**) an STO film on Si (001) taken along the Si <110> direction; and (**b**) for a 20 monolayer anatase film taken along the same direction (**b**). From [31]

reflection high energy electron diffraction (RHEED). Figure 7.1 shows a RHEED pattern for the STO/Si pseudo-substrate (Fig. 7.1a) and for a 20 monolayer anatase film taken along the <100> azimuth (Fig. 7.1b). The anatase surface shows a 4×1 reconstruction typically exhibited by well-ordered epitaxial anatase films [21]. Figure 7.2 shows a high resolution Z-contrast STEM image of the interface. The image was taken along the Si [110] zone axis. It shows a well-ordered epitaxial relationship between the Si substrate, STO and TiO$_2$. The most likely interface structure is shown by the overlay in Fig. 7.2. This interface structure was used as the starting model in the DFT calculations.

The main focus of the study [31] was to establish the band discontinuity at the STO/anatase (001) interface. The band offset is the fundamental physical parameter that largely affects the functionality of a heterostructure, owing to its profound effect on the carrier confinement and electronic transport along and across the interface. However, oxide heterointerfaces are not as well-understood as those between metals and semiconductors. The band alignment at oxide heterointerfaces has attracted considerable attention in the context of the high-k dielectric gate stack in field effect transistors [34–38]. More recently, the band alignment between complex oxides has been of great interest following the discovery of novel interfacial electronic phases emerging at the epitaxial complex oxide heterointerfaces [39–42]. To describe and control the band alignment between two materials, the interfacial chemistry of a given heterointerface has to be well understood [35, 36, 43–45].

Seo et al. used DFT within the local density approximation (LDA) as implemented in the VASP code [46]. They calculated the STO lattice parameter a to be 3.873 Å, and the TiO$_2$ lattice parameters a and c to be 3.766 Å and 9.456 Å ($c/a = 2.511$), respectively. These are in good agreement with low temperature experimental values of $a = 3.897$ Å for STO [47], and $a = 3.780$ Å and $c = 9.491$ Å ($c/a = 2.511$) for anatase TiO$_2$ [48] that are extrapolated based on the thermal expansion coefficients. Thus, theoretical biaxial tensile strain in the anatase film on STO (001) was calculated to be 2.8 %, in good agreement with the experimental value of 3.0 %. To model the evolution of the valence band with the anatase film

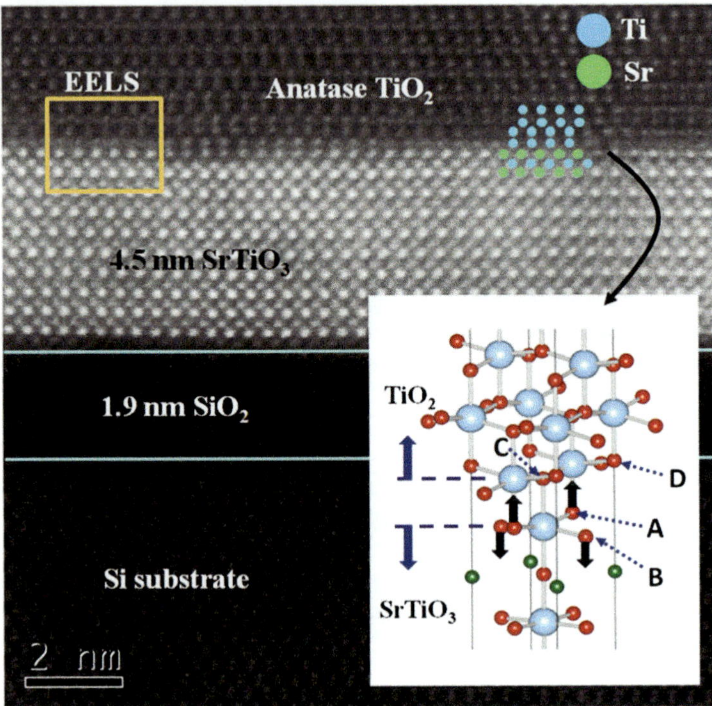

Fig. 7.2 Z-contrast HRTEM image of the TiO$_2$/STO/Si (001) structure. *Blue* (Ti) and *green* (Sr) balls are superimposed at the TiO$_2$/STO interface as a guide to the eye. Spatially resolved EELS measurement is performed at the region indicated by the *orange box*. *Inset* shows an interface model (relaxed) between anatase TiO$_2$ and STO (001) used in the DFT calculations. *Black arrows* show the relaxation pattern of the O ions at the interface. From [31]

thickness, supercell geometry was used. The substrate was modeled with a five-unit cell-thick TiO$_2$-terminated STO (001) slab, and four, six, and eight monolayers (MLs) of anatase (one anatase unit cell has four MLs) on both sides of STO were considered. To investigate whether the LDA caused the incorrect band offset prediction as suggested in [32], a quasi-particle (QP) correction within the G_oW_o approximation (implemented in the VASP code [49]) was used. We now shall discuss the major finding of this work in more detail.

7.1.1 Charge Transfer and Dielectric Screening at the Interface

To understand the mechanism of the band offset formation at the anatase/STO (001) interface, let us first discuss the Schottky limit when two oxide slabs are far apart so that there is no interaction between them. Seo et al. considered TiO$_2$-terminated

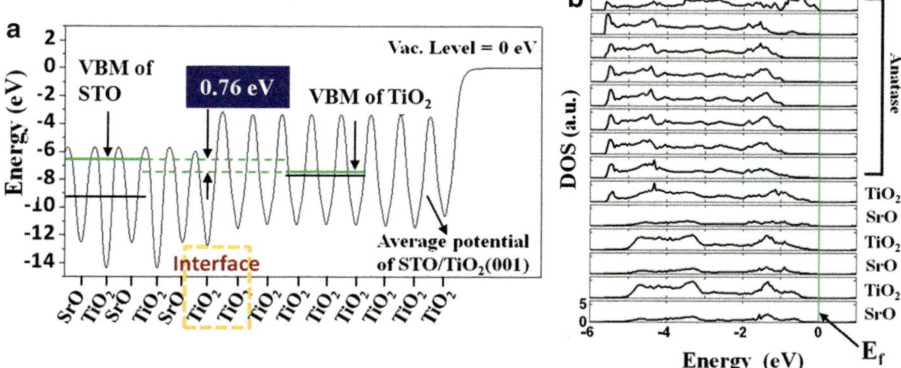

Fig. 7.3 (**a**) Planar-averaged electrostatic potential of the TiO$_2$/STO(001) heterostructure along the (001) direction. The *straight black lines* indicate the reference electrostatic energy positions with respect to the vacuum level (0 eV) in the bulk region of STO and TiO$_2$, respectively. The *green lines* indicate the relative positions of VBM of STO and TiO$_2$ with respect to their corresponding reference energy positions. (**b**) Layer-by-layer projected density of states (pDOS) of the TiO$_2$/SrTiO$_3$ (001) heterostructure. The *green line* is the Fermi level. From [31]

STO (001) and tensile-strained (2.8 %) anatase TiO$_2$ (001) slabs separately [31]. By using macroscopically averaged potentials and separate bulk calculations [50–52], they found the valence band offset ΔE_v ($=E_{VBM}$ (STO) $- E_{VBM}$ (TiO$_2$)) to be 0.94 eV. The offset changed to 0.89 eV if a relaxed anatase film ($a = b = 3.766$ Å) is considered, meaning that, in the Schottky limit, tensile strain plays only a minor role in determining the band offset.

When two oxide slabs are brought into contact, chemical bonds are created that result in charge redistribution in conjunction with a structural distortion at the interface. The physical interface model between anatase TiO$_2$ and STO (001) from [31] is shown in the inset of Fig. 7.2 and is based on the STEM image in Fig. 7.2. Note that half of O at the "STO surface" is bonded to Ti of anatase, leading to 3-fold coordinated O (O$_{3\text{-fold}}$, site A), while the other half of O remains 2-fold coordinated (O$_{2\text{-fold}}$, site B). Ti at the STO surface is also bonded to O of anatase at site C.

The valence band offset of the heterostructure can be written as $\Delta E_H = \Delta E_S + \Delta V$, where ΔE_H is the band offset of the actual heterostructure, ΔE_S is the band offset in the Schottky limit, and ΔV is the electrostatic potential drop at the interface due to the heterojunction formation. Using the macroscopic average of the electrostatic potential [50–52], Seo et al. calculated the valence band offset to be 0.76 eV, thus ΔV is -0.2 eV [31] as shown in Fig. 7.3a. This process is often described as the creation of an interface dipole or double layer involving charge transfer and subsequent dielectric screening at the interface [35, 36].

There are two channels for charge transfer: chemical bond-induced transfer [53] and transfer into evanescent gap states [54]. In Fig. 7.4a, we show the charge redistribution at the interface from [31], which was defined as

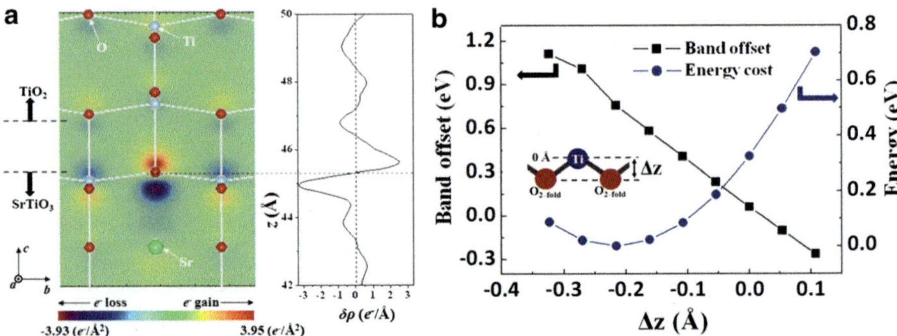

Fig. 7.4 (**a**) Two-dimensional (*left*) and one-dimensional (*right*) projections of the charge redistribution ($\delta\rho(x,y,z)$) at the TiO$_2$/SrTiO$_3$ (001) interface (see text). For the reference charge density of the free-standing STO and TiO$_2$ slabs, we use relaxed atomic geometry. (**b**) Band offset and total energy of the TiO$_2$/STO heterostructure as a function of the displacement of O$_{2\text{-fold}}$ ions at the interface. The inset is a schematic picture to show the lattice polarization by O$_{2\text{-fold}}$ at the interface. In the relaxed heterostructure, the optimal Δz is -0.22 Å, where the energy is minimum. From [31]

$$\delta\rho(x, y, z) = \rho(\text{heterostructure}) - \rho(\text{STO substrate}) - \rho(\text{TiO}_2 \text{ film}), \qquad (7.1)$$

where ρ is the valence charge density of a given structure. The two-dimensional (2D) and one-dimensional (1D) projections were defined as:

$$\delta\rho(y, z) = \int_0^a \delta\rho(x, y, z)dx \quad \text{and} \quad \delta\rho(z) = \int_0^a \int_0^a \delta\rho(x, y, z)dxdy \qquad (7.2)$$

respectively, where a is the STO lattice parameter. In Fig. 7.4a, it was observed that chemical bonds drive the charge transfer between O and Ti. Assuming that the chemical bonds locally modify the charge density, the density underneath the STO surface is shifted upward (towards anatase) and localized along the chemical bonds between O and Ti. The charge transfer to evanescent states of TiO$_2$ decays rapidly within 2 MLs of anatase as seen in Fig. 7.3b, where the finite density of states in the gap (between zero and -1 eV), can be only seen in the first two MLs of anatase.

For oxide interfaces, Sharia et al. have pointed out the importance of screening by O lattice polarization [35]. They have shown that as the coordination number of interfacial O increases, the Born effective charge of O increases as well and, as a result, the band offset is pushed back to the Schottky limit owing to the enhanced screening ability of interfacial O. The same argument was applied to the TiO$_2$/STO structure. In [31], it was found that the O$_{2\text{-fold}}$ ions at the interface shift by 0.22 Å from the Ti plane toward the STO bulk side as shown in Fig. 7.2. In Fig. 7.4b, the calculated band offset as a function of position of the O$_{2\text{-fold}}$ ions in the (001) direction from [31] is shown. Pushing the O$_{2\text{-fold}}$ ions back to the level of the Ti plane decreases the band offset from 0.76 to 0.06 eV. It suggests that lattice polarization by the O$_{2\text{-fold}}$ ions is indeed the main screening mechanism at the

Table 7.1 *GW* quasi-particle (QP) corrections for STO and anatase TiO$_2$

Material	CBM, δ^a (k-point/eV)	VBM, δ^a (k-point/eV)
SrTiO$_3$	Γ/1.9, R/2.1	Γ/−0.1, R/−0.06
TiO$_2$	Γ/1.97, X/1.93	Γ/−0.15, X/−0.02

$^a\delta$'s for the conduction band minimum (CBM) and the valence band maximum (VBM) are defined by $E^{LDA} - E^{QP}$ at the CBM and VBM, respectively

interface [35]. The potential change at the interface due to lattice polarization is given by [55, 56]:

$$\Delta V\left(O_{2-fold}\right) = \frac{1}{a^2} \frac{Z^*_{O_{2-fold}}{}^{(T)}}{\varepsilon_0 \varepsilon_\infty} \Delta u\left(O_{2-fold}\right), \qquad (7.3)$$

where a is the in-plane lattice parameter, $Z^*_{O_{2-fold}}{}^{(T)}$ is the Born effective charge of $O_{2\text{-fold}}$ along the (001) direction, $\Delta u(O_{2\text{-fold}})$ is the displacement of $O_{2\text{-fold}}$ with respect to the Ti plane in the (001) direction, and ε_∞ is the optical dielectric constant. Assuming $\varepsilon_\infty \approx 6.2$ [57–59], Seo et al. estimated the Z^* of $O_{2\text{-fold}}$ at the interface to be -1.66. This number is consistent with the model developed in [35].

To check the accuracy of the valence band offset computed within the LDA, quasi-particle (QP) corrections to the LDA eigenvalues were calculated using the *GW* method for bulk STO and TiO$_2$. The results are summarized in Table 7.1. The band offset within the *GW* formalism was calculated as:

$$\Delta E_{v-GW} = \Delta E_{v-LDA} - \left(\delta_{STO} - \delta_{TiO_2}\right), \qquad (7.4)$$

where $\Delta E_{v\text{-}LDA}$ is the valence band offset computed within DFT-LDA and δ_{STO} and δ_{TiO2} are the QP corrections to the LDA valence band maxima of STO and TiO$_2$, respectively. The conduction band offset can be corrected in a similar manner. Since the important quantities for computing band offsets are the valence band maxima and the conduction band minima, Seo et al. only reported the QP corrections at the R and Γ k-points for STO, and at the Γ and X points for anatase. Using the values of δ_{STO} and δ_{TiO2} at these k-points from Table 7.1 in (7.4), a *GW* correction of only 0.04 eV to the valence band offset calculated within LDA was found. Hence, although there is a significant QP correction to the band gaps of the individual bulk compounds, the valence band offset within the DFT-LDA is reliable as most of the correction comes from the conduction band for both the compounds.

7.1.2 EELS O K Edge Spectra Across the Interface

Experimentally, the evolution of the electronic structure across the interface can be monitored using electron energy loss spectroscopy (EELS) [60]. Seo et al. performed a similar measurement with atomic resolution across the region

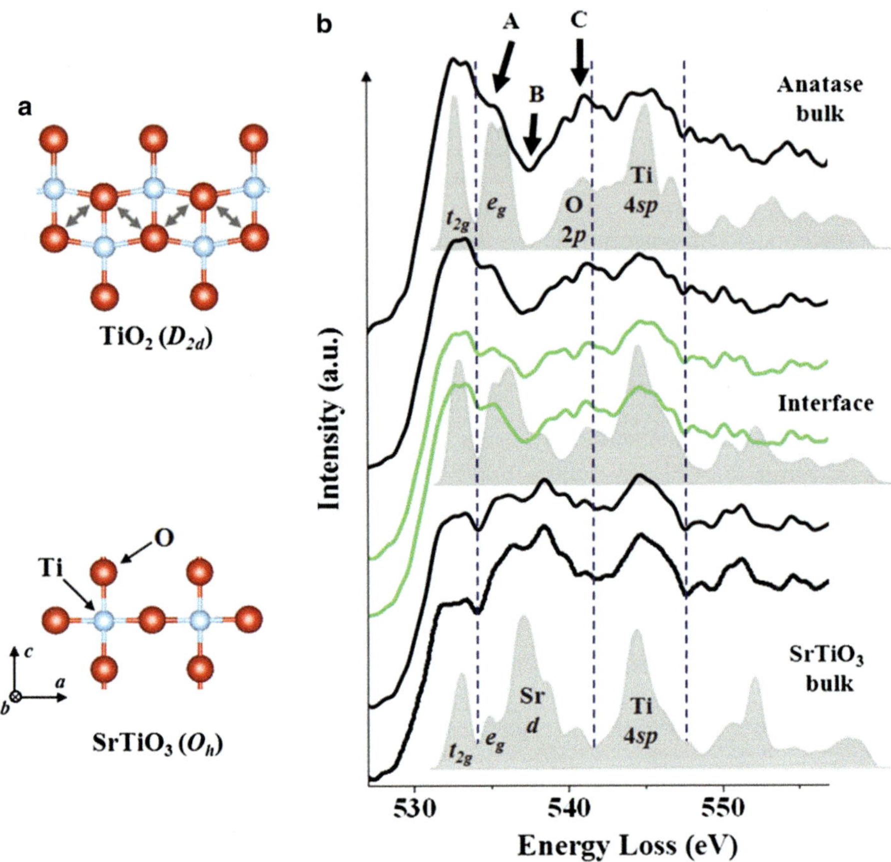

Fig. 7.5 (a) Schematic pictures of a TiO_2 plane in bulk STO (*left*) and bulk anatase TiO_2 (*right*) (b) EELS O K edge spectra taken at the interface from the STO side (two *bottom* spectra) to the anatase TiO_2 side (two *upper* spectra) through the interface (two *middle green* spectra). The corresponding 2p-projected DOS's are overlaid in *grey*. From [31]

indicated in Fig. 7.2 [31]. They focused on the O K edge spectra [61–63] rather than Ti $L_{2,3}$ edge spectra [64–66], as the O K edge in STO and TiO_2 better reflects the change in the local bonding environment (see Fig. 7.5a) [60, 67]. Theoretically, to include the effect of core holes generated in experiment, the so-called $Z + 1$ approximation [68, 69] was used. In Fig. 7.5b, we show the experimental EELS spectra along with the corresponding theoretical partial density of states (pDOS) reported in [31]. Both were broadened using the Gaussian convolution method with a full width at half maximum value of 0.7 eV.

At the bottom of Fig. 7.5b, the O K edge spectrum taken from the STO bulk region is compared to the 2p pDOS calculated at the O site in the STO bulk region of the supercell. In experiment, there are three main features between 530 and 550 eV, which are well reproduced in theory. Analyzing the entire set of pDOS's including contributions from the nearest neighbor Ti and Sr, Seo et al. identified that the first, second, and third main peaks are derived from the interaction with Ti t_{2g}, Sr d, and Ti $4sp$ states, respectively [31]. The Ti e_g peak is seen as a small peak in theory between the first t_{2g} and the second Sr d main peaks. However, it was noted that as the e_g band is largely broadened in STO with a bandwidth of about 5 eV, the e_g peak is not seen experimentally owing to the presence of the adjacent large Sr d peak.

Looking at the spectrum taken from the anatase bulk region (Fig. 7.5b), we observe quite different spectral features. First, the e_g and t_{2g} peaks become sharp and pronounced at position A compared to that of STO. Secondly, the Sr d peak is absent at position B, as there is no Sr on the TiO$_2$ side of the interface. Thirdly and most importantly, there is the appearance of a large spectral weight between 538 and 543 eV (position C), which is not found on the STO side. Seo et al. identified that this peak is derived from the interaction with the nearest neighbor O along the c-axis [31]. A schematic of the TiO$_2$ plane in STO and anatase is shown in Fig. 7.5a. Although the basic building block in both materials is a TiO$_6$ octahedron, the connecting geometry is different: the octahedra share corners in STO while they share four adjacent edges in anatase. In contrast to the straight Ti-O chain along the a or b axes in STO, it exhibits a zigzag pattern in anatase as shown in Fig. 7.5a. Seo et al. pointed out that this crystallographic feature leads to appreciable π-type overlap between O $2p$ orbitals along the c-direction as shown in Fig. 7.5a. Finally, Seo et al. noted that this spectral peak originally emerges at the interface as shown in the middle of Fig. 7.5b. By comparing with the $2p$ pDOS at the O$_{3\text{-fold}}$ site of the interfacial layer (site A in Fig. 7.2), Seo et al. inferred that this peak at the interface reflects the chemical bond formation between anatase and STO.

7.1.3   Oxygen Vacancy and Fluorine Impurity at the TiO$_2$/SrTiO$_3$ Interface

DFT calculations reported in [32] suggested a valence band offset of 0.5 eV, consistent with the results of [31] but in apparent variance with the measurement. To better understand this discrepancy, Seo et al. considered two types of interface defects: an oxygen vacancy and a substitutional fluorine impurity.

As described above, the interfacial O plays a crucial role in determining the dielectric response of the heterointerface. This brings a natural question: what would be the effect on the band alignment of an interfacial O vacancy? [70] Seo et al. considered a neutral vacancy and found that a vacancy at site B was the most stable configuration (see the inset of Fig. 7.2). The O vacancy formation energy at this site was calculated to be 3.8 eV while those of the A, C and D sites were 1.6, 1.1, and 1.4 eV higher in energy, respectively. The formation energy of 3.8 eV is significantly lower than that of a neutral O vacancy in bulk STO, which is larger than 6.0 eV [71]. Since oxygen at site B is responsible for the screening of the

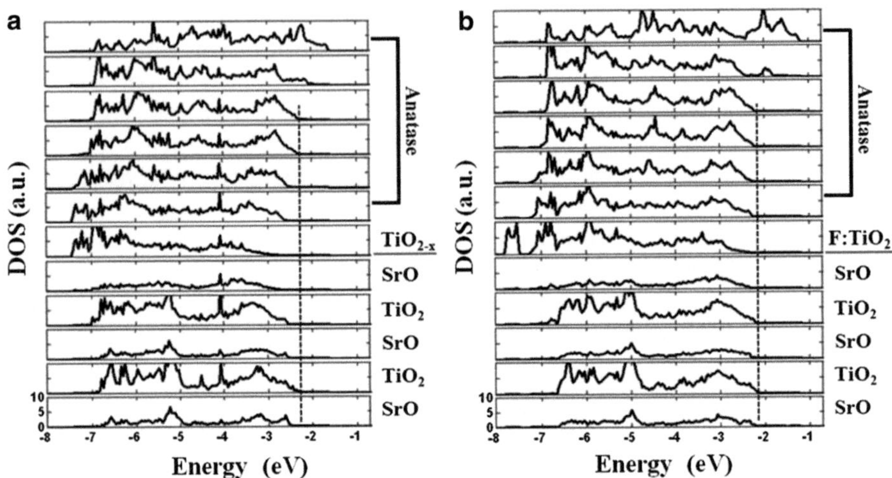

Fig. 7.6 Layer-by-layer valence band pDOS of the anatase TiO₂/STO(001) heterostructure in the presence of an interfacial O vacancy (**a**) and an interfacial F impurity (**b**). The *dotted straight line* is placed at the VBM in the STO bulk region and is extended into the TiO₂ side for comparison. From [31]

interfacial dipole, one can expect that the band offset would be significantly reduced by a vacancy at this site, as the charge transfer that tends to equilibrate the Fermi level of two oxides will not be fully screened. Using the average potential method, Seo et al. calculated the band offset for the interface with a vacancy to be 0.04 eV. The absence of a valence band offset in the presence of a vacancy is also seen in the pDOS of the heterostructure in Fig. 7.6a.

Seo et al. also considered a substitutional F impurity at the STO surface, which could be present when the STO substrate is etched in buffered HF solution in order to have a 1 × 1 TiO₂-terminated surface [32]. Considering all four interfacial O sites A, B, C, and D (see the inset of Fig. 7.2) for the F substitution, Seo et al. found that the most stable structure was obtained when F was substituted for 2-fold O at the B site. F at the A, C, and D sites was found to be higher in energy by 1.1, 0.7, and 0.8 eV, respectively. As mentioned earlier, 2-fold O at the B site is responsible for screening the interface dipole. Therefore, the dielectric screening at the interface is expected to be significantly reduced by the F substitution. The layer-by-layer pDOS in Fig. 7.6b reported in [31] indicates a negligible valence band offset of less than 0.1 eV between STO and TiO₂ in the presence of F at the interface.

7.1.4 Conclusions

In summary, high quality single crystal anatase was deposited using MBE on an STO layer that has been epitaxially grown on Si (001) [31]. By comparing the theoretical layer-by-layer pDOS and experimental EELS O *K* edge spectra, it was found that the evolution of the valence band across the STO/anatase interface is driven by the change in bonding configuration. It was shown that the charge transfer from STO to TiO_2 occurs mainly through the chemical bonds at the interface and equilibrates the Fermi level by forming a double layer. However, it was found that subsequent polarization of the $O_{2\text{-fold}}$ lattice largely screens the interfacial dipole, yielding a net valence band offset of 0.76 eV. In addition, theoretical analysis suggested that interfacial impurities such as an O vacancy or substitutional fluorine may be responsible for the experimentally observed absence of a valence band offset at the anatase TiO_2/STO interface [32]. It is worth noting that other mechanisms such as cationic exchange may also contribute to the disappearance of the valence band offset. For example, Ciancio et al. have observed a Sr-deficient epitaxial interphase at the TiO_2/STO interface grown by pulsed laser deposition (PLD) [72, 73]. This interphase formation has been attributed to the long range migration of Sr from the STO substrate to the anatase film.

7.2   Epitaxial Integration of Ferromagnetic Correlated Oxide LaCoO₃ with Si (100)

Cobaltates exhibit a wide variety of exciting electronic properties resulting from strong electron correlations including superconductivity, giant magnetoresistance, metal-insulator transitions, and strong thermoelectric effects, making them an excellent platform to study correlated electron physics, as well as being useful for various applications in electronics and sensors [74]. The prototypical cobaltate is $LaCoO_3$ (LCO), which exhibits interesting behavior in terms of its magnetism, conductivity, thermal expansion, and structural distortions [75]. LCO and its Sr-doped counterpart ($La_{1-x}Sr_xCoO_{3-\delta}$ or LSCO) are widely used in several applications. LSCO, which is metallic beyond a certain doping level, is often used as an electrode layer for epitaxial ferroelectric capacitor structures as well as a cathode material in solid oxide fuel cells [76]. They are also used in gas sensors [77] and thermoelectric applications [78].

Epitaxial growth of these materials on silicon potentially leads to device integration where multiple types of devices (e.g. a sensor and an actuator) are grown on a single wafer in combination with traditional Si-based CMOS functionality. Recent integration of $La_{0.7}Sr_{0.3}MnO_3$ on Si by Pradhan et al. is one example of integrating magnetism on Si [79]. The ability to grow various cobaltates using MBE, with its capability for precisely controlling the atomic layer sequence and interface structure in multi-layer heterostructures, is critical for complementary theoretical/experimental studies of the spin state transition in this material. It also allows for the exploration of phenomena that arise at the interface with other epitaxial layers. An additional benefit of being able to grow cobaltates by MBE is the ability to arbitrarily adjust the layer composition during deposition and be able to form compositionally graded layers needed for some device applications.

The difficulty of the MBE approach is the thermodynamics of LCO itself, which requires relatively high temperatures and oxygen pressures to form.

The minimal theoretical model to describe the basic electronic properties of LCO is a $[CoO_6]^{9-}$ octahedron within the ligand field theory [80]. In a cubic crystal field, the localized $3d$ orbitals are split into doubly degenerate e_g (d_{z^2} and $d_{x^2-y^2}$) and triply degenerate t_{2g} (d_{xy}, d_{yz} and d_{zx}) states separated by the crystal field splitting $10Dq$. Other important energy scales are the on-site Hubbard repulsion U, Hund's exchange coupling J_H, and the hopping matrix between the Co $3d$ and O $2p$ orbitals t. Due to all competing interactions being of the same order, Co^{3+} can be found in several different spin states: low-spin (LS, $t_{2g}^6 e_g^0$, S $= 0$), intermediate-spin (IS, $t_{2g}^5 e_g^1$, S $= 1$), or high-spin (HS, $t_{2g}^4 e_g^2$, S $= 2$). The ground state is insulating and nonmagnetic (NM) with Co^{3+} in the LS state. LCO undergoes a crossover to a paramagnetic insulating phase at about 100 K where electrons get promoted to the e_g level, and a metal-insulator transition above 500 K [81]. However, the spin structure at different temperatures has been highly debated. For example, the LS-HS [81, 82], LS-IS [83, 84], and LS-HS/LS crossover scenarios have been discussed in the literature [85–89].

An exciting example of the tremendous potential of cobaltate heterostructures is the recent demonstration of biaxial tensile strain stabilizing an insulating ferromagnetic (FM) ground state in LCO [90–99]. Fuchs et al. first demonstrated that the IS or HS state could be stabilized by epitaxial tensile strain resulting in a ferromagnetic ground state with a Curie temperature (T_C) of ~90 K when LCO is grown on STO using PLD [90, 91], which was later confirmed by Herklotz et al. [93]. Though LCO is a classic example of a correlated $3d$ transition metal perovskite oxide [80, 100], until the demonstration by Fuchs et al., FM correlation has never been observed for the ground state [100]. In addition to transport measurements showing insulating behavior for tensile-strained FM LCO [92], Fuchs and co-workers have also shown that both the population of higher spin states and the magnetization in LCO increase as tensile strain increases [91]. Using X-ray techniques, Merz and co-workers have suggested that the magnetic structure of tensile-strained LCO grown on STO is a mixture of Co^{3+} high spin (HS) and Co^{3+} LS states [98]. A recent report by Metha et al. also suggests that compressive strain by itself cannot produce a FM state in LCO [97], indicating the existence of an asymmetric orbital-lattice interaction [101]. Magnetization measurements of compressively strained LCO on LAO substrates show only weak to no ferromagnetism [91, 92, 97]. Most recently, Sterbinsky et al. have shown that inter-site hybridization involving Co and O states in LCO on STO is weaker than that in LCO on LAO by comparing the pre-edge structure of the Co K-edge X-ray absorption spectra [99].

Posadas et al. reported strain-induced ferromagnetism in epitaxial layers of LCO grown on STO/Si pseudosubstrate by MBE [95]. They found that MBE-grown, strained LCO on silicon also exhibited a ferromagnetic ground state with a T_C similar to PLD-grown films [91]. Epitaxial STO was first grown using a process described in Chap. 6 on epi-grade (100)-oriented silicon wafers to a total thickness of 20 unit cells. Atomic oxygen was then introduced by means of an RF plasma

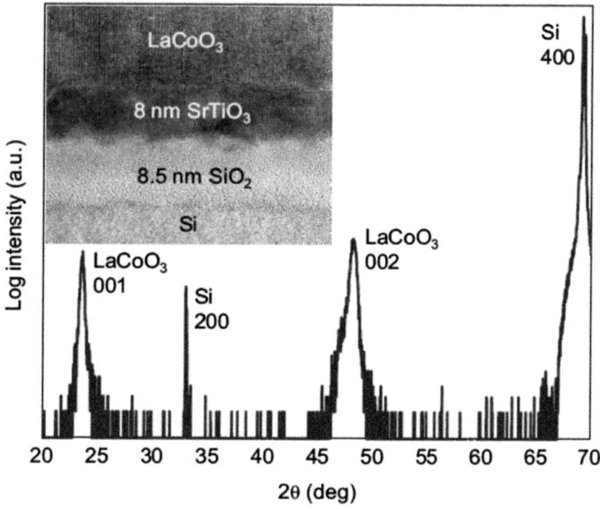

Fig. 7.7 X-ray diffraction 2θ − θ scan of LaCoO$_3$ on Si. The LaCoO$_3$ peaks are indexed using the pseudocubic notation. The *inset* shows a high-resolution cross-section transmission electron micrograph of LaCoO$_3$ on Si with the different layers labeled. Reprinted with permission from [95]. Copyright 2011, AIP Publishing LLC

source (with an oxygen background pressure of 1×10^{-5} Torr), then La and Co were co-deposited onto STO/Si at a substrate temperature of 725 °C. The LCO film was then cooled to room temperature in oxygen at a rate of 10 °C/min. In situ reflection high energy electron diffraction (RHEED) measurements indicate that the film is crystalline and epitaxially matched to the STO buffer layer.

The LCO film was measured using x-ray diffraction to determine lattice constants and overall crystalline quality [95]. A symmetric 2θ − θ scan of a 40-nm film (Fig. 7.7) shows only peaks from the silicon substrate and the 00*l* crystal planes of the LCO film. The *c* lattice constant was determined to be 3.77 Å from the symmetric scan, while measurement of the off-axis (103) Bragg peak yielded an in-plane lattice constant of 3.89 Å. These lattice constants are consistent with biaxially tensile-strained LCO with an in-plane lattice constant that is identical to that of the STO.

The inset of Fig. 7.7 shows a cross-sectional high-resolution electron micrograph of the composite LCO/STO/Si(100) material taken with a JEOL JEM-4000EX transmission electron microscope operated at 400 keV reported in [95]. The excellent crystal quality of the LCO layer, as well as the intermediary STO layer, are evident, even after high temperature and high oxygen pressure growth. However, a relatively thick (8.5 nm) SiO$_2$ layer has also formed at the interface between STO and Si as a result of high temperature growth in an oxygen plasma.

The magnetic properties of the LCO were measured using a Quantum Design superconducting quantum interference device (SQUID) magnetometer [95]. The magnetization as a function of temperature from 300 to 10 K was measured under an applied field of 1 kOe in both zero field-cooled and field-cooled conditions, after the film was first saturated at 10 K under a field of 40 kOe. The field was applied in the plane of the film. The results of the field-cooled measurement reported in [95] are shown in Fig. 7.8. The measurement showed a rapid increase in magnetization around 85 K, indicating that the strained LCO on Si is ferromagnetic, and with a T_C

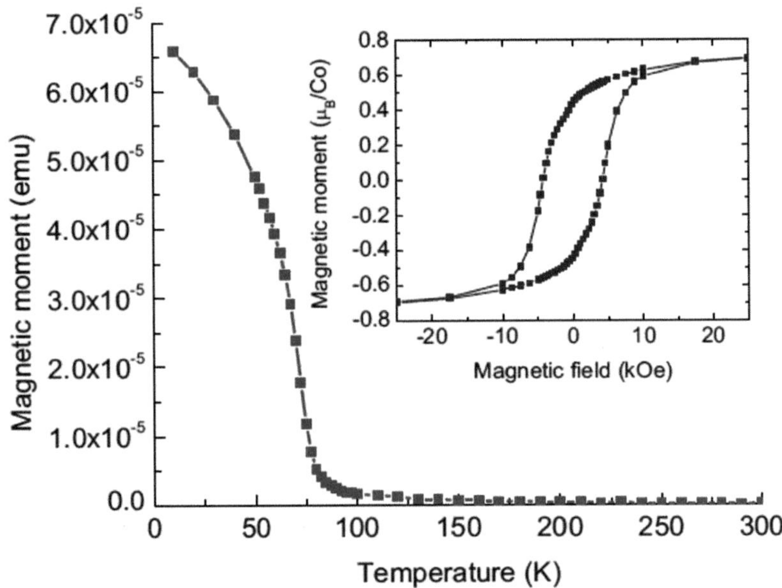

Fig. 7.8 Magnetization of LaCoO₃ as a function of temperature at a constant magnetic field of 1 kOe under field-cooled conditions. The film is ferromagnetic with a Curie temperature of 85 K. The inset shows magnetization of LaCoO₃ as a function of applied magnetic field measured at 10 K. The film has a coercive field of 3.8 kOe and a saturation moment equivalent to 0.7 μ_B per Co. Reprinted with permission from [95]. Copyright 2011, AIP Publishing LLC

that is consistent with the previous reports of strained films grown on single crystal STO using PLD [90, 91]. Magnetization as a function of applied magnetic field was also measured at 10 K, as shown in the inset of Fig. 7.8. The magnetic moment at 25 kOe was estimated to be about 0.7 μ_B per Co. The film also showed a very large coercive field of 3.8 kOe, which is similar to previous reports of strained LCO on bulk STO [91, 92] and could be an indication of strong magnetocrystalline anisotropy effects [100].

In conclusion, building on the seminal work of Fuchs et al. [90, 91] and Herklotz et al. [93], Posadas and co-workers have demonstrated monolithic integration of FM LCO on STO/Si pseudo-substrate using MBE [95].

7.2.1   Magnetism in Strained LaCoO₃: First Principles Theory

While a substantial body of experimental results for strained LCO has been reported, until recently there was a clear lack of theoretical understanding taking into account all the experimental observations [95, 103–105]. Using DFT, Gupta et al., have claimed that tensile strain is able to stabilize a FM ground state in

LCO [103] while Rondinelli et al., have suggested that strain by itself cannot produce a FM state [104]. However, Coulomb correlation effects for the localized $3d$ orbitals in LCO were not considered in [103] and proper structural optimization was not performed in [104]. Using local spin density approximation combined with the Hubbard U correction (LSDA + U), Posadas et al. have shown that a ferromagnetic state based on a *homogenous* intermediate spin (IS) state (S = 1) can be stabilized above 3.8 % tensile strain [95]. The ferromagnetic IS state is, however, inconsistent with two experimentally determined properties of strained LCO: the IS state is half-metallic while experiment shows that strained LCO is insulating [92], and a rather high critical strain of 3.8 % was required, which is somewhat higher than that in experiment (~2 %) [90–95]. Most recently, using a LDA + U approach, Hsu et al. have shown that a HS/LS mixed state has a lower energy than that of the IS state in LCO under a certain tensile strain [105].

A comprehensive first-principles analysis considering homogeneous IS states and inhomogeneous HS/LS mixed states as a function of biaxial strain from −4 to 4 % was recently performed by Seo and Demkov [106]. They showed that beyond a tensile strain of 2.5 %, LCO undergoes a spin state transition from LS to mixed HS/LS states [98], and explained why the higher concentration of HS Co^{3+} was preferred in tensile-strained LCO [91]. To understand the FM ordering in tensile-strained LCO found in experiment, they calculated the first and second nearest neighbor (n.n.) exchange parameters in the 1:1 HS/LS state [41, 107]. The qualitative feature of the exchange parameters was further verified within a model superexchange calculation showing that the first n.n. coupling was ferromagnetic. On the other hand, the second n.n. coupling was strongly antiferromagnetic (AFM). As a result, it was found that the most stable collinear magnetic structure of the HS/LS state is not a FM structure, but an AFM one with a ↑↑↓↓ order along the c-direction. The detailed description of calculations done using DFT could be found in [106]. Here we will discuss the main findings of this work.

7.2.2   *Strain-Induced Spin State Transition in LaCoO$_3$*

Seo and Demkov first tested homogeneous magnetic configurations with all Co^{3+} in either the IS or HS states [106]. They found that for the homogeneous HS state, only an antiferromagnetically (AFM) aligned solution (G-type) was stable, consistent with the Goodenough-Kanamori-Anderson rule [108–111]. At 3.5 % tensile-strain, however, the homogeneous IS and HS states were higher in energy than the NM state. They next considered mixed magnetic configurations with HS Co^{3+} ions embedded in a LS Co^{3+} matrix. The first important finding was that when LCO forms HS Co^{3+} ions, it was energetically favorable to separate them by LS Co^{3+} rather than having them to be the first nearest neighbors. Furthermore, tensile-strained LCO became more stable as the number of these second nearest neighbor HS pairs increased. Overall it was found that a 1:1 HS/LS mixed configuration was

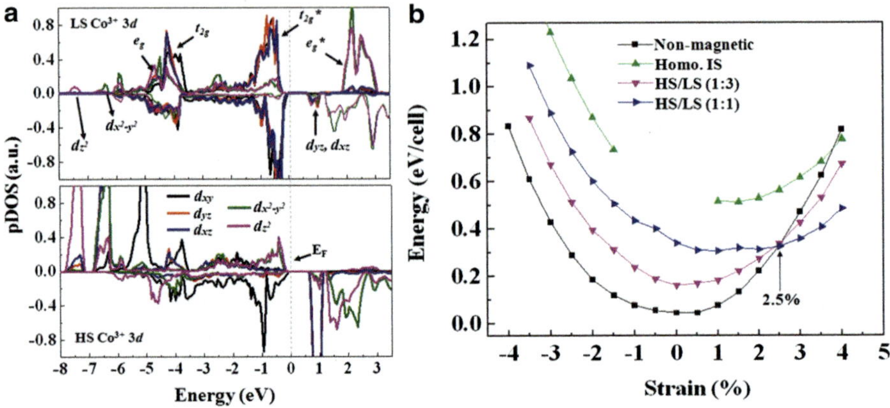

Fig. 7.9 (**a**) Projected density of states for $3d$ orbitals in the 1:1 HS/LS mixed at LS Co^{3+} site (*upper panel*) and out-of-plane HS Co^{3+} site (*lower panel*). The Fermi energy (*dashed vertical line*) is set to 0 eV. Positive (negative) DOS is for spin-up (spin-down). (**b**) Energy of LCO per $\sqrt{2} \times \sqrt{2} \times 2$ cell (four formula units) as a function of strain for non-magnetic (*filled squares*), homogeneous IS (*up triangles*), 1:3 HS/LS (*down triangles*), and 1:1 HS/LS states (*right triangles*). From [106]

the most stable magnetic solution for LCO under 3.5 % strain. In Fig. 7.9a, the $3d$-projected density of states at the LS and upper HS Co^{3+} sites along the c-axis direction for the 1:1 HS/LS state is shown [106]. There is an energy gap of 0.5 eV at the Fermi level defined by the t_{2g}* and e_g* splitting of the LS Co^{3+} sites. For the HS Co^{3+} site, the on-site U and J produce localized states from -8.0 to -5.0 eV in the spin-up channel, and empty d_{xz} and d_{yz} states in the spin down channel, consistent with the mean field picture of the HS state. The presence of the energy gap in the HS/LS mixed state is consistent with strained LCO being insulating [92].

To elucidate the effect of epitaxial strain on the magnetic state of LCO, Seo and Demkov compared the energy as a function of strain for mixed HS/LS configurations with 25 and 50 % concentration of HS Co^{3+}, the homogeneous IS state, and NM LCO for reference as shown in Fig. 7.9b. We see that the HS/LS mixed states are stable when compared to the homogeneous IS state at all strain levels, and that above a tensile strain of 2.5 %, the HS/LS states become more stable than NM LCO. Under zero strain, there is an energy cost to excite LS Co^{3+} to HS Co^{3+}. However, as a function of tensile strain, the energy of the mixed HS/LS states increases more slowly than that of NM LCO, inducing a spin state transition at 2.5 %. It is also evident from Fig. 7.9b that LCO with higher concentration of HS Co^{3+} is softer against tensile strain [91], and that compressive strain does not stabilize a magnetic state [91, 92, 97].

To shed more light on the mechanism of the strain-induced spin state transition in LCO, Seo and Demkov considered the energy gap in NM LCO as a function of strain. The energy gap in LCO forms between the t_{2g}* and e_g* bands and is given by $10Dq - [W(e_g*) + W(t_{2g}*)]/2$, where $10Dq$ is the crystal field splitting and W's are

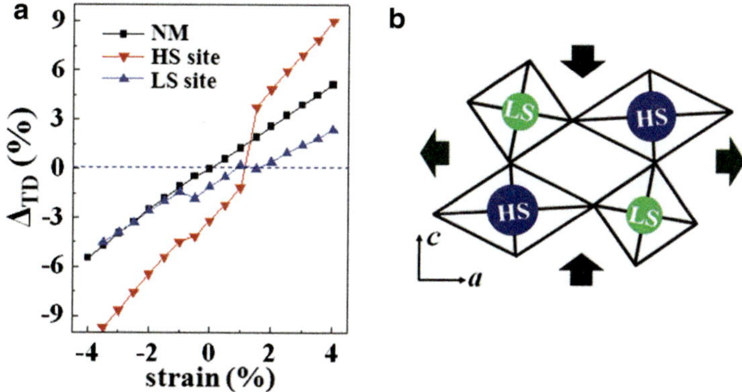

Fig. 7.10 (a) Local tetragonality ($\Delta_{TD} = 2 \times (b_{in} - b_{out})/|b_{in} + b_{out}|$) as a function of strain for CoO$_6$ octahedra in NM LCO (*squares*), for HS Co^{3+} sites (*down triangles*) and LS Co^{3+} sites (*up triangles*) in 1:1 HS/LS FM LCO. (b) Schematic of octahedral distortion in 1:1 HS/LS FM LCO above 1.5 % strain. *Lateral arrows* stand for the epitaxial constraint in the *ab* plane imposed by biaxial tensile strain while *vertical arrows* stand for the contraction of LCO in the *c*-direction due to the Poisson effect. From [106]

the bandwidths of corresponding bands. They found that the energy gap becomes less than 58 meV above a strain level of 2.5 %, thus allowing for the spin state transition [86]. They noted, however, that the band gap also narrows for compressively strained LCO, but this does not result in a magnetic solution as shown in Fig. 7.9b. This result suggested that the standard picture in terms of the competition between the crystal field splitting and the Hund's rule coupling is not sufficient to consistently describe magnetism in strained LCO. Instead, they found an important structural transition in LCO under tensile strain that accompanies the spin state transition.

When biaxially strained, LCO responds in the out-of-plane *c* direction due to the Poisson effect [91, 95]. Since the LS CoO$_6$ unit is rigid due to covalency of the Co-O bond (bond stretching costs a large amount of energy), strain is mainly accommodated by tilting and rotation of CoO$_6$ octahedra [112]. Microscopically, this is achieved by changes in the in-plane (θ_{in}) and out-of-plane (θ_{out}) Co-O-Co angles accompanied by slight changes in the Co-O bond length or local tetragonality ($\Delta_{TD} = 2 \times (b_{in} - b_{out})/|b_{in} + b_{out}|$, where b_{in} and b_{out} are the in-plane and out-of-plane Co-O bond lengths, respectively.). As shown in Fig. 7.10a, in NM LCO, local tetragonality in NM LCO increases almost linearly as tensile or compressive strain is applied. They noted that the energy curve for NM LCO in Fig. 7.9b could be thought of as $\frac{1}{2}k\Delta_{TD}^2$, where k is a spring constant determined by the covalent mixing between Co 3d and O 2p states. Therefore, to minimize the bond stretching or Δ_{TD} under tensile strain, the octahedral rotation is largely suppressed (θ_{in} greater than the bulk value) while the tilting is enhanced (θ_{out} smaller than the bulk value). The opposite is true for compressive strain: θ_{in} becomes smaller than the bulk value in conjunction with the disappearance of the tilting mode ($\theta_{out} = 180°$).

Interestingly, Seo and Demkov found that strained 1:1 HS/LS LCO undergoes an unusual structural transition above 1.5 % tensile strain [106]. It manifests itself as a substantial increase in Δ_{TD} of the HS CoO$_6$ clusters as shown in Fig. 7.10a, b. On the other hand, Δ_{TD} of the LS clusters in 1:1 HS/LS LCO drops by more than a factor of two compared to NM LCO. This suggests that above 1.5 %, tensile strain is accommodated mainly by the HS CoO$_6$ units through bond length changes, allowing the LS octahedra to be less distorted thus relieving their elastic energy. This is possible because HS Co^{3+} has a softer Co-O bond under stretch [113–115]. As a result, both bond angles θ_{in} and θ_{out} almost recover their bulk values since octahedral rotation and tilting are no longer needed for strain accommodation.

Mapping the exchange coupling between the local moments on an effective Heisenberg Hamiltonian, Seo and Demkov found that the first n.n. coupling $J_{1,in}$ and $J_{1,out}$ were ferromagnetic and 2.5 and 2.7 meV/pair, respectively [106, 116]. However, the second n.n. couplings were found to be strongly antiferromagnetic and $|J_{2,out}|$ was larger than $|J_1|$'s by more than a factor of two. To explore the effect of $J_{2,out}$ in the 1:1 HS/LS configuration of tensile-strained (3.5 %) LCO, they also performed several calculations based on the unconstrained non-collinear spin density functional formalism [117]. One of the lowest energy canted spin structures that they have found was that the spin moments rotate by 90° (for φ) as it goes to the next upper *ab*-plane according to the second n.n. AFM coupling. However, half of the local moments were slightly canted toward the *c*-axis, yielding a small magnetic moment of about 0.26 μ_B/Co^{3+} in the system [106]. These results suggested that the relatively low magnetic moment of 0.7 μ_B/Co^{3+} in experiment [95] may be due to the presence of the strong AFM coupling screening the FM ordering in the system.

7.2.3 Conclusions

In conclusion, considering various high-spin/low-spin configurations, Seo and Demkov showed that high-spin Co^{3+} ions in LCO prefer to be separated by low-spin Co^{3+} ions [106]. They demonstrated that above a tensile strain of 2.5 %, the ground state of LCO was an insulator with a 1:1 HS/LS mixed state. In contrast, compressive strain was not able to produce a magnetic state. They attributed the stabilization of the HS/LS state to increased compliance of LCO when it has a higher concentration of HS Co^{3+} ions. They examined the exchange parameters in the 1:1 HS/LS state of tensile-strained LCO by considering various collinear magnetic structures. Due to the strong antiferromagnetic coupling, they found that the lowest energy collinear structure is one with an up-up-down-down order along the *c*-direction. However, the results suggested that the competition between the FM and AFM couplings in the system may lead to a canted (non-collinear) spin structure with a finite net magnetization.

7.3 Cobalt-Substituted SrTiO$_3$ Epitaxially Integrated with Silicon

The ability to manipulate the spin degrees of freedom in electronic devices, in addition to the traditional control of charge, can potentially lead the way to advanced device structures that have higher speed and lower energy consumption [118]. One such proposed device is the spin field effect transistor (spin-FET) [119]. The hardest challenge in realizing spin-FETs is the efficient injection of a spin-polarized current from a ferromagnetic metal into a semiconductor such as silicon. Among the issues are the large density of states mismatch between a metal and semiconductor resulting in an injection efficiency of less than 1 % [120], and the fact that the interface between a typical ferromagnetic metal and silicon is chemically unstable leading to silicide pockets and other defects such as stacking faults [121]. The use of tunnel barriers such as ultrathin MgO or Al$_2$O$_3$ layers has partially alleviated the issue, raising injection efficiency to approximately 30 % [122, 123]. Using dilute magnetic semiconductors as spin injection contacts is a natural choice, as these materials have high quality interfaces to regular semi-conductors and do not have the conductivity mismatch problem. However, mag-netic semiconductors have Curie temperatures well below room temperature (less than 180 K for the best reported magnetic transition temperature) limiting their practical application [124]. Transition metal oxides doped with small amounts of magnetic ions have been found to exhibit room temperature ferromagnetism [21, 27, 124–133]. There has also been an extensive effort to increase the Curie temperature of these systems as well as to look for new candidates for room temperature ferromagnetism [134]. One issue limiting the application potential of most ferromagnetic doped metal oxides is the difficulty of integrating them with a suitable semiconductor spin host such as silicon. Most of the known dilute ferro-magnetic oxides are not thermodynamically stable in contact with silicon [135]. For example, ferromagnetic cobalt-doped anatase TiO$_2$ is known to result in substantial SiO$_2$ formation and TiO$_2$ reduction when grown epitaxially on silicon using a STO buffer [136].

Because it can be epitaxially grown directly on silicon, STO is a good candidate for the integration a ferromagnetic oxide onto silicon. A limited amount of work has been reported on using STO as a host material for magnetic ion doping with contradictory results. Bulk synthesis methods generally do not show ferromagne-tism in cobalt-doped STO except when a significant number of oxygen vacancies are present [137–141]. Thin films of insulating/semiconducting cobalt-doped STO grown by pulsed laser deposition (PLD) on the other hand, are reported to exhibit ferromagnetism at very high doping (> ~20 %) but not at low doping [142, 143]. Cobalt-doped metallic (La,Sr)TiO$_3$ thin films, however, do show ferromagnetism and highly spin-polarized carriers even at very low doping (~2 %) [144].

In the work described in [145], the epitaxial integration of Co-substituted STO on Si by MBE was demonstrated. The films exhibited room temperature insulating ferromagnetic behavior in the composition range of 30–40 % cobalt. In situ compositional analysis of the films using XPS indicated cobalt ions in the +2 valence state substituting for titanium, along with a significant amount of oxygen vacancies. Using first principles DFT calculations, the magnetic behavior of Co-substituted STO, with particular emphasis on the role of oxygen vacancies in promoting magnetism was also studied. The results suggested that a cobalt (II) ion–oxygen vacancy complex is responsible for the observed ferromagnetic insulating behavior in this system. Such a room temperature ferromagnetic insulator that can be integrated on silicon is potentially useful for spin filtering type injection contacts in spin-FETs.

7.3.1 Film Growth and Characterization

Cobalt-substituted STO was grown by MBE on a thin (three unit cells) undoped STO/Si pseudo-substrate [145]. The cobalt-substituted STO was grown at 550 °C under a constant background oxygen partial pressure of 2×10^{-7} Torr with a nominal sample stoichiometry of SrTi$_{1-x}$Co$_x$O$_{3-y}$. Samples with target cobalt compositions of $x = 10, 20, 30, 40,$ and 50 % were grown. During the growth, the Sr shutter was kept open while the Co and Ti shutters were opened alternately for a length of time corresponding to the target cobalt composition. The films were crystalline and epitaxial as-deposited with no further annealing treatment necessary.

After growth, the films were characterized in situ using a VG Scienta XPS system with R3000 electron analyzer [145]. High resolution scans of the Co *2p*, O *1s*, Ti *2p*, and Sr *3d* core levels were performed for each sample to determine stoichiometry. All films were also characterized ex situ using x-ray diffraction (Bruker D8 Advance) and magnetometry (Quantum Design MPMS). Magnetization vs. magnetic field measurements were performed at 300 K and at 10 K in a magnetic field range of ±10 kOe. Selected samples were prepared and imaged in cross-section using a JEOL JEM-4000EX transmission electron microscope operated at 400 keV. Resistivity vs. temperature down to liquid nitrogen temperature using a van der Pauw configuration was also measured on samples exhibiting ferromagnetism to determine whether they are metallic or insulating.

7.3.2 Film Crystalline Structure

RHEED patterns for a 10-nm Co-substituted STO film grown on 1.5 nm undoped STO/Si reported in [145] are shown in Fig. 7.11. The RHEED pattern for 10 % cobalt substitution (Fig. 7.11a) shows very sharp streaks similar to that obtained for undoped STO on Si. When the amount of cobalt was increased to 20–30 % (Fig. 7.11b), the streaks became broader, indicative of a greater degree of crystalline disorder, but still well-defined and single-phase. For 40 % Co substitution

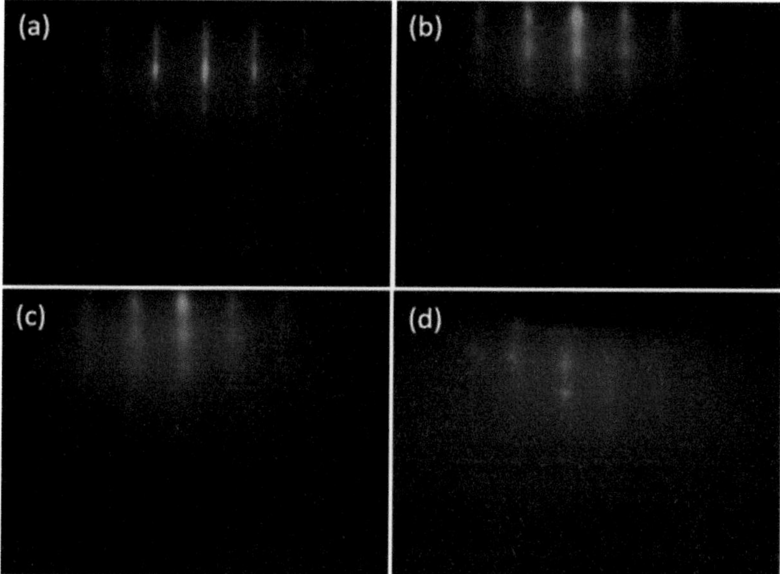

Fig. 7.11 Reflection high energy electron diffraction patterns taken along the (110) azimuth of Co-substituted SrTiO₃ for (**a**) 10 %, (**b**) 30 %, (**c**) 40 %, and (**d**) 50 % nominal cobalt composition. The diffraction patterns are taken using 18 keV electron energy at a grazing angle of 4°. From [145]

(Fig. 7.11c), RHEED shows the appearance of weak extra spots that indicate the presence of incommensurate secondary phases. The appearance of secondary phases becomes even clearer at 50 % cobalt composition where polycrystalline arcs begin to emerge (Fig. 7.11d).

The Co-substituted STO films were measured using x-ray diffraction to determine lattice constants and bulk crystalline quality [145]. For compositions 10–40 %, the symmetric $2\theta - \theta$ scan showed only peaks from the silicon substrate and the $00l$ crystal planes of STO, confirming the 001-orientation of the Co-substituted films. A small extra peak at ~42° observed for 50 % cobalt composition was attributed to CoO precipitates. A rocking curve measurement about the (002) STO peak for 10 % cobalt yielded a full width at half maximum of 0.4°, while a value of 0.7° was obtained for 30 % cobalt.

Figure 7.12 is a cross-sectional high-resolution electron micrograph of a 5-nm 30 % cobalt-substituted film grown on 1.5 nm undoped STO/Si reported in [145]. The uniformity of the cobalt-substituted layer is evident. A wider area low resolution scan of the same sample does not show the presence of precipitates. Figure 7.12 also shows that the initial undoped STO layer appears to have become amorphized and that a thin (~1.5 nm) SiO$_x$ layer has also formed. The reason for the loss of crystallinity in the undoped STO layer are still unclear and may be due to either preferential amorphization as a result of the TEM sample preparation [146] or to reaction of the thin STO layer with silicon forming silicates and silicides [147, 148].

Fig. 7.12 High resolution cross-section transmission electron microscopy images for 30 % Co-substituted STO grown on four unit cells undoped STO/Si. The uniform crystallinity of the Co-substituted layer is evident but the image also shows amorphization of the undoped layer as well as a thin (~1 nm) SiO$_2$ layer. From [145]

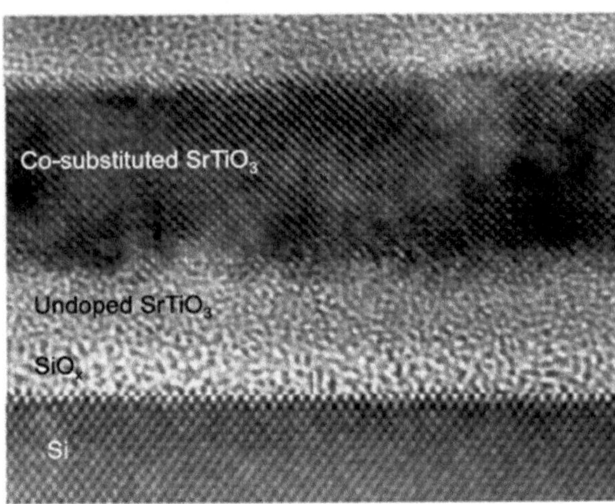

Table 7.2 Composition of Co-substituted SrTiO$_3$ films for different nominal Co compositions as measured by high resolution X-ray photoelectron spectroscopy

Nominal Co composition	Co/(Co + Ti)	Ti/(Co + Ti)	Sr/(Co + Ti)	O/(Co + Ti)
10 %	0.12	0.88	1.06	2.90
20 %	0.19	0.81	0.98	2.75
30 %	0.32	0.68	0.95	2.65
40 %	0.41	0.59	0.86	2.50
50 %	0.42	0.58	0.80	2.45

The integrated areas of the Sr *3d*, Ti *2p*, Co *2p*, and O *1s* spectra were utilized using appropriate sensitivity factors and corrected for sampling depth. From [145]

7.3.3   *Film Composition*

The Co 2*p* spectrum measured by in situ XPS showed a peak at 780.5 eV (2$p_{3/2}$) with a strong shake-up satellite feature at about 6.0 eV higher binding energy [145]. This spectrum is consistent with previously reported spectra for Co in the +2 valence state [149, 150] and confirmed by comparison to an epitaxial CoO thin film grown separately [145]. There was no signal at 778 eV, confirming that there was no detectable free Co metal in the sample [151]. The Sr *3d*, Ti *2p*, and O *1s* spectra had qualitative features similar to in situ spectra for undoped STO [145]. The XPS high resolution spectra for Sr *3d*, Ti *2p*, Co *2p*, and O *1s* were also used to determine the stoichiometry of all films [145]. Calculated stoichiometries for samples with various target cobalt compositions reported in [145] are shown in Table 7.2. The measured Co concentrations are within ~10 % of the target cobalt composition. XPS also showed that the amount of oxygen in the Co-substituted films are all significantly less than the ideal O/(Co + Ti) ratio of 3 and correlated roughly with the amount of cobalt substitution [145]. In particular, for 20 % cobalt substitution, an oxygen ratio of 2.75 was observed indicating as much as 25 % oxygen deficiency in the film. This observation implies that cobalt substitution facilitates the removal of a nearly identical amount of oxygen (compensated doping).

7.3.4   Magnetic and Transport Properties

The magnetic properties of the cobalt-substituted films were measured using a Quantum Design SQUID magnetometer [145]. The magnetization as a function of magnetic field was measured at 10 and 300 K. The field was applied in the plane of the film. The results of the 300 K measurement reported in [145] are shown in Fig. 7.13 for various cobalt compositions. The measurement shows paramagnetic behavior for low doping (10 %). A small hysteresis opens up at 20 % cobalt and a well-defined hysteresis loop is observed at 30–40 % cobalt. Based on the film volume used in the magnetic measurement, the saturation magnetic moment was calculated to be about 3.1 μ_B per Co with a remnant moment equivalent to 0.6 μ_B per Co. The 30 % cobalt film also shows a coercive field of 95 Oe at room temperature, slightly larger than typical values reported for cobalt-doped anatase [133] and PLD-grown thin films of cobalt-doped STO [143]. For very high cobalt concentrations (50 %), magnetic ordering in the film is lost and only diamagnetic behavior from the substrate was observed at room temperature. Measurement of magnetization vs. temperature between 10 and 300 K showed no phase transition and only a gradual increase in the magnetic moment as temperature was decreased. This indicates that the Curie temperature of the films is above room temperature. It has been reported by Bi et al. that the Curie temperature for PLD-grown cobalt-substituted STO is higher than 1,000 K [143]. Resistivity measurement of a ferromagnetic sample at room temperature showed a high sheet resistance of >1.1 GΩ/□ [145]. The sheet resistance increases as the temperature of the sample was decreased towards 77 K, indicating that the Co-substituted STO is insulating.

7.3.5   Electronic Structure of Co in STO

In order to study the electronic structure and magnetic moment of an isolated Co atom, a dilute concentration of Co is needed. For this, a $3 \times 3 \times 3$ supercell of STO was employed where a single Ti atom is replaced by Co resulting in a stoichiometry of SrCo$_x$Ti$_{1-x}$O$_3$ with $x = 3.7$ % [145]. Substituting a Ti atom in STO imparts a valency of +4 to Co meaning that it has five electrons to fill its valence states. Depending on the relative strength of the crystal field splitting and Hund's coupling of the Co atom, these five electrons can either occupy only t_{2g} levels forming a low-spin $[(t_{2g}\uparrow)^3(t_{2g}\downarrow)^2]$ (1μ_B) state, or occupy both e_g and t_{2g} levels forming a high-spin $[(t_{2g}\uparrow)^3(e_g\uparrow)^2]$ (5μ_B) state. The calculations in [145] indicated that Co stabilizes in the low spin state, where all the majority t_{2g} states are

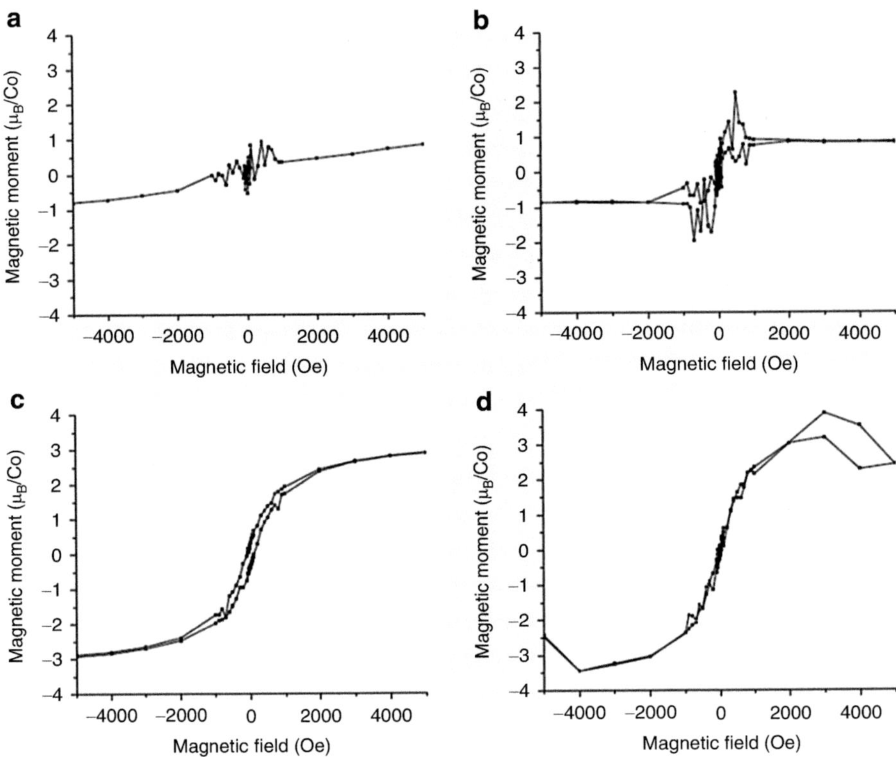

Fig. 7.13 Room temperature magnetization vs. magnetic field measurements for Co-substituted SrTiO$_3$ with different compositions: (**a**) 10 %, (**b**) 20 %, (**c**) 30 %, and (**d**) 40 % Co. The magnetic field is applied in the plane of the film and ranges from -5 to $+5$ kOe. At 10 % Co, the film is paramagnetic. A small hysteresis loop opens up at 20 %, becoming very well defined at 30–40 %. The coercive field for 30 % Co is 95 Oe with a saturation magnetic moment equivalent to ~3.1 μ_B/Co. At 50 % Co (not shown), only a diamagnetic signal from is observed. From [145]

occupied while one of the minority t_{2g} states is unoccupied. On the other hand, both the majority and minority spins of the e_g states were found to be unoccupied. It was also found that the influence of Co on Ti is short-ranged. At a site located two lattice constants away, the Ti local density of states (DOS) was found to be essentially identical to that of the bulk STO. This implies that any Co-Co interaction mediated is short-ranged.

To study the magnetic coupling between Co atoms, two Ti atoms were substituted in the same supercell (the cell is sufficiently large to treat up to third nearest neighbors) of STO with Co [145]. The magnetic interaction for both first nearest neighbor Co atoms as well as second nearest neighbors was calculated. For each case, the total energy of a ferromagnetic ($E_{\uparrow\uparrow}$) and an antiferromagnetic ($E_{\uparrow\downarrow}$) configuration was compared. The calculations showed that for nearest neighbor Co atoms, the ferromagnetic configuration is favored over the antiferromagnetic one by

60 meV/cell. However, when two Co atoms were arranged as second nearest neighbors the ferromagnetic interaction became zero, confirming that the magnetic interaction between Co atoms in STO is short-ranged. This suggests that, to induce ferromagnetism, Co atoms need to cluster close to each other. However, comparing the total energy of the first and second nearest neighbor configurations, it was found that clustering of Co atoms is not energetically favored. Hence, extrinsic sources such as point defects are needed to stabilize the experimentally observed room-temperature ferromagnetism in Co-substituted STO.

7.3.6   Role of Oxygen Vacancies in Ferromagnetism

Because the presence of a significant number of oxygen vacancies in as-deposited MBE-grown Co-substituted STO was observed experimentally, they are likely involved in the origin of ferromagnetism in this system [145]. The role of oxygen vacancies has been previously investigated by Griffin Roberts et al. for anatase [152] and by Florez et al. for STO [153]. Griffin Roberts et al. showed that in cobalt-doped anatase, a Co^{2+} interstitial coupled with Ti^{3+} via an oxygen vacancy are responsible for the observed ferromagnetism with cobalt doping at the level of 3 % [152]. In the study of Florez et al., mixed spin states of Co^{3+}, in the presence of an oxygen vacancy, were found to produce ferromagnetic interactions between Co atoms in cobalt-doped STO at a doping level of 12.5 % [153]. However, it was found experimentally that robust magnetic ordering in cobalt-doped STO only occurs at higher concentrations of Co (~30 %) and that Co is in the +2 valence state [145].

Theoretically, $SrTi_{1-x}Co_xO_{3-\delta}$ was modeled with a $2 \times 2 \times 2$ supercell to simulate the higher Co concentration [145]. The effect of a single oxygen vacancy on a neighboring Co atom (Fig. 7.14a) was first examined. It was first checked whether or not vacancies are likely to occur in the vicinity of Co atoms. In order to do this, the formation energy of an oxygen vacancy in the vicinity of a Co atom was computed. This was done following the Zhang-Northrup formalism [154]. The formation energy was defined by

$$E_{form} = E_{total}(SrTi_{1-x}Co_xO_{3-\delta}) - E_{total}(SrTi_{1-x}Co_xO_3) - \mu_{O_2} \qquad (7.5)$$

where the first two terms are the total energies with and without an oxygen vacancy, respectively, and the last term is the chemical potential of oxygen taken as half of the binding energy of an oxygen molecule [155]. The formation energy of an oxygen vacancy in $SrTi_{1-x}Co_xO_{3-\delta}$ was calculated to be 4.56 eV when it was placed adjacent to a Co atom at a distance of 1.93 Å [145]. However, when the vacancy was placed 7 Å away from the Co atom, the formation energy increased to 5.84 eV. Hence, it is clear that oxygen vacancies are likely to occur near Co atoms. Compared to the neutral vacancy formation energy in undoped STO, which is about

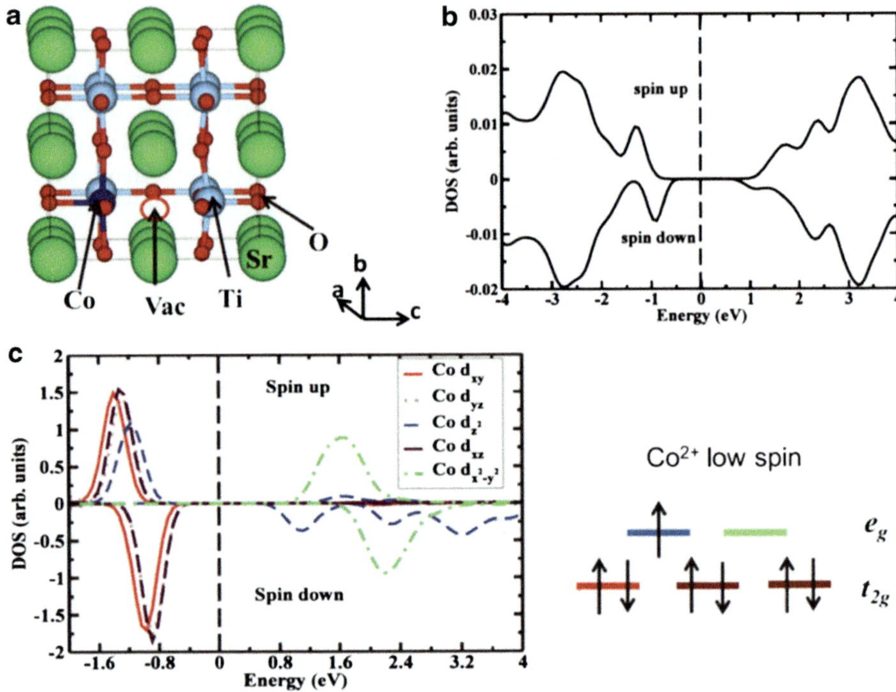

Fig. 7.14 (a) Calculation supercell for one Co and one oxygen vacancy in SrTiO$_3$, with the oxygen vacancy next to the Co atom. The vacancy is directly on the right of Co (*dark blue*) in the figure. (b) Total density of states plot showing the existence of an energy gap resulting in insulating behavior. (c) Calculated local density of states for Co with an oxygen vacancy next to it for spin up (*upper half*) and spin down (*lower half*) electrons. The calculations indicate that Co is in the low spin Co^{2+} configuration. From [145]

6.5 eV [71], the presence of a Co atom substantially (~2 eV) lowers the formation energy because Co provides empty low energy states for the two extra electrons associated with the vacancy to occupy. From this point on, it was assumed that the vacancy is located adjacent to substitutional Co [145].

Before examining the electronic structure, it is interesting to point out the structural changes brought by the introduction of a Co-oxygen vacancy complex. The equilibrium lattice constant of bulk STO was found to be 3.92 Å within an LDA + U approximation [145]. This means there is an overestimation of 0.5 % compared to the experimental value of 3.90 Å. Upon the introduction of a Co-oxygen vacancy complex, the equilibrium lattice constant of SrTi$_{1-x}$Co$_x$O$_{3-\delta}$ (x = δ = 0.125) was reduced to 3.90 Å [145]. As can be seen from the total DOS plot (Fig. 7.14b), the system was found to be an insulator. An oxygen vacancy is an *n*-type defect and results in the donation of two electrons to the neighboring Co atom. This changes the cobalt valence state from Co^{4+} to Co^{2+}. With seven electrons in its outermost valence shell, Posadas et al., found that Co stabilizes in the low spin state with a magnetic moment of 1μ$_B$ [145]. The orbital resolved DOS

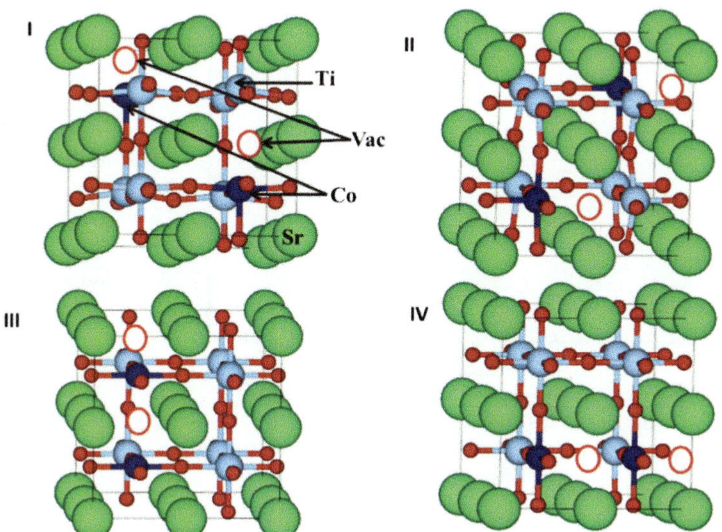

Fig. 7.15   Four lowest energy configurations for two Co atoms with adjacent oxygen vacancies in a 2 × 2 × 2 supercell. From [145]

plot of Co (in the presence of an oxygen vacancy) shows that while all its t_{2g} orbitals are completely filled, one of its e_g orbitals, the d_z^2 orbital is half filled and the other, the $d_{x^2-y^2}$ is completely empty (Fig. 7.14c). This results in stabilization of the low spin state.

Experimentally, a well-defined hysteresis loop and a high-spin state were observed at a Co concentration of more than 20 % [145]. Thus, Posadas et al. have theoretically considered the 25 % concentration of Co with two Ti atoms being substituted by Co in the 2 × 2 × 2 supercell [145]. The occurrence of Co in the +2 valence state suggested that for every Co atom introduced, one needs a compensating oxygen vacancy. At this concentration, various configurations of Co and oxygen vacancies are possible. As described earlier, oxygen vacancies are more likely to be found near a Co atom.

From in-plane x-ray diffraction measurements [145], the 20 % Co-substituted STO sample on Si was under an in-plane tensile strain (1 %) and an overall tetragonal distortion was present with the ratio between the out-of plane and in-plane lattice constants c/a being 0.989. This was modeled theoretically by fixing the in-plane lattice constant to 3.963 Å, which implies a 1 % tensile strain in a theoretical structure, and then optimizing the out-of plane lattice constant c. There are many configurations possible for the Co atoms and vacancies at this concentration. In Fig. 7.15, the four lowest energy structures reported in [145] are shown. In structures I and II, Co atoms are placed along the body diagonal with vacancies along the c axis in I, and along the a axis in II. In structures III and IV,

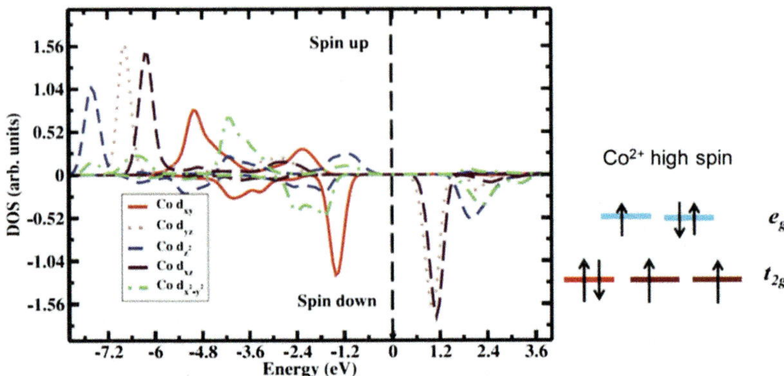

Fig. 7.16 Local density of states for Co for the lowest energy configuration of two Co atoms plus two oxygen vacancies (configuration IV in Fig. 7.15). The Co stabilizes in the high spin state with a magnetic moment of 3μ_B/Co. From [145]

the cobalt-vacancy complexes are aligned along the c and a axes, respectively. After relaxation, the resulting c/a ratios were found to be 0.973, 0.974, 0.969 and 0.975, respectively, for structures I, II, III and IV. Configuration IV was found to have the lowest energy, with the difference between E_{III} and E_{IV} being 164 meV/Co. Structures I and II are 359 and 320 meV/Co, respectively, higher in energy than structure IV. The in-plane tensile strain appears to stabilize a linear arrangement of Co-vacancy complexes along one of the in-plane directions. Structure IV was also found to be insulating in agreement with experiment. An orbital-resolved local DOS (Fig. 7.16) of one of the Co atoms in structure IV shows one of the e_g orbitals and two of the t_{2g} orbitals to be half-filled, indicating that Co is in the high-spin state with a magnetic moment of $3\mu_B$/Co. It should be noted that the stabilization of one of the e_g states (a combination of $d_{x^2-y^2}$ and d_{z^2} orbitals) originates from its local hybridization with the Co $4s$ state in the presence of an oxygen vacancy [156].

The theoretical investigations imply that a combination of an oxygen vacancy and strain are crucial in order to explain the experimental observations in [145]. First, a Co ion and oxygen vacancy tend to form a complex and the two electrons provided by a vacancy are trapped at Co site, consistent with the observed Co^{2+} state. The presence of Co lowers the formation energy of an oxygen vacancy by as much as 2 eV, consistent with the observation of the presence of an approximately equal number of oxygen vacancies as the amount of cobalt substitution. Second, a high concentration of Co (~25 %) is needed to stabilize the high spin state, in agreement with the occurrence of ferromagnetism only at cobalt concentrations greater than 20 %. Finally, strain appears to stabilize the in-plane orientation of Co-vacancy complexes that result in an insulating state as measured experimentally.

7.3.7 Conclusions

Posadas et al. have successfully demonstrated integration of cobalt-substituted STO on silicon using MBE. Films with 30–40 % cobalt showed room temperature ferromagnetism with a saturation moment of ~3 μ_B/Co. X-ray photoelectron spectroscopy indicated that Co is in the +2 valence state and that an approximately equal number of oxygen vacancies are created by the cobalt substitution. Resistivity measurements showed that the combination of cobalt substitution and oxygen vacancy creation result in an insulating material. First principles calculations revealed that oxygen vacancies are crucial in stabilizing ferromagnetism in this system and that a cobalt-oxygen vacancy complex is responsible for the observed insulating and magnetic behavior.

7.4 Ferroelectric BaTiO$_3$ Epitaxially Integrated with Silicon

The epitaxial integration of BaTiO$_3$ (BTO) on Si (001) presents several challenges, particularly for applications requiring that the ferroelectric polarization be pointing out of plane. The lattice mismatch between BTO and Si is 4 % (and even larger at typical growth temperatures) with the BTO compressed in-plane. While the direction of the mismatch is favorable for producing c-axis oriented BTO, the relatively large mismatch usually results in a very high defect density that can degrade the electrical properties of the film. Another challenge is the large thermal expansion mismatch between BTO and Si, with BTO having a thermal expansion coefficient about three times larger than Si [157]. While BTO films have been grown on Si using various buffers [158–162], these typically produce a-axis oriented films as a result of the thermal expansion mismatch, which causes the BTO to experience tensile stress while cooling down through the Curie temperature. To resolve this problem, a combination of a suitable buffer that reduces both lattice and thermal expansion mismatch, and slow cool down to reduce stress is necessary to obtain c-axis oriented films on Si. The epitaxial growth of tetragonal BTO on Si with the c-axis being out-of-plane has been reported by several groups [157, 163, 164].

Figure 7.17 (from [157]) shows the stability of various ferroelectric phases and domain structures in (100)-oriented BTO thin films as a function of temperature and biaxial strain. The temperature-strain stability diagram was calculated using phase-field simulations with an eighth-order Landau-Devonshire thermodynamic potential describing the bulk free energy of BTO. Positive strain values are for films in a biaxial tensile strain state while negative strain values are for films in a biaxial compressive strain state. The letters **T**, **O**, and **M** in the graph indicate tetragonal, orthorhombic, and monoclinic crystallographic symmetries, respectively, under a constraint. The paraelectric and ferroelectric natures of the phases are indicated by the superscript P and F, respectively. $M_1^F + O_2^F$ implies a mixture of M_1^F and O_2^F

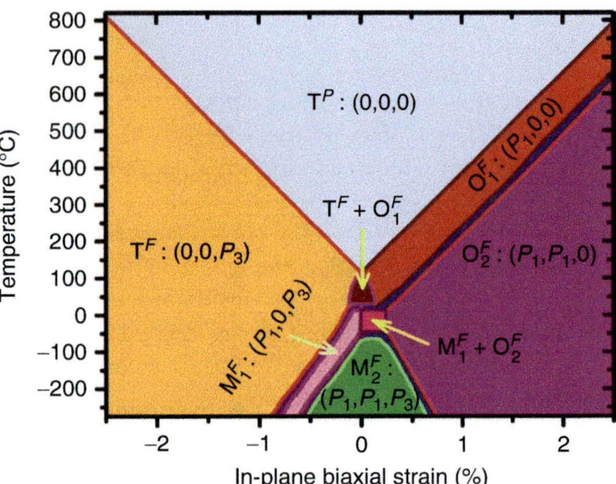

Fig. 7.17 Stability phase diagram for epitaxial (100) oriented BaTiO₃ thin films as a function of temperature and total in-plane biaxial strain (resulting from lattice mismatch and thermal expansion mismatch), predicted using phase-field simulations. Reprinted with permission from [157]. Copyright 2006, AIP Publishing LLC

phases. The components of the polarization vector P corresponding to the phases (along the crystallographic directions of the pseudocubic BTO) are indicated within the parentheses following the phase notation. From Fig. 7.17, it is evident that by tailoring the total strain (lattice mismatch plus thermal expansion mismatch) between the film and the substrate, the resulting orientation of a BTO film can be varied from a purely *a*-axis oriented (where the *a*-axis lies perpendicular to the plane of the substrate), to a mixture of *c* and *a* axes oriented, to purely *c*-axis oriented (where the *c*-axis lies perpendicular to the plane of the substrate). Large biaxial tensile strain results in *a*-axis films, while large biaxial compressive strain results in *c*-axis films. This suggests that using appropriate buffer layers for strain affords certain control over the polarization state of the BTO film.

Vaithyanathan et al. have reported *c*-oriented BTO grown on Si by reactive molecular beam epitaxy [157]. Their strategy was to use a relaxed buffer layer of Ba$_x$Sr$_{1-x}$TiO₃ between the BTO film and silicon substrate. Such a buffer accomplishes two important requirements. Its lattice constant is sufficiently close to BTO, but the biaxial compressive strain it exerts on BTO is sufficient to overcome the tensile strain of +0.4 % caused by the thermal mismatch. Using a buffer with composition x = 0.7, high quality *c*-oriented BTO was grown. The rocking curve widths of the BTO films were as narrow as 0.4°. X-ray diffraction and second harmonic generation experiments reveal the *out-of-plane c*-axis orientation of the film. Piezoresponse atomic force microscopy was used to write ferroelectric domains, corroborating the orientation of the ferroelectric film. In Fig. 7.18 we show the piezoresponse for a set of domains written into the BTO film of the 100 Å BaTiO₃/300 Å Ba$_{0.7}$ Sr$_{0.3}$TiO₃/Si (001) ferroelectric heterostructure at a *dc* bias of +2 V on the AFM tip. The line profiles demonstrate the sharpness of the domain boundaries obtained. Subsequent experiments confirmed the reversibility of the film polarization induced in this manner.

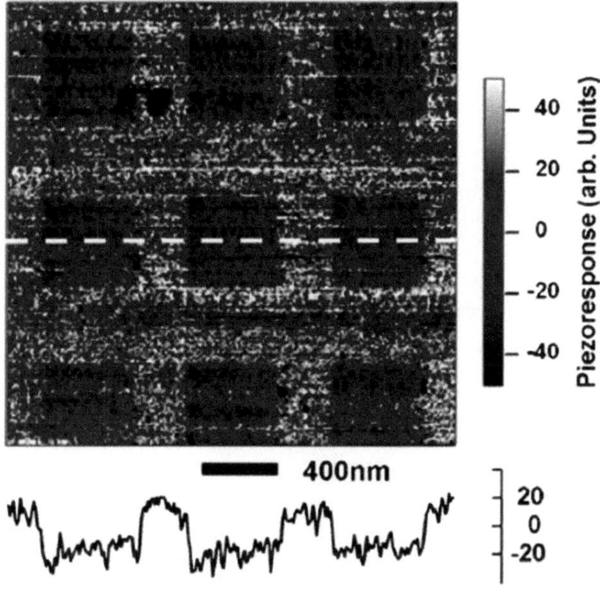

Fig. 7.18 Ferroelectric domain patterns written on the 100 Å BTO/300 Å Ba$_{0.7}$Sr$_{0.3}$TiO$_3$/Si (001) heterostructure using piezoresponse AFM, corroborating the c-axis orientation of the BTO film. Reprinted with permission from [157]. Copyright 2006, AIP Publishing LLC

Niu and co-workers reported the epitaxial growth and electrical characterization of single crystalline, (001)-oriented BTO film on STO/Si(001) template using both molecular beam epitaxy (MBE) and pulsed laser deposition (PLD) methods [164]. MBE is well known for its unique advantages of surface, interface and stoichiometry control. However, the oxygen pressure in an MBE chamber is limited to the order of 10^{-5} Torr, which likely leads to the formation of oxygen vacancies that dramatically impact the electrical properties of the oxide films [165]. PLD enables the growth of oxide thin films under much higher oxygen partial pressure of ~1 Torr, which should suppress oxygen vacancy formation in the film. Due to the limited oxygen pressure during growth, the BTO films prepared by MBE show no ferroelectric properties but only typical dielectric behavior despite a post-deposition rapid thermal annealing (RTA) being performed. In Fig. 7.19 we show the I–V and C–V curves (f = 1 MHz) of as-deposited MBE-grown BTO film (black) and that annealed by RTA process under O$_2$ ambient (red). The I–V curve of the as-deposited BTO sample exhibits no ferroelectric characteristic but a typical dielectric behavior and the leakage current is -2.5 μA at V $= -5$ V. Though the RTA process under oxygen ambient decreased the leakage current in the film by a factor of 10, ferroelectricity was still absent.

The much higher oxygen pressure used in the PLD growth resulted in a BTO film with a hysteretic C-V curve characteristic of a ferroelectric. The C-V and G-V curves are shown in Fig. 7.20, showing hysteresis loops in C-V and an asymmetric feature in the G-V curve corresponding to the switching current of the ferroelectric domain at the coercive field. The measured C-V memory window has a width of 0.75 V. This shows that epitaxial BTO films on Si can be possibly used in the non-volatile memory applications as long as they are sufficiently oxygenated.

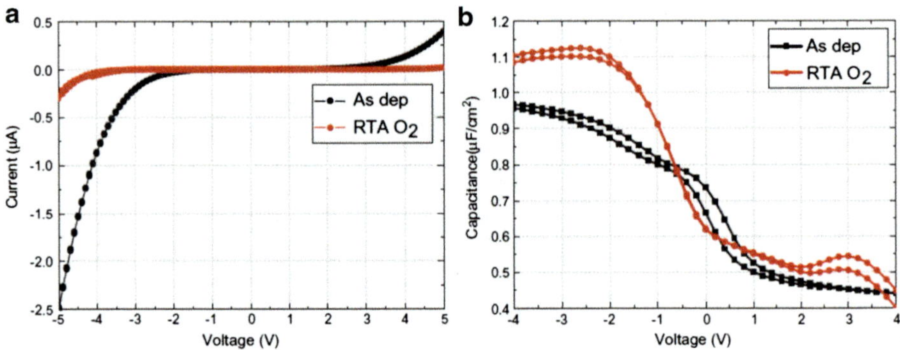

Fig. 7.19 (**a**) I–V characteristics of the BTO/STO/SiO$_2$/Si(001) samples of as-deposited one and O$_2$-RTA annealed one; (**b**) corresponding C–V characteristics at f = 1 MHz for the samples. The electrode area is 100 × 100 μm^2. Reprinted from [164], Copyright 2011, with permission from Elsevier

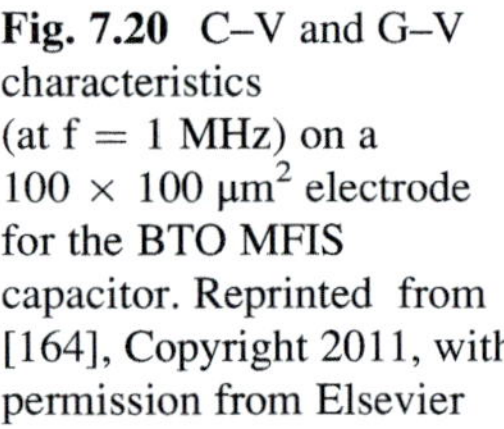

Fig. 7.20 C–V and G–V characteristics (at f = 1 MHz) on a 100 × 100 μm^2 electrode for the BTO MFIS capacitor. Reprinted from [164], Copyright 2011, with permission from Elsevier

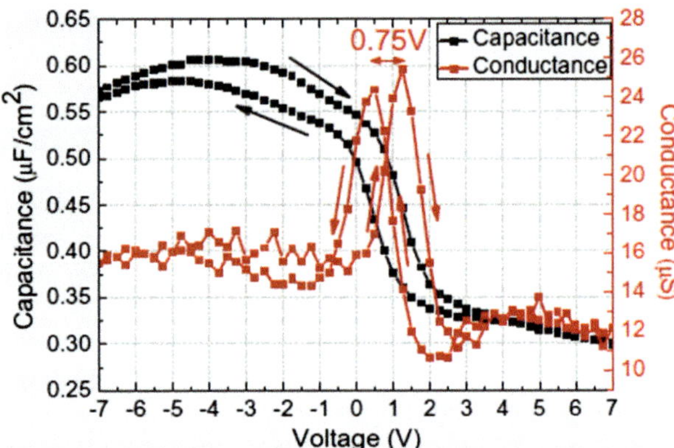

Most recently, another two successful attempts at growing out-of-plane polarized BTO on Si(001) have been reported [166, 167] both using an STO buffer grown epitaxially on Si. Dubourdieu et al. were interested in realizing a ferroelectric field-effect transistor (FETs) for non-volatile memory [168] and logic applications [169, 170]. It has been recently proposed that introducing a ferroelectric as a gate oxide could decrease the sub-threshold slope of FETs below the intrinsic thermodynamic limit of 60 mV/decade at room temperature thereby enabling low voltage operation of logic devices [171, 172] and thus reducing power consumption. Abel et al. pursued the integration of electro-optical active films on silicon, which could pave the way towards power-efficient, ultra-compact integrated devices, such as modulators, tuning elements and bistable switches [167, 171].

The strategy of Dubourdieu et al. was to use a fully relaxed layer of STO grown on Si as a pseudo-substrate for the subsequent BTO deposition. The lattice mismatch between BTO and STO still induces compressive strain that stabilizes

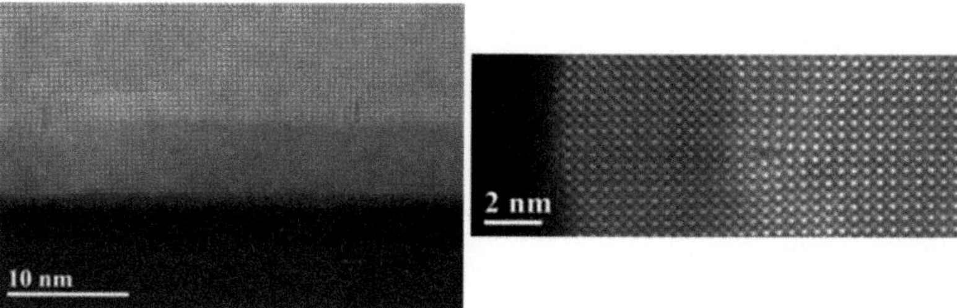

Fig. 7.21 TEM and in-plane X-ray diffraction of BTO/STO/silicate/SiO$_2$ stacks. Transmission electron micrographs emphasizing the interface between STO and BTO films. Edge dislocations are observed at the interface. From [166]

out of plane orientation of the BTO film with a significantly reduced lattice mismatch of 2.2 %, as well as a reduced thermal expansion mismatch such that BTO now experiences favorable compressive stress during cool down. To further reduce the lattice mismatch between BTO and STO, the STO pseudo-substrate was annealed in oxygen prior to BTO deposition to allow for plastic relaxation of the STO on Si as a result of thin SiO$_x$ interlayer formation [33]. After the STO growth, the substrate was heated at 20 °C/min to 700 °C while simultaneously ramping the O$_2$ pressure to 5×10^{-6} Torr. When the target O$_2$ pressure and substrate temperature were reached, BTO growth by means of alternating monolayer dosing of Ba and Ti was initiated, with Ba first and Ti last. After the desired thickness of BTO was grown, the sample was cooled at 5 °C/min in O$_2$ to room temperature. BTO films with thicknesses ranging from 1.6 to 40 nm were grown. Thus by tailoring the heterostructure layers and their interfaces, it was indeed possible overcome both lattice and thermal expansion mismatch issues and grow c-axis oriented BTO films on Si.

High resolution TEM images such as those shown in Fig. 7.21 confirm the high crystalline quality of the stacks at a local scale with a sharp structural and chemical interface between STO and BTO. Misfit dislocations appear close to the STO/BTO interface. Relaxation of the in-plane parameter with increasing BTO thickness is clearly observed using grazing incidence x-ray diffraction, as illustrated in Fig. 7.22 on the 200 reflection. The ultrathin 1.6 nm film is quasi-pseudomorphic with STO while for thicker films distinct contributions are observed for BTO and STO.

Though negative capacitance effects were not demonstrated, Dubourdieu et al. have demonstrated ferroelectric switching of perpendicular polarization in epitaxial BTO films on STO-buffered silicon in the absence of a bottom metallic electrode for BTO film thicknesses of 8–40 nm. Shown in Fig. 7.23 are hysteresis loop measurements on the BTO films as function of BTO thickness. For the 40 nm film, the characteristic bulk hysteresis loop shape with well-saturated response is observed. For the thinner 16 and 8 nm BTO films, the hysteresis loops become more elongated and the remnant responses become lower. This behavior is consistent with that predicted for thin ferroelectric films as driven by depolarization field effects [172]. However, the hysteresis loops are still well defined and variability of response between dissimilar locations is small compared to the loop opening.

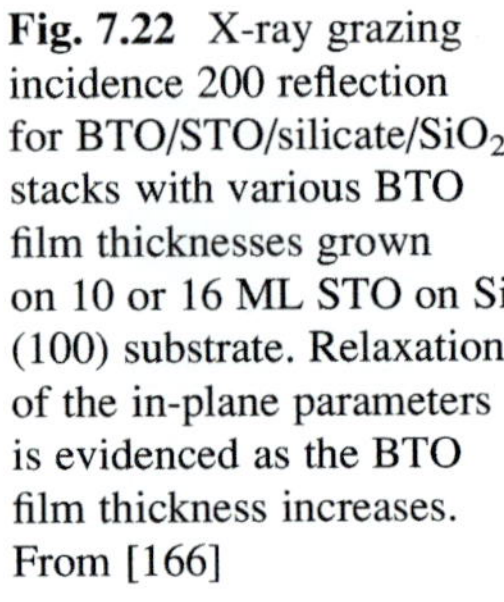

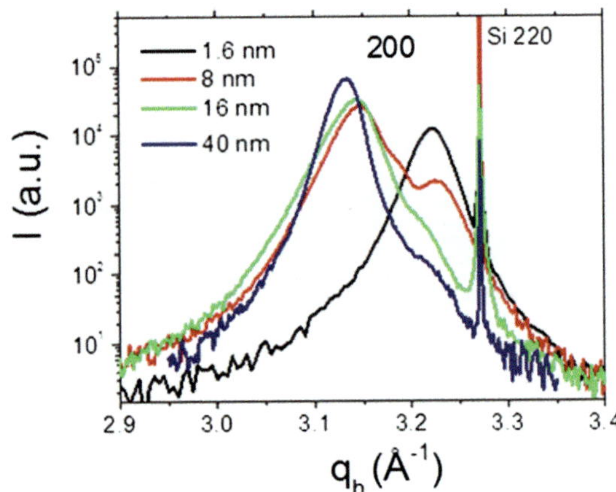

Fig. 7.22 X-ray grazing incidence 200 reflection for BTO/STO/silicate/SiO$_2$ stacks with various BTO film thicknesses grown on 10 or 16 ML STO on Si (100) substrate. Relaxation of the in-plane parameters is evidenced as the BTO film thickness increases. From [166]

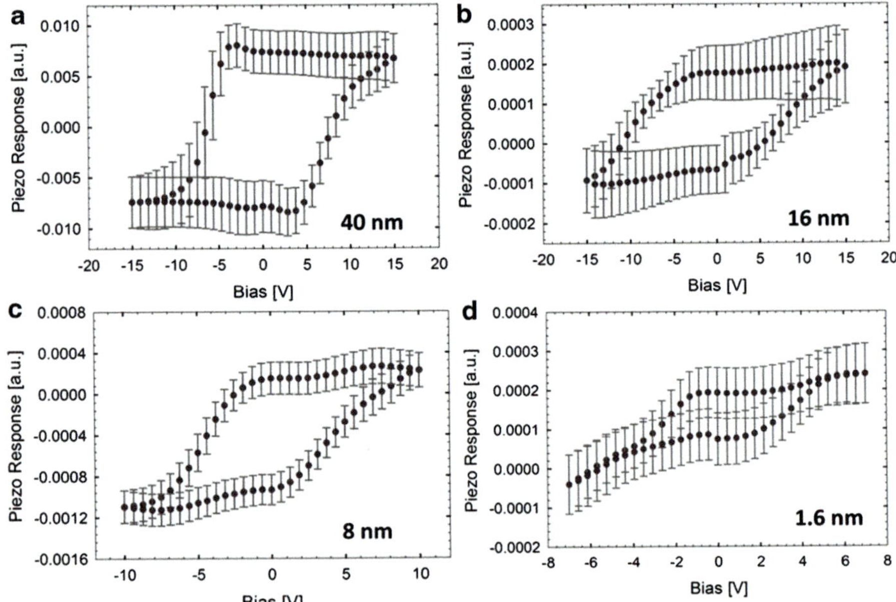

Fig. 7.23 Hysteresis loops measured by PFM for BaTiO$_3$ films of different thicknesses (**a**) 40 nm, (**b**) 16 nm, (**c**) 8 nm and (**d**) 1.6 nm. The error bars represent the dispersion of the signal measured at different locations of the sample surface. Open, saturated loops are measured for 40, 16 and 8 nm thick films. The 1.6 nm film exhibits a much noisier signal, which does not allow concluding on the ferroelectricity of such thin films. From [166]

Finally, for the ultra-thin 1.6 nm film, the hysteresis loops at individual locations are highly irregular and response variability exceeds loop opening, precluding definitive conclusion on the existence of the ferroelectric state. For field-effect devices, a ferroelectric film thickness below ~10 nm is best suited to ensure mostly c-domain orientation.

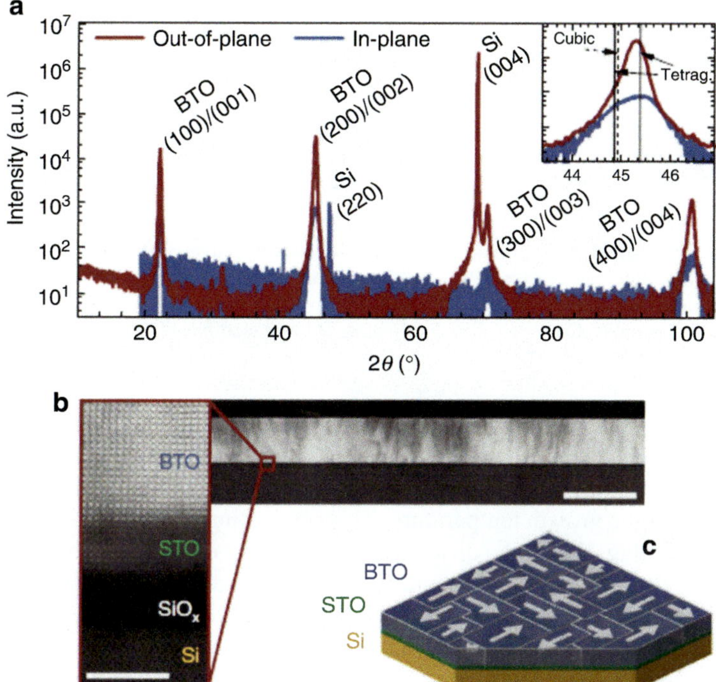

Fig. 7.24 (a) XRD diagram of 130 nm BTO on 4 nm STO on Si; the *inset* shows a magnification of the BTO (200)/(002) peak obtained with out-of-plane and in-plane geometries. The vertical lines correspond to bulk values of tetragonal (*solid*) and cubic (*dashed*) BTO. (**b**) Cross-sectional high-angle annular dark-field STEM image of the interface region showing Si, SiOx, STO and BTO (scale bar, 5 nm), and a corresponding low-magnification cross-section of the sample in dark-field mode (scale bar, 200 nm). The *red square* indicates the area from which the high-resolution micrograph was recorded. Defects penetrate mainly perpendicularly to the interface. (**c**) Schematic of the domain structure in the BTO film with randomly oriented spontaneous polarization associated with the c-axis direction (*white arrows* in *blue cuboids*). *a.u.* arbitrary unit. Reprinted from [167] by permission from Macmillan Publishers Ltd, Copyright 2013

The IBM group in Zurich recently reported on the electro-optical (EO) properties of thin BTO films epitaxially grown on silicon substrates [167, 171]. They extracted a very large effective Pockels coefficient of $r_{eff} = 148$ pm V^{-1}, which is five times larger than that of the current standard material for electro-optical devices, lithium niobate LiNbO$_3$. The non-vanishing linear Pockels effect in BTO can only be observed in a non-centrosymmetric tetragonal phase (space group P4mm). Unfortunately, BTO thin films tend to stabilize in a cubic symmetry [173], which originates from the formation of small grains, without Pockels effect. Abel et al. suppressed such grain formation by depositing a 130-nm-thick BTO film epitaxially on a Si substrate covered by a 4-nm-thick STO buffer layer [167]. Out-of-plane and grazing-incidence in-plane X-ray diffraction (XRD) measurements (Fig. 7.24a) confirmed the epitaxial relationship and showed that the crystal symmetry is tetragonal with its long tetragonal c-axis

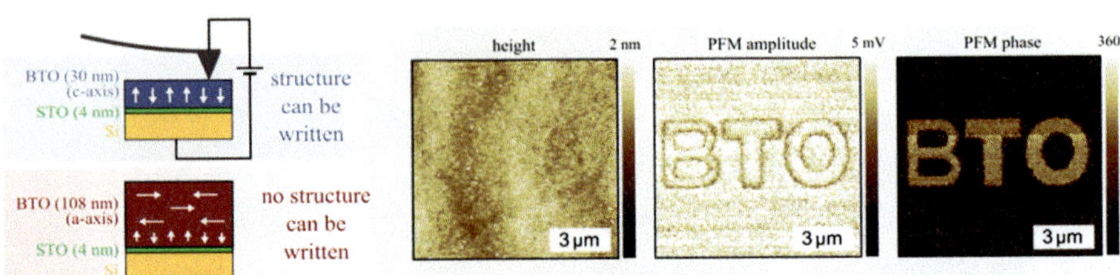

Fig. 7.25 On the *left*: sketches indicating the sample structures investigated by PFM. On the *right*: topography, amplitude and phase of the PFM signal measured on a mainly *c*-axis oriented BTO thin film after initially writing the letters "BTO". No PFM contrast was observable for thicker, *a*-axis oriented samples. Copyright IOP Publishing. Reproduced from [171] by permission of IOP Publishing. All rights reserved

parallel to the sample surface. While an out-of-plane oriented c-axis can be obtained by growing highly compressively strained films, for example by reducing their thickness, an in-plane-oriented *c*-axis is generally expected in thick films that are fully relaxed at the growth temperature. This orientation results from biaxial tensile strain after cooling, due to the different thermal expansion coefficients of BTO and Si [157]. As the template for the epitaxial growth is cubic, two equivalent orientations of the tetragonal *c*-axis exist. Strain originating from the lattice mismatch can be compensated by forming domains in which the *c*-axis is rotated by 90° within the plane of the film. The existence of two orthogonal domains is evidenced by the presence of the convoluted (200) BTO and (002) BTO diffraction peaks in the grazing-incidence XRD data (Fig. 7.24a, inset). The epitaxial relationship between BTO, STO and Si is visualized in high-resolution scanning transmission electron microscopy (STEM) analysis (Fig. 7.24b, left). Low-resolution images reveal that defects in the BTO develop mainly perpendicular to the interface (Fig. 7.24b, right), likely because of strain compensation at the interfaces of orthogonal domains. The XRD and STEM analysis suggests that the different domains penetrate the film to form a structure as shown in Fig. 7.24c. The arrows indicate the spontaneous polarization present in tetragonal BTO along the crystalline *c*-axis. The ferroelectric response of the grown layers was investigated by piezo-force response microscopy [171]. Figure 7.25 shows domains with up/down orientation written on the *c*-axis BTO film. While the height signal indicates a flat topography, the structure written into the film is clearly visible in the PFM amplitude and phase image. The phase between two different domain states changed by 180°, as expected for domains oriented in opposing directions. Also, the vanishing amplitude of the PFM signal at the edges of the written structure was in agreement with the expected behaviour at domain walls.

The electro-optical properties of the film were determined by investigating the polarization changes at the standard telecommunications wavelength of $\lambda = 1.55\,\mu m$ of a laser beam transmitted between a pair of electrodes separated by a small, micrometer scale gap. Applying a potential difference between two electrodes results in an electric field E within the gap, modifying the refractive index n(E) of BTO. The modification of the refractive index due to the applied field leads to a rotation of

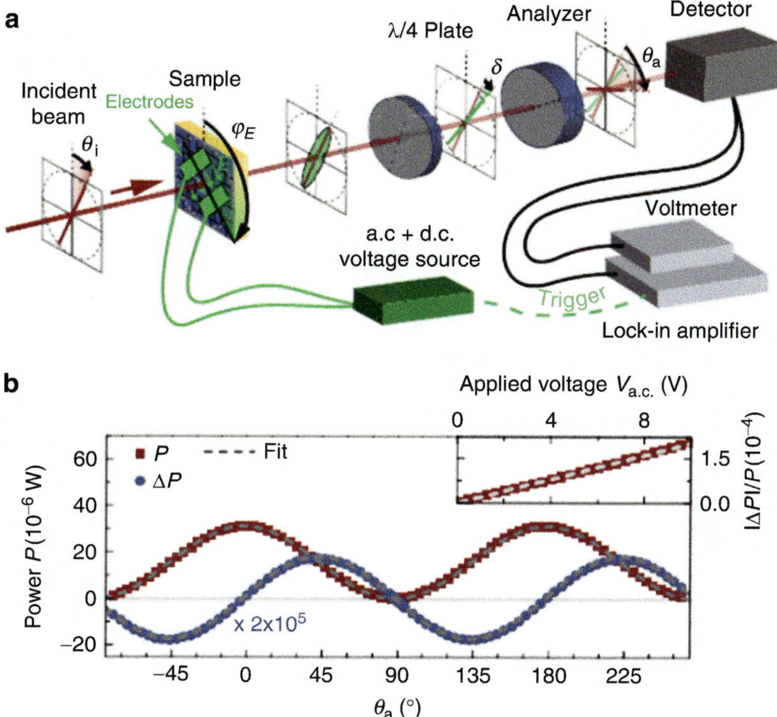

Fig. 7.26 (**a**) Schematics of measurement setup indicating the variation of the polarization states when an electric field. (**b**) Transmitted power P and variation in power ΔP due to the EO effect as a function of the analyzer position θ_a. Reprinted from [167] by permission from Macmillan Publishers Ltd, Copyright 2013

the linear polarization of the transmitted beam by an angle δ (Fig. 7.26a). After passing through an analyzer oriented at an angle θ_a relative to the polarization of the incident beam, the transmitted power $P = \cos^2(\theta_a - \delta)$ depends on the induced rotation of the polarization [174]. For small angles δ, the power variations are expected to be proportional to the derivative $dP/d\theta_a$ [167]. The measured response of the sample is in excellent agreement with such a dependence (Fig. 7.26b), confirming the existence of an electro-optical effect. The development of silicon photonics could greatly benefit from the electro-optical properties of ferroelectric oxides as a novel way to seamlessly connect the electrical and optical domain.

7.5   Integration of Epitaxial High-k Dielectric LaAlO$_3$ on Si(001)

Because of the lack of a conduction band offset with Si, STO has not been utilized as a gate dielectric on Si-based transistors even though it has a very high dielectric constant of ~300 at room temperature. For this reason, other higher band gap perovksite oxides have been looked at as a possible replacement gate dielectric. LaAlO$_3$ (LAO) is one potential candidate for use as an epitaxial gate dielectric on silicon with large band offsets to both valence and conduction bands. However, direct epitaxy on Si has proven to be difficult so researchers have used an ultrathin buffer of STO as a transition layer between LAO and Si. Four groups have successfully integrated epitaxial LAO on Si using this method.

Wang et al. at the Institute of Materials Research and Engineering in Singapore first reported crystalline LAO on Si using STO [175]. Using the Motorola process, they first deposited an ultrathin STO layer on Si. Afterwards, they formed LAO using co-deposition of La and Al metal sources under an oxygen pressure of 10^{-5} mbar and 600 °C substrate temperature initially, lowering it to 400 °C later in the growth. This process resulted in a small but unspecified amount of interfacial reaction between STO and Si as observed in cross-section TEM. A more careful growth using a similar method but resulting in no interfacial reaction was reported by Reiner et al. at Yale University [176]. In this work, 2.5 unit cells of STO were first grown on Si, which was determined to be the smallest thickness of STO with a bulk-like surface. La and Al were then co-deposited on the STO/Si at a substrate temperature of 400 °C and oxygen pressure of 10^{-8} mbar. At these conditions, the LAO was amorphous. Two unit cells of amorphous LAO were deposited and then crystallized by heating in vacuum at 700 °C. As soon as LAO started crystallizing, more La and Al were deposited in the presence of 10^{-7} mbar oxygen at the same temperature. Using this process, LAO films up to 100 nm thick could be grown with no interfacial SiO$_2$. Capacitance measurements of this stack showed a series capacitance equivalent to less than a monolayer of SiO$_x$ and an effective dielectric constant of 24 [177]. An alternative means of depositing LAO was reported by Merckling et al. [178]. Instead of using La and Al metal sources, they used electron beam evaporation of LAO crystals. LAO was evaporated onto a three-unit cell STO/Si at 700 °C. This process, however, results in the formation of a 2 nm-thick amorphous silicate layer, which causes the crystallinity of LAO to become progressively worse as the film is grown thicker.

Epitaxial LAO films were also grown epitaxially on STO/Si using atomic layer deposition (ALD) [179]. The ALD growth of LAO was done at 250 °C by using tris (N, N'-diisopropylformamidinate)-lanthanum, trimethylaluminum, and water as co-reactants. The as-deposited LAO films were amorphous and became crystalline after vacuum annealing at 600 °C for 2 h. Figure 7.27 shows the high degree of crystallinity of LAO films and shows the sharp interface between LAO and STO, as well as the existence of a ~1 nm amorphous interlayer between STO and Si. It also shows an amorphous LAO column where the STO also appears amorphous.

By keeping the annealing temperature relatively low (compared to other works where epitaxial LAO was grown on Si), the interfacial amorphous layer at the STO/Si interface was minimized to about 1.0 nm. The result demonstrates a method to form epitaxial LAO films on STO-buffered Si(001) by ALD with a minimal amorphous interfacial layer between STO/Si by maintaining the annealing temperature as low as possible. The ability to obtain high crystalline quality epitaxial LAO films on Si using ALD provides an alternative chemical route for fabricating complex oxide heterostructures and superlattices on silicon.

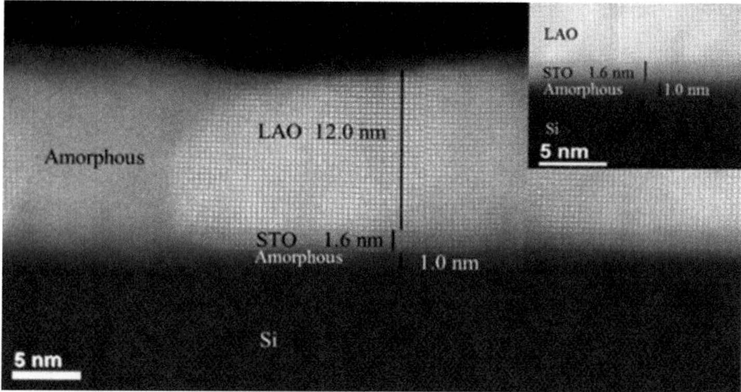

Fig. 7.27 Cross-sectional Z-contrast TEM image of a 12 nm LAO film on STO-buffered Si(001) after annealing at 600 °C for 2 h that illustrates the Si/STO/LAO interfaces. The *inset* is an expanded image of Si/STO/LAO interface region. Reprinted from [179], Copyright 2012, with permission from Elsevier

7.6   Multiferroic BiFeO$_3$ on Si

BiFeO$_3$ (BFO) is a multiferroic material, exhibiting both robust ferroelectricity with a large spontaneous polarization, and antiferromagnetic ordering. Both ferroelectric and magnetic properties are present at room temperature making BFO an attractive material for magnetoelectric device applications.

BFO was deposited on STO/Si by Wang et al. [180]. The STO on Si was fabricated by Motorola and transferred ex situ to a PLD growth chamber. On top of the STO on Si, 20 nm of conducting SrRuO$_3$ was first deposited by PLD to serve as a bottom electrode. BFO with a thickness of 200 nm was then deposited, also by PLD, on top of the SrRuO$_3$. Both SrRuO$_3$ and BFO depositions are done under relatively high temperatures (~650 °C) and high oxygen pressures compared to MBE. For SrRuO$_3$, the oxygen pressure used was 100 mTorr, while for BFO, a pressure of 40 mTorr was used. After deposition, the films were annealed at 390 °C under 1 atm of oxygen. This process is expected to result in complete uncoupling of the functional properties of BFO from the Si substrate as a result of the high oxygen pressures used, which would produce a very thick SiO$_2$ interfacial layer. Such heterostructures can still be used for making devices where the Si does not provide functionality [181], or for forming free-standing BFO membranes where the Si is a sacrificial substrate [182].

A gentler method of forming BFO on Si was recently reported by Laughlin et al. [183]. Using the Motorola STO on Si process, pseudosubstrates consisting of 20 nm of STO on Si with no interfacial SiO$_2$ were fabricated. BFO was then grown by MBE in the same growth system on the STO/Si using elemental Bi and Fe sources and atomic oxygen to ensure complete oxidation of Fe to the +3 state. For BFO growth, a substrate temperature of 650–700 °C was used with the oxygen plasma source operated at 300 W and background pressure of 1×10^{-6} mbar. The growth of BFO is adsorption-limited, being controlled by the Fe flux. For their growth, an overpressure of Bi flux was utilized. The grown BFO films exhibit a sixfold surface reconstruction with a 2D growth front, as observed by in situ RHEED. The use of atomic oxygen results in the oxidation of the Si-STO interface, forming a SiO_2 layer about 2.5 nm in thickness. This SiO_2 layer may, however, be sufficiently thin for coupling of the multiferroic properties of BFO to Si.

7.7 Summary

In this chapter, we have reviewed selected key developments in the integration of functional complex oxides on silicon by means of an MBE-grown epitaxial STO buffer. Materials that are high-k dielectric, ferroelectric, ferromagnetic, photo-catalytic, and multiferroic have been integrated on silicon. The properties exhibited by these materials are similar in quality to bulk, showing the high degree of crystallinity and robustness of the STO/Si pseudo-substrate even under somewhat high temperatures and oxygen pressures needed to grow some of these materials. We expect that many more kinds of functional oxide materials will eventually be integrated on silicon that may ultimately result in novel device architectures for future applications.

References

1. B. O'Regan, M. Grätzel, Nature **353**, 737 (1991)
2. A. Fujishima, X. Zhang, D.A. Tryk, Surf. Sci. Rep. **63**, 515 (2008)
3. X. Chen, S. Shen, L. Guo, S.S. Mao, Chem. Rev. **110**, 6503 (2010)
4. N.S. Lewis, D.G. Nocera, Proc. Natl. Acad. Sci. **103**, 15729 (2006)
5. A.L. Linsebigler, G. Lu, J.T. Yates, Chem. Rev. **95**, 735 (1995)
6. H.G. Yang, C.H. Sun, S.Z. Qiao, J. Zou, G. Liu, S.C. Smith, H.M. Cheng, G.Q. Lu, Nature **453**, 638 (2008)
7. A. Selloni, Nat. Mater. **7**, 613 (2008)
8. L. Forro, O. Chauvet, D. Emin, L. Zuppiroli, H. Berger, F. Lévy, J. Appl. Phys. **75**, 633 (1994)
9. H. Tang, K. Prasad, R. Sanjinès, F. Lévy, J. Appl. Phys. **75**, 2042 (1994)
10. M. Xu, Y. Gao, E.M. Moreno, M. Kunst, M. Muhler, Y. Wang, H. Idriss, C. Wöll, Phys. Rev. Lett. **106**, 138302 (2011)
11. F.E. Osterloh, B.A. Parkinson, MRS Bull. **36**, 17 (2011)
12. R. Asahi, T. Morikawa, T. Ohwaki, K. Aoki, Y. Taga, Science **293**, 269 (2001)
13. Y. Gai, J. Li, S. Li, J. Xia, S. Wei, Phys. Rev. Lett. **102**, 036402 (2009)
14. W. Zhu, X. Qiu, V. Iancu, X. Chen, H. Pan, W. Wang, N.M. Dimitrijevic, T. Rajh, H.M. Meyer III, M. Parans Paranthaman, G.M. Stocks, H.H. Weitering, B. Gu, G. Eres, Z. Zhang, Phys. Rev. Lett. **103**, 226401 (2009)
15. X. Chen, L. Liu, P.Y. Yu, S.S. Mao, Science **331**, 746 (2011)
16. T. Ohno, K. Sarukawa, K. Tokieda, M. Matsumura, J. Catal. **203**, 82 (2001)

17. D.C. Hurum, A.G. Agrios, K.A. Gray, J. Phys. Chem. B **107**, 4545 (2003)
18. N. Siedl, M.J. Elser, J. Bernardi, O. Diwald, J. Phys. Chem. C **113**, 15792 (2009)
19. D.L. Liao, C.A. Badour, B.Q. Liao, J. Photochem. Photobiol. A **194**, 11 (2008)
20. N. Nilius, T. Risse, S. Schauermann, S. Shaikhutdinov, M. Sterrer, H.–.J. Freund, Top. Catal. **54**, 4 (2011)
21. S.A. Chambers, C.M. Wang, S. Thevuthasan, T. Droubay, D.E. McCready, A.S. Lea, V. Shutthanandan, C.F. Wndisch Jr., Thin Solid Films **418**, 197 (2002)
22. Z. Wang, W. Zeng, L. Gu, M. Saito, S. Tsukimoto, Y. Ikuhara, J. Appl. Phys. **108**, 113701 (2010)
23. N.V. Burbure, P.A. Salvador, G.S. Rohrer, Chem. Mater. **22**, 5823 (2010)
24. D. Kazazis, S. Guha, N.A. Bojarczuk, A. Zaslavsky, H.-C. Kim, Appl. Phys. Lett. **95**, 064103 (2009)
25. R. Shao, C. Wang, D.E. McCready, T.C. Droubay, S.A. Chambers, Surf. Sci. **601**, 1582 (2007)
26. H. Ohta, S. Kim, Y. Mune, T. Mizoguchi, K. Nomura, S. Ohta, T. Nomura, Y. Nakanishi, Y. Ikuhara, M. Hirano, H. Hosono, K. Koumoto, Nat. Mater. **6**, 129 (2007)
27. Y. Matsumoto, M. Murakami, T. Shono, T. Hasegawa, T. Fukumura, M. Kawasaki, P. Ahmet, T. Chikyow, S. Koshihara, H. Koinuma, Science **291**, 854 (2001)
28. M. Katayama, S. Ikesaka, J. Kuwano, H. Koinuma, Y. Matsumoto, Appl. Phys. Lett. **92**, 132107 (2008)
29. T.C. Kaspar, T. Droubay, V. Shutthanandan, S.M. Heald, C.M. Wang, D.E. McCready, S. Thevuthasan, J.D. Bryan, D.R. Gamelin, A.J. Kellock, M.F. Toney, X. Hong, C.H. Ahn, S.A. Chambers, Phys. Rev. B **73**, 155327 (2006)
30. Y. Yamada, K. Ueno, T. Fukumura, H.T. Yuan, H. Shimotani, Y. Iwasa, L. Gu, S. Tsukimoto, Y. Ikuhara, M. Kawasaki, Science **332**, 1065 (2011)
31. H. Seo, A.B. Posadas, C. Mitra, J. Ramdani, A.V. Kvit, A.A. Demkov, Phys. Rev. B **86**, 075301 (2012)
32. S.A. Chambers, T. Ohsawa, C.M. Wang, I. Lyubinetsky, J.E. Jaffe, Surf. Sci. **603**, 771 (2009)
33. M. Choi, A. Posadas, R. Dargis, C. Shih, A.A. Demkov, J. Appl. Phys. **111**, 064112 (2012)
34. A.A. Demkov, L.R.C. Fonseca, E. Verret, J. Tomfohr, O.F. Sankey, Phys. Rev. B **71**, 195306 (2005)
35. O. Sharia, A.A. Demkov, G. Bersuker, B.H. Lee, Phys. Rev. B **75**, 035306 (2007)
36. X. Luo, G. Bersuker, A.A. Demkov, Phys. Rev. B **84**, 195309 (2011)
37. J. Junquera, M. Zimmer, P. Ordejón, P. Ghosez, Phys. Rev. B **67**, 155327 (2003)
38. K. Kita, A. Toriumi, Appl. Phys. Lett. **94**, 132902 (2009)
39. A. Ohtomo, H.Y. Hwang, Nature **427**, 423 (2004)
40. J.K. Lee, A.A. Demkov, Phys. Rev. B **78**, 193104 (2008)
41. J.K. Lee, N. Sai, A.A. Demkov, Phys. Rev. B **82**, 235305 (2010)
42. M. Nakamura, A. Sawa, J. Fujioka, M. Kawasaki, Y. Tokura, Phys. Rev. B **82**, 201101 (R) (2010)
43. R.A. McKee, F.J. Walker, M.B. Nardelli, W.A. Shelton, G.M. Stocks, Science **300**, 1726 (2003)
44. Y. Hikita, M. Nishikawa, T. Yajima, H.Y. Hwang, Phys. Rev. B **79**, 073101 (2009)
45. J.D. Burton, E.Y. Tsymbal, Phys. Rev. B **82**, 161407 (2010)
46. G. Kresse, J. Furthmüller, Phys. Rev. B **54**, 11169 (1996)
47. R. Loetzsch, A. Lübcke, I. Uschmann, E. Förster, V. Große, M. Thuerk, T. Koettig, F. Schmidl, P. Seidel, Appl. Phys. Lett. **96**, 071901 (2010)
48. D.R. Hummer, P.J. Heaney, J.E. Post, Powder Diffract. **22**, 352 (2007)
49. M. Shishkin, G. Kresse, Phys. Rev. B **74**, 035101 (2006)
50. L. Kleinman, Phys. Rev. B **24**, 7412 (1981)
51. D.M. Bylander, L. Kleinman, Phys. Rev. B **36**, 3229 (1987)
52. C.G. Van de Walle, Phys. Rev. B **39**, 1871 (1989)
53. R.T. Tung, Phys. Rev. Lett. **84**, 6078 (2000)

54. J. Tersoff, Phys. Rev. B **32**, 6968 (1985)
55. R.M. Martin, K. Kunc, Phys. Rev. B **24**, 2081 (1981)
56. P. Ghosez, J.–.P. Michenaud, X. Gonze, Phys. Rev. B **58**, 6224 (1998)
57. M. Mikami, S. Nakamura, O. Kitao, H. Arakawa, Phys. Rev. B **66**, 155213 (2002)
58. R.J. Gonzalez, R. Zallen, H. Berger, Phys. Rev. B **55**, 7014 (1997)
59. C.J. Fennie, K.M. Rabe, Phys. Rev. B **68**, 184111 (2003)
60. D.A. Muller, Nat. Mater. **8**, 263 (2009)
61. F.M.F. de Groot, J. Faber, J.J.M. Michiels, M.T. Czyżyk, M. Abbate, J.C. Fuggle, Phys. Rev. B **48**, 2074 (1993)
62. K. van Benthem, C. Elsässer, M. Rühle, Ultramicroscopy **96**, 509 (2003)
63. D.A. Muller, T. Sorsch, S. Moccio, F.H. Baumann, K. Evans-Lutterodt, G. Timp, Nature **399**, 758 (1999)
64. R. Laskowski, P. Blaha, Phys. Rev. B **82**, 205104 (2010)
65. K. Ogasawara, T. Iwata, Y. Koyama, T. Ishii, I. Tanaka, H. Adachi, Phys. Rev. B **64**, 115413 (2001)
66. P. Krüger, Phys. Rev. B **81**, 125121 (2010)
67. D.A. Muller, D.J. Singh, J. Silcox, Phys. Rev. B **57**, 8181 (1998)
68. C. Elsässer, S. Köstlmeier, Ultramicroscopy **86**, 325 (2001)
69. G. Duscher, R. Buczko, S.J. Pennycook, S.T. Pantelides, Ultramicroscopy **86**, 355 (2001)
70. X. Weng, P. Fisher, M. Skowronski, P.A. Salvador, O. Maksimov, J. Cryst. Growth **310**, 545 (2008)
71. C. Mitra, C. Lin, J. Robertson, A.A. Demkov, Phys. Rev. B **86**, 155105 (2012)
72. R. Ciancio, E. Carlino, C. Aruta, D. Maccariello, F.M. Granozio, U.S. di Uccio, Nanoscale **4**, 91 (2012)
73. M. Radović, M. Salluzzo, Z. Ristić, R. di Capua, N. Lampis, R. Vaglio, F.M. Granozio, J. Chem. Phys. **135**, 034705 (2011)
74. N.B. Ivanova, S.G. Ovchinnikov, M.M. Korshunov, I.M. Eremin, N.V. Kazak, Physics – Uspkehi **52**, 789 (2009)
75. K. Knízek, Z. Jirák, J. Hejtmánek, M. Veverka, M. Marysko, G. Maris, T.T.M. Palstra, Eur. Phys. J. B **47**, 213 (2005)
76. X. Chen, N.J. Wu, L. Smith, A. Ignatiev, Appl. Phys. Lett. **84**, 2700 (2004)
77. J.W. Fergus, Sensors Actuators B **123**, 1169 (2007)
78. W. Kobayashi, Y. Teraoka, I. Terasaki, Appl. Phys. Lett. **95**, 171905 (2009)
79. A.K. Pradhan, J.B. Dadson, D. Hunter, K. Zhang, S. Mohanty, E.M. Jackson, B. Lasley-Hunter, K. Lord, T.M. Williams, R.R. Rakhimov, J. Zhang, D.J. Sellmyer, K. Inaba, T. Hasegawa, S. Methews, B. Joseph, B.R. Sekhar, U.N. Roy, Y. Cui, A. Burger, J. Appl. Phys. **100**, 033903 (2006)
80. S. Maekawa, T. Tohyama, S.E. Barnes, S. Ishihara, W. Koshibae, G. Khaliullin, *Physics of Transition Metal Oxides* (Springer, Berlin, 2004)
81. P.M. Raccah, J.B. Goodenough, Phys. Rev. **155**, 932 (1967)
82. A. Podlesnyak, S. Streule, J. Mesot, M. Medarde, E. Pomjakushina, K. Conder, A. Tanaka, M.W. Haverkort, D.I. Khomskii, Phys. Rev. Lett. **97**, 247208 (2006)
83. M.A. Korotin, S.Y. Ezhov, I.V. Solovyev, V.I. Anisimov, D.I. Khomskii, G.A. Sawatzky, Phys. Rev. B **54**, 5309 (1996)
84. P.G. Radaelli, S.W. Cheong, Phys. Rev. B **66**, 094408 (2002)
85. M. Zhuang, W. Zhang, N. Ming, Phys. Rev. B **57**, 10705 (1998)
86. M.W. Haverkort, Z. Hu, J.C. Cezar, T. Burnus, H. Hartmann, M. Reuther, C. Zobel, T. Lorenz, A. Tanaka, N.B. Brookes, H.H. Hsieh, H.-J. Lin, C.T. Chen, L.H. Tjeng, Phys. Rev. Lett. **97**, 176405 (2006)
87. K. Knížek, Z. Jirák, J. Hejtmánek, P. Novák, W. Ku, Phys. Rev. B **79**, 014430 (2009)
88. J. Kuneš, V. Křápek, Phys. Rev. Lett. **106**, 256401 (2011)
89. H. Hsu, P. Blaha, R.M. Wentzcovitch, C. Leighton, Phys. Rev. B **82**, 100406(R) (2010)

90. D. Fuchs, C. Pinta, T. Schwarz, P. Schweiss, P. Nagel, S. Schuppler, R. Schneider, M. Merz, G. Roth, H.v. Löhneysen, Phys. Rev. B **75**, 144402 (2007)
91. D. Fuchs, E. Arac, C. Pinta, S. Schuppler, R. Schneider, H.b. Löhneysen, Phys. Rev. B **77**, 014434 (2008)
92. J.W. Freeland, J.X. Ma, J. Shi, Appl. Phys. Lett. **93**, 212501 (2008)
93. A. Herklotz, A.D. Rata, L. Schultz, K. Dörr, Phys. Rev. B **79**, 092409 (2009)
94. S. Park, P. Ryan, E. Karapetrova, J.W. Kim, J.X. Ma, J. Shi, J.W. Freeland, W. Wu, Appl. Phys. Lett. **95**, 072508 (2009)
95. A. Posadas, M. Berg, H. Seo, A. de Lozanne, A.A. Demkov, D.J. Smith, A.P. Kirk, D. Zhernokletov, R.M. Wallace, Appl. Phys. Lett. **98**, 053104 (2011)
96. V. Mehta, M. Liberati, F.J. Wong, R.V. Chopdekar, E. Arenholz, Y. Suzuki, J. Appl. Phys. **105**, 07E503 (2009)
97. V. Mehta, Y. Suzuki, J. Appl. Phys. **109**, 07D717 (2011)
98. M. Merz, P. Nagel, C. Pinta, A. Samartsev, H. Löhneysen, M. Wissinger, S. Uebe, A. Assmann, D. Fuchs, S. Schuppler, Phys. Rev. B **82**, 174416 (2010)
99. G.E. Sterbinsky, P.J. Ryan, J.-W. Kim, E. Karapetrova, J.X. Ma, J. Shi, J.C. Woicik, Phys. Rev. B **85**, 020403(R) (2012)
100. M.A. Senaris-Rodriguez, J.B. Goodenough, J. Solid State Chem. **116**, 224 (1995)
101. J. Chakhalian, J.M. Rondinelli, L. Jian, B.A. Gray, M. Kareev, E.J. Moon, N. Prasai, J.L. Cohn, M. Varela, I.C. Tung, M.J. Bedzyk, S.G. Altendorf, F. Strigari, B. Dabrowski, L.H. Tjeng, P.J. Ryan, J.W. Freeland, Phys. Rev. Lett **107**, 116805 (2011)
102. R.J. Zeches, M.D. Rossell, J.X. Zhang, A.J. Hatt, Q. He, C.-H. Yang, A. Kumar, C.H. Wang, A. Melville, C. Adamo, G. Sheng, Y.-H. Chu, J.F. Ihlefeld, R. Erni, C. Ederer, V. Gopalan, L.Q. Chen, D.G. Schlom, N.A. Spaldin, L.W. Martin, R. Ramesh, Science **326**, 977 (2009)
103. K. Gupta, P. Mahadevan, Phys. Rev. B **79**, 020406 (2009)
104. J.M. Rondinelli, N.A. Spaldin, Phys. Rev. B **79**, 054409 (2009)
105. H. Hsu, P. Blaha, R.M. Wentzcovitch, Phys. Rev. B **85**, 140404(R) (2012)
106. H. Seo, A. Posadas, A.A. Demkov, Phys. Rev. B **86**, 014430 (2012)
107. G. Fischer, M. Däne, A. Ernst, P. Bruno, M. Lüders, Z. Szotek, W. Temmerman, W. Hergert, Phys. Rev. B **80**, 014408 (2009)
108. J.B. Goodenough, Phys. Rev. **100**, 564 (1955)
109. J.B. Goodenough, J. Phys. Chem. Solids **6**, 287 (1958)
110. J. Kanamori, J. Phys. Chem. Solids **10**, 87 (1959)
111. P.W. Anderson, Solid State Phys. **14**, 99 (1963)
112. S.J. May, J.-W. Kim, J.M. Rondinelli, E. Karapetrova, N.A. Spaldin, A. Bhattacharya, P.J. Ryan, Phys. Rev. B **82**, 014110 (2010)
113. T. Vogt, J.A. Hriljac, N.C. Hyatt, P. Woodward, Phys. Rev. B **67**, 140401(R) (2003)
114. J.-S. Zhou, J.-Q. Yan, J.B. Goodenough, Phys. Rev. B **71**, 220103 (2005)
115. The bulk modulus of LCO with 0%, 25%, 50% and 100% concentrations of HS Co^{3+} was calculated to be 203, 186, 181, and 176 GPa, respectively in [106], showing significant softening as concentration of HS Co^{3+} increases.
116. To ensure that the FM coupling is robust in a reasonable range of U_{eff}, Seo and Demkov tested three U_{eff} values of 3.0, 3.5, and 4.0 eV, yielding coupling strengths of 3.0, 2.8 and 2.6 meV/pair, respectively in [106].
117. D. Hobbs, G. Kresse, J. Hafner, Phys. Rev. B **62**, 11556 (2000)
118. S.A. Wolf, D.D. Awschalom, R.A. Buhrman, J.M. Daughton, S. von Molnar, M.L. Roukes, A.Y. Chtchelkanova, D.M. Treger, Science **294**, 1488 (2001)
119. S. Datta, B. Das, Appl. Phys. Lett. **56**, 665 (1990)
120. G. Schmidt, D. Ferrand, L.W. Molenkamp, A.T. Filip, B.J. van Wees, Phys. Rev. B **62**, R4790 (2000)
121. R. Klasges, C. Carbone, W. Eberhardt, C. Pampuch, O. Rader, T. Kachel, W. Gudat, Phys. Rev. B **56**, 10801 (1997)
122. S.P. Dash, S. Sharma, R.S. Patel, M.P. de Jong, R. Jansen, Nature **462**, 491 (2009)

123. B.T. Jonker, G. Kioseoglou, A.T. Hanbicki, C.H. Li, P.E. Thompson, Nat. Phys. **3**, 542 (2007)
124. A.H. MacDonald, P. Schiffer, N. Samarth, Nat. Mater. **4**, 195 (2005)
125. S. Chambers, Surf. Sci. Rep. **61**, 345 (2006)
126. S.J. Pearton, W.H. Heo, M. Ivill, D.P. Norton, T. Steiner, Semicond. Sci. Technol. **19**, R59 (2004)
127. C. Song, K.W. Geng, F. Zeng, X.B. Wang, Y.X. Shen, F. Pan, Y.N. Xie, T. Liu, H.T. Zhou, Z. Fan, Phys. Rev. B **73**, 024405 (2006)
128. K. Ueda, H. Tabata, T. Kawai, Appl. Phys. Lett. **79**, 988 (2001)
129. H.S. Kim, L. Bi, G.F. Dionne, C.A. Ross, H.J. Paik, Phys. Rev. B **77**, 214436 (2008)
130. S.B. Ogale, R.J. Choudhary, J.P. Buban, S.E. Lofland, S.R. Shinde, S.N. Kale, V.N. Kulkarni, J. Higgins, C. Lanci, J.R. Simpson, N.D. Browning, S. Das Sarma, H.D. Drew, R.L. Greene, T. Venkatesan, Phys. Rev. Lett. **91**, 077205 (2003)
131. J. Philip, A. Punnoose, B.I. Kim, K.M. Reddy, S. Layne, J.O. Holmes, B. Satpati, P.R. Leclair, T.S. Santos, J. Moodera, Nat. Mater. **5**, 298 (2006)
132. Y.K. Yoo, Q. Xue, H.-C. Lee, S. Cheng, X.D. Xiang, G.F. Dionne, S. Xu, J. He, Y.S. Chu, S.D. Preite, S.E. Lofland, I. Takeuchi, Appl. Phys. Lett. **86**, 042506 (2005)
133. K.A. Griffin, A.B. Pakhomov, C.M. Wang, S.M. Heald, K.M. Krishnan, Phys. Rev. Lett. **94**, 157204 (2005)
134. J.M.D. Coey, Curr. Opinion Solid State Mater. Sci. **10**, 83 (2006)
135. K.J. Hubbard, D.G. Schlom, J. Mater. Res. **11**, 2757 (1996)
136. T.C. Kaspar, T. Droubay, C.M. Wang, S.M. Heald, A.S. Lea, S.A. Chambers, J. Appl. Phys. **97**, 073511 (2005)
137. C. Pascanut, N. Dragoe, P. Berthet, J. Magn. Magn. Mater. **305**, 6 (2006)
138. C. Decorse-Pascanut, J. Berthon, L. Pinsard-Gaudart, N. Dragoe, P. Berthet, J. Magn. Magn. Mater. **321**, 3526 (2009)
139. S. Malo, A. Maignan, Inorg. Chem. **43**, 8169 (2004)
140. P. Galinetto, A. Casiraghi, M.C. Mozazti, C.B. Azzoni, D. Norton, L.A. Boatner, V. Trepakov, Ferroelectrics **368**, 120–130 (2008)
141. D. Yao, X. Zhou, S. Ge, Appl. Surf. Sci. **257**, 9233 (2011)
142. S.X. Zhang, S.B. Ogale, D.C. Kundaliya, L.F. Fu, N.D. Browning, S. Dhar, W. Ramadan, J.S. Higgins, R.L. Greene, T. Venkatesan, Appl. Phys. Lett. **89**, 012501 (2006)
143. L. Bi, H.-S. Kim, G.F. Dionne, C.A. Ross, New J. Phys. **12**, 043044 (2010)
144. G. Herranz, M. Basletić, M. Bibes, R. Ranchal, A. Hamzić, H. Jaffrès, E. Tafra, K. Bouzehouane, E. Jacquet, J.P. Contour, A. Barthélémy, A. Fert, J. Magn. Magn. Mater. **310**, 2111 (2007)
145. A.B. Posadas, C. Mitra, C. Lin, A. Dhamdhere, D.J. Smith, M. Tsoi, A.A. Demkov, Phys. Rev. B **87**, 144422 (2013)
146. E. Eberg, A.F. Monsen, T. Tybell, A.T.J. Van Helvoort, R. Holmestad, J. Electron Microsc. **57**, 175 (2008)
147. G.J. Yong, R.M. Kolagani, S. Adhikari, W. Vanderlinde, Y. Liang, K. Muramatsu, S. Friedrich, J. Appl. Phys. **108**, 033502 (2010)
148. J.Q. He, C.L. Jia, V. Vaithnayathan, D.G. Schlom, J. Schubert, A. Gerber, H.H. Kohlstedt, R.H. Wang, J. Appl. Phys. **97**, 104921 (2005)
149. N.S. McIntyre, M.G. Cook, Anal. Chem. **47**, 2208 (1975)
150. T.J. Chuang, C.R. Brundle, D.W. Rice, Surf. Sci. **59**, 413 (1976)
151. C.D. Wagner, W.M. Riggs, L.E. Davis, J.F. Moulder, G.E. Mullenberg, *Handbook of X-ray Photoelectron Spectroscopy* (Perkin-Elmer Corp., Physical Electronics Division, Eden Prairie, MN, 1979)
152. K. Griffin Roberts, M. Varela, S. Rashkeev, S.T. Pantelides, S.J. Pennycook, K.M. Krishnan, Phys. Rev. B **78**, 014409 (2008)
153. J.M. Florez, S.P. Ong, M.C. Onbaşli, G.F. Dionne, P. Vargas, G. Ceder, C.A. Ross, Appl. Phys. Lett. **100**, 252904 (2012)
154. S.B. Zhang, J.E. Northrup, Phys. Rev. Lett. **67**, 2339 (1991)

155. R.O. Jones, O. Gunnarsson, Rev. Mod. Phys. **61**, 689 (1989)
156. C. Lin, C. Mitra, A.A. Demkov, Phys. Rev. B **86**, 161102(R) (2012)
157. V. Vaithyanathan, J. Lettieri, W. Tian, A. Sharan, A. Vasudevarao, Y.L. Li, A. Kochhar, H. Ma, J. Levy, P. Zschack, J.C. Woicik, L.Q. Chen, V. Gopalan, D.G. Schlom, J. Appl. Phys. **100**, 024108 (2006)
158. R.A. McKee, F.J. Walker, J.R. Conner, E.D. Specht, D.E. Zelmon, Appl. Phys. Lett. **59**, 782 (1991)
159. M.-B. Lee, M. Kawasaki, M. Yoshimoto, H. Koinuma, Appl. Phys. Lett. **66**, 1331 (1995)
160. Z. Yu, J. Ramdani, J.A. Curless, C.D. Overgaard, J.M. Finder, R. Droopad, K.W. Eisenbeiser, J.A. Hallmark, W.J. Ooms, V.S. Kaushik, J. Vac. Sci. Technol. B **18**, 2139 (2000)
161. M. Kondo, K. Maruyama, K. Kurihara, Fujitsu Sci. Tech. J. **38**, 46 (2002)
162. A.R. Meier, F. Niu, B.W. Wessels, J. Cryst. Growth **294**, 401 (2006)
163. F. Niu, B.W. Wessels, J. Vac. Sci. Technol. B **25**, 1053 (2007)
164. G. Niu, S. Yin, G. Saint-Girons, B. Gautier, P. Lecoeur, V. Pillard, G. Hollinger, B. Vilquin, Microelectron. Eng. **88**, 1232 (2011)
165. J. Hiltunen, D. Seneviratne, H.L. Tuller, J. Lappalainen, V. Lantto, J. Electroceram. **22**, 395 (2009)
166. C. Dubourdieu, J. Bruley, T. M. Arruda, A.B. Posadas, J. Jordan-Sweet, M. M. Frank, E. Cartier, D. J. Frank, S. V. Kalinin, A.A. Demkov, and V. Narayanan, Nature Nanotechnol. **8**, 748 (2013)
167. S. Abel, T. Stööferle, C. Marchiori, C. Rossel, M.D. Rossell, R. Erni, D. Caimi, M. Sousa, A. Chelnokov, B.J. Offrein, J. Fompeyrine, Nat. Commun. **4**, 1671 (2013)
168. J.F. Scott, *Ferroelectric Memories* (Springer, Berlin, 2000) (chapter 2 and 12)
169. S. Salahuddin, S. Datta, Nano Lett. **8**, 405–410 (2008)
170. V.V. Zhrinov, R.K. Cavin, Nat. Nanotechnol. **3**, 77–78 (2008)
171. S. Abel, M. Sousa, C. Rossel, D. Caimi, M.D. Rossell, R. Erni, J. Fompeyrine, C. Marchiori, Nanotechnology **24**, 285701 (2013)
172. J. Paul, T. Nishimatsu, Y. Kawazoe, U.V. Waghmare, Phys. Rev. Lett. **99**, 077601 (2007)
173. J.W. Jang, S.J. Chung, W.J. Cho, T.S. Hahn, S.S. Choi, J. Appl. Phys. **81**, 6322 (1997)
174. A. Yariv, P. Yeh, *Optical Waves in Crystals* (John Wiley & Sons, New York, 1984)
175. Y.Y. Mi, Z. Yu, S.J. Wang, P.C. Lim, Y.L. Foo, A.C.H. Huan, C.K. Ong, Appl. Phys. Lett. **90**, 181925 (2007)
176. J.W. Reiner, A. Posadas, M. Wang, T.P. Ma, C.H. Ahn, Microelectron. Eng. **85**, 36 (2008)
177. J.W. Reiner, A. Posadas, M. Wang, M. Sidorov, Z. Krivokapic, F.J. Walker, T.P. Ma, C.H. Ahn, J. Appl. Phys. **105**, 124501 (2009)
178. C. Merckling, G. Delhaye, M. El-Kazzi, S. Gaillard, Y. Rozier, L. Rapenne, B. Chenevier, O. Marty, G. Saint-Girons, M. Gendry, Y. Robach, G. Hollinger, Microelectron. Reliab. **47**, 540–543 (2007)
179. T.Q. Ngo, A. Posadas, M.D. McDaniel, D.A. Ferrer, J. Bruley, C. Breslin, A.A. Demkov, J.G. Ekerdt, J. Cryst. Growth **363**, 150 (2013)
180. J. Wang, H. Zheng, Z. Ma, S. Prasertchoung, M. Wuttig, R. Droopad, J. Yu, K. Eisenbeiser, R. Ramesh, Appl. Phys. Lett. **85**, 2574 (2004)
181. L.W. Martin, Y.-H. Chu, Q. Zhan, R. Ramesh, S.-J. Han, S.X. Wang, M. Warusawithana, D.G. Schlom, Appl. Phys. Lett. **91**, 172513 (2007)
182. H.W. Jang, S.H. Baek, D. Ortiz, C.M. Folkman, C.B. Eom, Y.H. Chu, P. Shafer, R. Ramesh, V. Vaithyanathan, D.G. Schlom, Appl. Phys. Lett. **92**, 062910 (2008)
183. R.P. Laughlin, D.A. Currie, R. Contreras-Guererro, A. Dedigama, W. Priyantha, R. Droopad, N. Theodoropoulou, P. Gao, X. Pan, J. Appl. Phys. **113**, 17D919 (2013)

Chapter 8
Other Epitaxial Oxides on Semiconductors

While $SrTiO_3$ on Si(100) is the most extensively studied epitaxial oxide on semiconductor system both experimentally and theoretically, there has been significant effort into the epitaxial growth of other oxides on Si(100), oxides on Si (111), as well as oxides on other semiconductor substrates. In this chapter, we will give an account of the major developments in the research on epitaxial growth of oxides on semiconductors other than $SrTiO_3$ on Si(100). We will also describe some recent developments in the opposite stacking sequence, namely the growth of semiconductor layers on oxide surfaces.

8.1 Other Oxides on Si(100)

8.1.1 MgO

MgO is more thermodynamically stable than SiO_2 [1] making it well-suited for epitaxial growth on Si without forming an amorphous SiO_2 interlayer. Epitaxial growth of MgO on Si(100) without SiO_2 formation was first reported by Fork et al. [2] by starting with hydrogen-terminated Si(100), as achieved by an HF-last procedure and using reactive PLD with a Mg metal sputtering target and a low oxygen partial pressure (5×10^{-6} Torr) to deposit the MgO at a substrate temperature of 400 °C. Under these conditions, growth of MgO is in the so-called adsorption-controlled regime where unoxidized Mg metal will re-evaporate from the substrate while stoichiometric MgO will stick. Despite the large lattice mismatch (22.5 %) between MgO and Si, a coincidence lattice match of three Si unit cells to every four MgO unit cells occurs that still results in epitaxy. Electron beam evaporation from an MgO source has also been demonstrated at a substrate temperature of 300 °C but with the formation of a thin (<0.5 nm) disordered interlayer (Fig. 8.1) [3].

A.A. Demkov and A.B. Posadas, *Integration of Functional Oxides with Semiconductors*, DOI 10.1007/978-1-4614-9320-4_8, © The Author(s) 2014

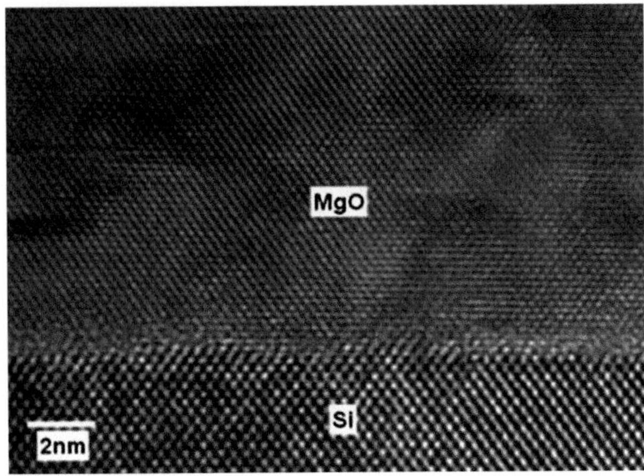

Fig. 8.1 Cross-section TEM of electron beam evaporated MgO on Si(100) showing epitaxy with a thin disordered interfacial layer. Reprinted with permission from [3]. Copyright 2008, AIP Publishing LLC

8.1.2 γ-Al_2O_3

Al_2O_3 is another binary oxide that is well-established to be more thermodynamically stable than SiO_2. Growing Al_2O_3 on a clean Si(100) surface results in the formation of the cubic variant of Al_2O_3 known as γ-Al_2O_3, which has a defective spinel structure [4]. For the first two monolayers, the γ-Al_2O_3 grows in the cubic 100 orientation. Beyond this critical thickness, the orientation transitions to the cubic 111 direction even on Si(100) [5]. Growth of γ-Al_2O_3 on Si(100) has been demonstrated by metal-organic MBE using trimethyl aluminum and N_2O at a substrate temperature of 760 °C [6], and also by electron beam evaporation from an alumina source at a substrate temperature of 850 °C and very low oxygen partial pressure (below 1×10^{-8} Torr) [7]. γ-Al_2O_3 on Si has been utilized as a pseudo-substrate for the growth of single crystalline Pt [8], Si [9] and GaN [10] overlayers on Si(100).

8.1.3   *Yttria-Stabilized ZrO_2*

Epitaxial growth of yttria-stabilized ZrO_2 (YSZ) has been demonstrated by several groups using PLD from a ceramic target [11, 12]. The growth was done using a background oxygen partial pressure of ~4×10^{-4} Torr at a growth temperature of 730–800 °C. Epitaxy apparently occurs whether one starts with a hydrogen-terminated Si surface or even with the native SiO_2 layer still present. Because Zr has a higher oxygen affinity than Si, exposure of SiO_2 to Zr metal results in the reduction of SiO_2 during the deposition. This has been confirmed by XPS depth profiling and cross-sectional TEM showing little to no SiO_x at the interface between Si and YSZ (Fig. 8.2) [13].

Fig. 8.2 XPS spectrum
of Si 2p core level for
PLD-grown YSZ on Si(100)
as a function of ion
sputtering time. Shorter
times mean the scan is
closer to the surface while
longer times mean the scan
is closer to the bulk Si
substrate. The absence of a
feature at 103.5 eV is an
indication of a SiO_2-free
interface. Reprinted with
permission from [13].
Copyright 2001, AIP
Publishing LLC

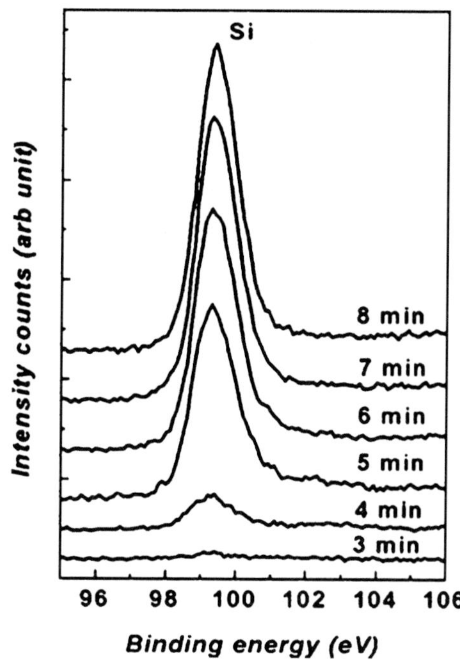

8.1.4   MgAl₂O₄

The complex oxide $MgAl_2O_4$ with normal spinel structure is actually the earliest
reported epitaxial oxide growth on Si(100) with no interfacial amorphous layer.
Ihara et al. [14], using halide vapor phase epitaxy (a variant of CVD), was able to
grow $MgAl_2O_4$ from $AlCl_3$, $MgCl_2$, and CO_2 vapor sources at a substrate temperature of 900–1,000 °C. They also demonstrated the growth of epitaxial Si on the
$MgAl_2O_4$. Other groups later utilized this $MgAl_2O_4$ on Si as a buffer for the growth
of $YBa_2Cu_3O_x$ high-T_c superconductors [15] as well as ferroelectric $BaTiO_3$ [16]
on Si. The growth of BTO and YBCO both induce the formation of an amorphous
SiO_x interlayer between Si and $MgAl_2O_4$.

8.1.5   SrHfO₃

Due to the practically zero conduction band offset between $SrTiO_3$ and Si [17], a
new candidate high-k epitaxial oxide on Si was proposed by IBM Zurich. The
material is $SrHfO_3$, which is a cubic perovskite with lattice constant of 4.08 Å. The
material has a band gap of 6.1 eV [18] and a conduction band offset with Si of
2.3 eV. Utilizing a process that is analogous to the method of forming $SrTiO_3$
directly on Si [19, 20], $SrHfO_3$ to a thickness of 4 nm was grown on Si(100) using

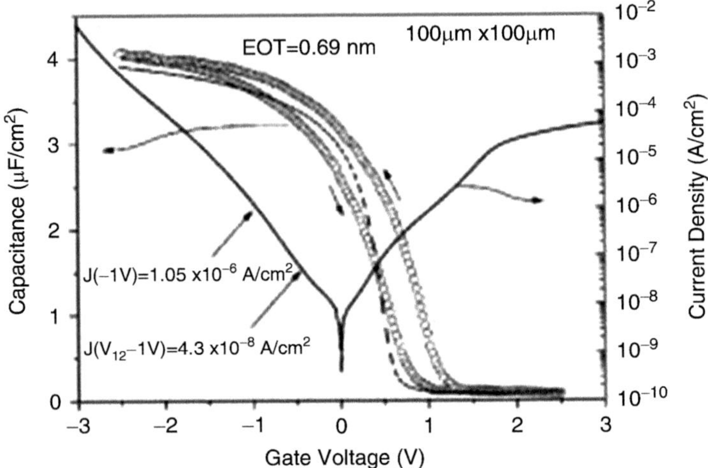

Fig. 8.3 Capacitance vs. voltage and Current vs. voltage characteristics of 10 unit cell thick $SrHfO_3$ transistor. Reprinted from [21], Copyright 2007, with permission from Elsevier

MBE. One-half monolayer of Sr was deposited followed by a low temperature (~100 °C) deposition of one monolayer of epitaxial SrO and ten unit cells of amorphous $SrHfO_3$ at ~3 × 10^{-8} Torr of oxygen. The amorphous layer was then crystallized at 500–600 °C in vacuum. A post-deposition anneal in the presence of atomic oxygen was performed at 120 °C to ensure full oxygenation of the film. The crystallization and post-deposition anneal result in an interfacial layer consisting of Sr, Si, and O with a thickness of ~0.9 nm as determined from XPS measurements. The relatively large lattice mismatch to Si of about 6 % results in a broad x-ray diffraction rocking curve of 4°. Transistor devices fabricated out of $SrHfO_3$ show an EOT of 0.7 nm with a leakage of 1.6 × 10^{-3} A/cm^2 (Fig. 8.3) [21, 22].

8.1.6   Gd$_2$O$_3$(011)

The Osten group in Hanover has shown that Gd_2O_3 can be grown on surfactant-mediated grown Ge layers (900 nm thick) that have been grown on Si(001). The Gd_2O_3 grows in the 011 orientation out of plane. This growth direction on a square symmetric substrate results in the formation of two in-plane domains. Gd_2O_3 films were deposited by evaporating Gd_2O_3 from an electron beam evaporator under a 5 × 10^{-7} Torr oxygen partial pressure ambient. To achieve a flat surface, the substrate temperature was initially at 400 °C at the start of the growth and later increased to 670 °C while growing. Growing at a constant temperature of 670 °C resulted in three-dimensional island growth [23].

8.2 Epitaxial Oxides on Si(111)

8.2.1 $Gd_2O_3(111)$ and Other Bixbyite Oxides

The Osten group has also investigated epitaxy of crystalline Gd_2O_3 on Si (111) [24]. This material is of interest in CMOS technology and other nanoelectronics applications owing to its thermodynamic stability, large 5.9 eV band gap, large band offsets to Si, and high dielectric constant ($k \sim 20$–25). These properties are shared by other lanthanide oxides such as Y_2O_3, Sc_2O_3, Pr_2O_3 and Nd_2O_3. The Gd_2O_3 films were grown on n-type Si(111) substrates using MBE. The Si(111) substrates were wet chemically cleaned using diluted (1:100) HF as the last step. Commercially available granular Gd_2O_3 was evaporated using electron beam heating. To prevent the oxidation of Si during Gd_2O_3 growth, the substrate was exposed to Gd_2O_3 at 300 °C for 7 min at low source power compared to that during the actual deposition. This resulted in the formation of a passivation layer. The substrate temperature during deposition was 675 °C. During the main deposition step, additional molecular oxygen at a partial pressure of 4×10^{-7} Torr was supplied into the growth chamber to ensure complete oxidation during growth. Detailed structural examinations by RHEED and XRD reveal that the Gd_2O_3 layer has a high-quality cubic bixbyite structure with a single domain orientation. The orientation relationship is [111]Gd_2O_3//[111]Si and [1–10]Gd_2O_3//[–110]Si (Fig. 8.4). A 10.9 nm Gd_2O_3 layer was found to have a mismatch of only -0.1 % to the Si substrate in the in-plane direction. It is suggested that the 10.9 nm Gd_2O_3 layer is partially strain relaxed with a small residual compressive strain in the out-of-plane direction and tensile strain in the in-plane direction.

Other bixbyite oxides have been epitaxially grown on Si(111), including Y_2O_3 by PLD [25], La_2O_3-Y_2O_3 by MBE [26], cubic Pr_2O_3 by MBE [27], Nd_2O_3 [28, 29], and Sc_2O_3 by MBE [30]. Y_2O_3 was first utilized as a buffer layer on Si(111) by Park et al. in 1998 [31]. Park deposited Y_2O_3 using electron beam evaporation of Y_2O_3 under an oxygen ambient of 5×10^{-6} Torr. Y_2O_3 has been successfully utilized as a buffer for interfacing ferroelectrics on Si(111), including PZT [31], SBT [32], and $YMnO_3$ [33]. Y_2O_3 has also been deposited using PLD [25]. Films deposited at

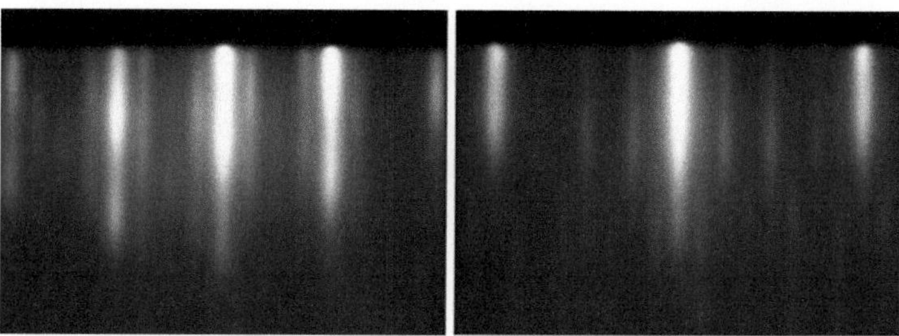

Fig. 8.4 RHEED patterns for Gd_2O_3 on Si(111) along the $<110>$ (*left*) and $<211>$ (*right*) azimuths showing single domain epitaxy. Copyright IOP Publishing. Reproduced from [24] by permission of IOP Publishing. All rights reserved

600–750 °C show no amorphous interlayer but post-annealing at any higher temperature results in the formation of ~3 nm thick amorphous layers. To get better lattice matching with Si, Guha et al. deposited a solid solution of Y_2O_3 and La_2O_3 using MBE from elemental sources [26]. The oxygen pressure used was 2×10^{-5} Torr at a substrate temperature of 700–750 °C. Flat layer by layer growth is observed for this process using RHEED. Silicon has been grown on the $(La,Y)_2O_3$ layer but the silicon has a large number of stacking faults [34]. A related approach was demonstrated by Schroeder using Y_2O_3 and Pr_2O_3 instead [35]. In this case, however, a layer approach was used instead of forming a solid solution. The Pr_2O_3 grows pseudomorphically to Si while the thickness of the Y_2O_3 controls the resulting surface in-plane lattice constant over a fairly wide range. Si(111) overlayers have been grown on this buffer structure [36]. In the case of Sc_2O_3, the deposition was done by MBE at 400 °C using Sc metal flux in the presence of 5×10^{-8} Torr O_2. The native SiO_2 was thermally desorbed at 950 °C prior to growth [30]. Hexagonal ZnO and GaN have both been grown on Sc_2O_3/Si(111) [37, 38]. Nd_2O_3 has also been epitaxially grown on Si(111) using electron beam evaporation and can be utilized for strain control by mixing with Gd_2O_3 [39–41].

8.2.2 Hexagonal Pr_2O_3

Epitaxial Pr_2O_3 on Si(111) was first reported by Tarsa et al. in 1993 [42] using PLD with a growth temperature of 600 °C. The growth was done on a hydrogen-terminated Si surface. The target used had a nominal composition of Pr_6O_{11} and the deposition was done without introducing any oxygen gas into the growth chamber. The resulting surface turns out to be rough as determined both by RHEED and cross-sectional TEM. Osten et al. later showed that hexagonal Pr_2O_3 can also be grown on Si(111) using electron beam evaporation, also using source material with composition Pr_6O_{11} and with no additional oxygen supply. Cross-section TEM shows a sharp interface between Pr_2O_3 and Si with no interfacial layer. It is also possible to grow epitaxial Si on top of this Pr_2O_3 layer [43].

8.2.3 $CeO_2(111)$

CeO_2 has a fluorite structure with a lattice constant of 5.41 Å, making it lattice and atomically matched to Si. Epitaxy of CeO_2 on Si(111) has been reported even at room temperature using PLD [44]. The Si substrates were hydrogen-terminated via an HF-last processing step. To minimize SiO_2 formation, nucleation of CeO_2 was performed in vacuum. Later growth was performed under an oxygen partial pressure of 5×10^{-5} Torr. If the growth is done at higher substrate temperature, an interfacial SiO_2 layer forms [45]. PrO_2, also a fluorite structure rare earth oxide, has also been epitaxially grown on Si(111) by PLD [46].

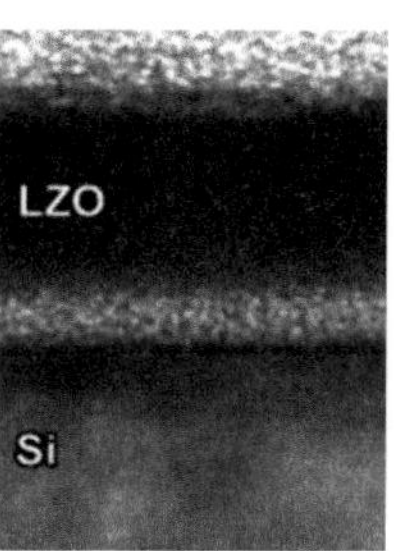

Fig. 8.5 Epitaxial
$La_2Zr_2O_7$ grown on Si(111)
showing ~2 nm SiO_2
interfacial layer. Reprinted
with permission from [47].
Copyright 2003, AIP
Publishing LLC

8.2.4 $La_2Zr_2O_7$

The pyrochlore structure $La_2Zr_2O_7$ (LZO) has a cubic lattice parameter of 10.79 Å, less than 1 % smaller than twice the cubic lattice parameter of Si. Further, both La_2O_3 and ZrO_2 are both known to be thermodynamically stable in contact with Si [1]. Seo et al. [47] have demonstrated that it is possible to achieve direct epitaxy of LZO(111) on Si(111) using MBE. There is a strong competition between SiO_2 formation, La silicide formation, and LZO crystallization during the deposition of LZO on Si. By carefully choosing the growth temperature, a window for LZO growth with only minimal SiO_2 formation is possible. The optimized growth temperature (650 °C) produces a SiO_2 layer of not more than 2 nm thick with about 10 % of the areas having almost no SiO_2 (Fig. 8.5).

8.3   Epitaxial Oxides on Non-silicon Semiconductors

The use of semiconductors other than silicon has also seen some development although not as extensive. We discuss some of the more important developments in the epitaxial integration of oxides on non-silicon semiconductors, specifically, GaN, SiC, Ge, GaAs, and InP.

8.3.1   GaN

GaN appears to be amenable to oxide epitaxial growth without any special surface preparation necessary. There have been several groups working on the integration of dielectric and ferroelectric oxides in epitaxial form on GaN. One common issue is that most ferroelectric materials are cubic or rhombohedral resulting in at best a trigonal layer on a hexagonal substrate. This symmetry mismatch typically results in two trigonal domains that are rotated by 180° from each other. The lack of high quality, large area single crystals of GaN also means that most epitaxial oxide studies use thick GaN films grown on sapphire. This limits the overall quality of oxide films grown on GaN as the GaN thick films themselves are of relatively poor quality compared to single crystal substrates of Si or GaAs.

8.3.1.1 Ferroelectrics on GaN

$(Ba,Sr)TiO_3$ or BST is a non-linear dielectric material used for voltage tunable capacitance applications [48]. BST in bulk ceramic form is known to have large tunability and low loss but suffers from large losses in thin film form when grown as a polycrystal. The integration of BST thin films in epitaxial form with a wide bandgap semiconductor, such as GaN, for frequency agile microwave applications has been a goal for the past decade. The first breakthrough in this area came from the work of C.-R. Cho et al. from the Korea Basic Science Institute in 2004 [49]. 111-oriented BST films were grown by PLD on thick (~2 μm) epitaxial 0001 GaN grown on sapphire substrates. The growth was done under high oxygen partial pressures of 100–500 mTorr with no special surface treatment or templating of the GaN. Because GaN(0001) is hexagonal while BST(111) is trigonal, the film grows in two domains related by a 180° in-plane rotation. Additionally, the large lattice mismatch (12 %) results in columnar growth with a reported crystallite diameter of ~50 nm. Dielectric measurements show decent permittivity and loss tangent with a memory window of 2 V [49]. Improved crystalline quality of BST grown directly on GaN was achieved in 2008 by the group of J.-P. Maria from North Carolina State University [50]. This was achieved by depositing the BST using rf sputtering under 10 mTorr Ar atmosphere using a low plasma power (75 W), resulting in a slow growth rate of 1 nm/min. An analysis of the in-plane epitaxial relationship shows that the BST grows in two domains and that the oxygen sublattice of BST prefers to line up with the Ga surface atoms even though this results in a large lattice mismatch. This is analogous to the case for $YMnO_3$ on GaN described below. Cross-sectional STEM images of the BST/GaN interface still shows columnar growth with crystallite sizes on the order of 20 nm [50].

Another ferroelectric material that has been epitaxially grown on GaN is $YMnO_3$. $YMnO_3$ belongs to the class of materials known as hexagonal manganites that show both ferroelectric and magnetic ordering. Epitaxial $YMnO_3$ on GaN was first reported in 2005 by the Ahn group at Yale University using rf sputtering under low pressure Ar atmosphere [51]. Using low power (75 W) to reduce the growth rate and off-axis geometry to obtain a particulate free surface, direct growth of $YMnO_3$ on GaN was demonstrated. In-plane x-ray diffraction analysis shows that the film and substrate have a 30° relative in-plane rotation to a configuration with a larger lattice mismatch (~10 %) (Fig. 8.6). This has been attributed to the stronger chemical bonding gained from lining up the oxygen sublattice of $YMnO_3$ with the Ga surface atoms of GaN. Ferroelectric measurements show robust ferroelectricity with a somewhat reduced polarization of ~3 μC/cm^2 vs. 5.5 μC/cm^2 in the bulk [51]. The Ahn group has also subsequently demonstrated the growth of $YMnO_3$ on ZnO with similar structural and electrical characteristics [52].

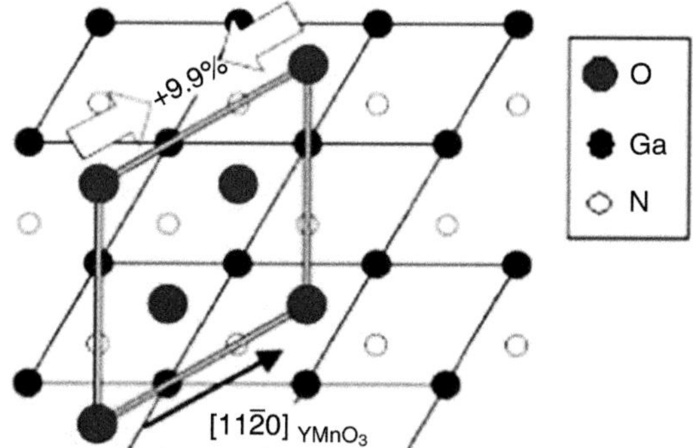

Fig. 8.6 Schematic of the in-plane epitaxial relationship observed for YMnO₃ on GaN showing a situation where chemical bonding considerations overcoming strain considerations. Reprinted with permission from [51]. Copyright 2005, AIP Publishing LLC

The group of D. Lederman at West Virginia University has also succeeded in the epitaxial growth of $YMnO_3$ on GaN using MBE showing the same epitaxial relationship observed for the sputtered films [53].

Sai and co-workers reported the details of the electronic structure of the hexagonal $YMnO_3$–GaN heterojunction [54]. Calculations were done using DFT within the LSDA + U formalism. Using the interface structure proposed by Posadas et al. [51], they considered two inequivalent Ga–O terminated interfaces that can be found in $YMnO_3$ films grown on (0001) and (000$\bar{1}$) oriented GaN substrates and two possible orientations of the $YMnO_3$ polarization with respect to that of the GaN substrate. The main finding was that the band offsets for spin-up and spin-down components are different, with a larger variance at the (000$\bar{1}$) interface. The band offset also depends on the orientation of the polarization in $YMnO_3$ layer with respect to that of the substrate. The spin-dependent interface barriers suggest that these heterostructures may be applicable in spin filtering tunneling devices. These results may be relevant not only to $YMnO_3$ films but also to other multiferroic thin films with coexisting antiferromagnetic and ferroelectric structures.

In 2005, the group of J.S. Speck at UC Santa Barbara reported the epitaxial growth of 100-oriented rutile TiO_2 on 0001 GaN by MBE [55]. The pseudo-two-fold symmetry of 100 rutile results in the growth of three rotational variants on the six-fold symmetric GaN surface. This, as expected, also results in columnar growth, with observed grain sizes on the order of a few tens of nm. This epitaxial TiO_2 layer worked reasonably well as a passivation layer for AlGaN/GaN high electron mobility transistor structures, reducing gate leakage by four orders of magnitude although the band offsets are not favorable for device applications [55]. This epitaxial rutile layer was later utilized by Schlom and Ramesh in 2007 to deposit 111-oriented STO on the rutile-buffered GaN using MBE, and then further depositing the multiferroic material $BiFeO_3$ on top by means of MOCVD or PLD [56]. The 111 STO layer grows in two twin variants which is transmitted to the BFO layer. Ferroelectric measurements of the BFO show very large polarization of ~90 µC/cm².

$LiNbO_3$ is a non-linear optical and ferroelectric material with rhombohedral crystal structure [57]. It has been used as a substrate for the growth of GaN [58]

as well, making it crystallographically compatible with GaN. Not surprisingly, there have been several efforts in the epitaxial growth of $LiNbO_3$ on GaN as well as on SiC. The Speck group at UC Santa Barbara first reported the epitaxial growth of $LiNbO_3$ on GaN in 2005 using rf sputtering from a $LiNbO_3$ target containing 5 mol% excess Li_2O [59]. No other substrate preparation or templating was done other than a standard degrease process. Because of the existence of several Li-Nb oxides, it is not trivial to find the sputtering process parameters to achieve single phase growth. The optimal parameters found for homoepitaxy of $LiNbO_3$ were used to grow $LiNbO_3$ on cleaned GaN/sapphire substrates. A relatively high gas flow rate (100 sccm) at low background pressure (30 mTorr) of a 40 % O_2 in Ar process gas was used in conjunction with a plasma power of 140 W and a substrate temperature of 500 °C. These growth conditions produce single phase, highly oriented films but with a modest x-ray rocking curve width of 1.38°. In-plane x-ray diffraction scans reveal a 30° rotational offset between the unit cells of $LiNbO_3$ and GaN and the existence of two rotational domains. Cross-section TEM images of the samples reveal a fairly thick (~5 nm) amorphous interlayer between the film and substrate, which is detrimental for ferroelectric field effect transistor applications [59]. Slight improvement in the interface quality was reported in 2012 by Hao et al. [60] They utilized pulsed laser deposition of $LiNbO_3$ at higher background pressure (150 mTorr). They also obtained two rotational domains (as expected from the point group mismatch) and the same 30° relative rotation between the unit cells of $LiNbO_3$ and GaN. The amorphous interlayer, however, had been reduced to ~2–3 nm and the C-V characteristics of the heterostructure show a robust memory window of 5.8 V for thick (120 nm) $LiNbO_3$ films.

8.3.1.2   Rocksalt Oxides on GaN

Epitaxial growth of 111-oriented rocksalt oxides on GaN was pioneered by the group of J.-P. Maria at North Carolina State University. In 2006, they reported a study on the direct growth of MgO by MBE on thick GaN (0001) films grown on sapphire substrates. They used a technique of adsorption-controlled growth where the substrate temperature is high enough that the pure Mg metal would evaporate faster than the arrival rate from the effusion cell. The growth rate of MgO would then be controlled by the arrival rate of oxygen. The MgO films, however, exhibited RHEED patterns that were spotty [61]. As the (111) surface of rocksalt has a very high surface energy, it typically facets to expose the (100) surface [62], making flat layer-by-layer growth impossible without additional surface treatment. This problem is further exacerbated by the fairly large lattice mismatch between MgO and GaN of ~7 %, as well as the symmetry mismatch (trigonal versus hexagonal) resulting in twinned growth of MgO. In 2012, the same group pioneered a method of growing (111) MgO in layer by layer fashion by controlling the surface energy of the growing layer. The technique involves utilizing water vapor as the oxidant rather than molecular oxygen. The water vapor produces a hydroxylated surface that enables the (111) surface to remain flat [63]. However, it was observed that

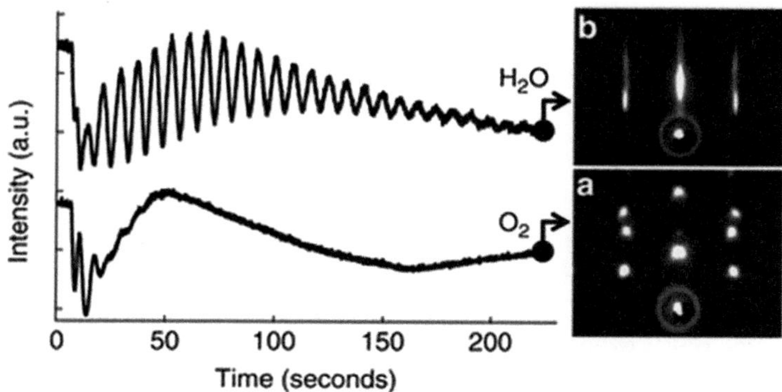

Fig. 8.7   Comparison of RHEED intensity oscillations and RHEED pattern for CaO grown on GaN using water vapor and molecular oxygen as oxidants. The flatness of the film using water vapor is evident. Reprinted from [64] by permission from Macmillan Publishers Ltd., copyright 2011

accumulation of MgO stops after three to four layers, when the surface was saturated by hydroxy groups. To work around this problem, they utilize PLD instead of MBE to deposit the MgO in the presence of a water vapor atmosphere. The smooth (111) MgO films made using the PLD/water vapor process showed more than an order of magnitude improvement in leakage current for 4.5-nm-thick capacitor structures. This technique of using water vapor as a surfactant/oxidant for growth of smooth (111) rocksalt films on GaN was first demonstrated for CaO (Fig. 8.7) [64]. The growth process relies on forming an initial monolayer of CaO by reaction of Ca metal with the native oxide of GaN.

The North Carolina State University group has also succeeded in growing YbO on GaN [65]. YbO is a metastable phase where Yb exhibits a +2 oxidation state rather than the normal +3. This is achieved by carefully matching both the ytterbium and oxygen fluxes. A similar approach was used by the Schlom group to grow EuO on GaN [66]. EuO is a ferromagnetic semiconductor that can be used for certain spintronics applications and suffers from a similar issue with YbO where there is a delicate balance needed between the europium and oxygen fluxes to obtain EuO rather than the non-ferromagnetic Eu_2O_3.

8.3.2   SiC

Hexagonal SiC is another wide band gap semiconductor that is being studied for use in high temperature, high power, and high frequency electronic devices. In this sense, it occupies the same area of application as GaN. High quality SiC substrates, however, are available commercially and should yield higher quality epitaxial oxide films. One drawback is the necessity of removing the native amorphous oxide of SiC and the extensive surface preparation needed to obtain a flat SiC surface prior to growth. There have been two notable epitaxial oxide on SiC systems that have been reported: LiNbO$_3$, which can work as a ferroelectric and optical material, and MgO, which can be used as a dielectric and also as a buffer layer for the growth of other functional oxides.

8.3.2.1   LiNbO$_3$ on SiC

The Doolittle group at the Georgia Institute of Technology has pioneered a method of depositing LiNbO$_3$ on SiC using NbCl$_5$ as the niobium source in an MBE system [67]. The NbCl$_5$ was evaporated from an effusion cell operated at 35 °C. The NbCl$_5$ is stable up to 1,100 °C and is not decomposed on the substrate in the absence of oxygen. To obtain a smooth SiC surface, the substrates are vacuum annealed at 850–950 °C for several hours, followed by annealing under exposure to a Ga flux for several more hours. This process reduces the mean surface roughness of SiC from 1.2 nm as received to about 0.4 nm. The LiNbO$_3$ is grown under an oxygen plasma at 850–950 °C by co-depositing Li metal and NbCl$_5$. A chemical reaction at or near the surface of the substrate produces LiNbO$_3$ and Cl$_2$ gas. Chemical analysis of the LiNbO$_3$ shows no chlorine incorporation in the film. A drawback of this technique, however, is the need for special equipment to handle the corrosive chlorine gas being produced by the chemical reaction. By observing the growth process in RHEED, the growth was found to proceed with multiple island formation that coalesces into a single film at a thickness of 5 nm. X-ray diffraction reveals the LiNbO$_3$ films to have a rocking curve width of 0.7° or better. The Doolittle group has also demonstrated the epitaxial growth of the related material LiNbO$_2$, a layered material that has shown potential for memristor-type applications [68].

8.3.2.2   MgO on SiC

MBE-growth of (111)-oriented MgO on SiC was demonstrated by the group of K. S. Ziemer at Northeastern University [69, 70]. The key to the high quality MgO film lies in the substrate preparation prior to film growth. After the standard degreasing process, the SiC was loaded into a high temperature furnace and exposed to flowing H$_2$ gas (11.4 slpm) at a temperature of 1,700 °C for 30 min. This simultaneously removes the native oxide and smoothens out the scratches in the as received substrate. AFM analysis shows a mean roughness of less than 0.45 nm with a stepped surface, while RHEED shows the characteristic $\sqrt{3} \times \sqrt{3}$ R30 surface reconstruction of Si-terminated SiC surface with an ordered silicate adlayer. The presence of this silicate adlayer turns out to be important for subsequent growth of MgO, both for wetting and lattice matching considerations. The in-plane surface oxygen spacing of the silicate adlayer reconstruction is found to be 5.3 Å. With a 30° relative rotation between reconstructed SiC and the MgO (111) plane, the difference in oxygen spacing is modest, about ~3.3 % with the MgO in tension. The MgO deposition was performed using MBE with an rf oxygen plasma source operated at 90 W power,

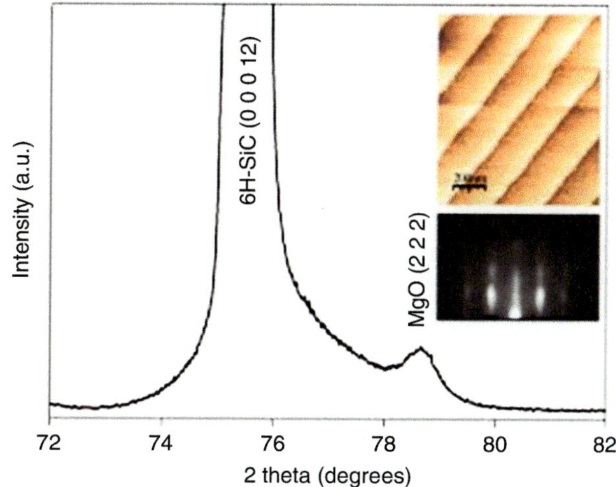

Fig. 8.8 XRD, AFM, and RHEED scans of MgO(111) on 6H-SiC. Reprinted with permission from [71]. Copyright 2008, AIP Publishing LLC

and a Mg effusion cell. The substrate was held at ~140 °C during growth. Growth was also performed using molecular oxygen but this resulted in cluster growth rather than the conformal growth obtained when using an oxygen plasma. Smooth growth of the (111) surface is obtained for thicknesses up to 5 nm. Thicker films start exhibiting evidence of 3D growth with the RHEED becoming spotty. This was attributed to a twin along the growth direction [69]. This Stranski-Krastanov growth mode was confirmed using high resolution cross-section STEM imaging. The thin flat MgO on SiC was shown to have excellent dielectric properties suitable for use as a gate oxide [71]. Thin MgO on SiC was also utilized as a template layer for the growth of functional oxides on SiC. The Ziemer group has demonstrated the growth of both ferroelectric $BaTiO_3$ [72] and ferrimagnetic insulating barium hexaferrite [73] on the MgO-buffered SiC, showing the suitability of MgO as a template layer for functional oxide integration on SiC (Fig. 8.8).

Depositing $BaTiO_3$ directly on SiC results in an amorphous material with a rough surface. Using PLD to deposit BST directly on hydrogen-cleaned SiC results in a mean surface roughness of ~16 nm. However, when the BST was deposited on a 2.5-nm MgO buffer layer grown on SiC, the resulting BST films come out smoother with surface roughness within a terrace as low as 0.3 nm [74]. The use of MBE to deposit the $BaTiO_3$ was also studied [72]. Similar to PLD-grown BST, direct growth on SiC resulted only in amorphous films at the optimum BTO growth condition, while (111)-oriented $BaTiO_3$ forms when the MgO buffer layer is present. RHEED shows that the growth mode is three-dimensional with an overall surface roughness of 0.82 nm.

The Ziemer group has also demonstrated the epitaxial integration of the hexagonal ferrite $BaO-(Fe_2O_3)_6$ both on MgO/SiC and on BTO/MgO/SiC. Barium hexaferrite is a ferrimagnetic insulator with high anisotropy and high permeability. High crystalline quality barium hexaferrite films on SiC are desirable for their potential use in monolithic integrated circuits for high frequency and high power applications.

Direct growth of hexaferrite films on SiC had been reported [75] but the formation of native oxides of SiC during hexaferrite deposition by PLD resulted in loss of epitaxy. Through the use of the MBE-grown MgO buffer, a high quality barium hexaferrite film was grown on SiC with a ferromagnetic resonance linewidth of 100 Oe (i.e. $f = 53$ GHz) [76]. To achieve this, barium hexaferrite was deposited by PLD at 915 °C with a low pressure oxygen background of 20 mTorr on MgO/SiC [77, 78]. It was determined from EDX line scans of a cross-section TEM image that there is an interface reaction that occurs between the MgO and the hexaferrite during deposition. The reacted layer is crystalline and is believed to consist of spinel structure $MgFe_2O_4$, which actually facilitates the epitaxy of the barium hexaferrite. A very thick amorphous SiO_x interlayer (10–20 nm) also forms although this does not disrupt the epitaxy of the overlying layers. MBE was also utilized to grow the barium hexaferrite on MgO/SiC [73]. When grown directly on MgO/SiC, the hexaferrite surface showed a roughness of about 1.4 nm. However, when grown on the BTO(111) surface, the surface roughness increased to 2.9 nm, even though the underlying BTO layer was measured to have a roughness of 0.4 nm. The hexaferrite growth mode also appears to be strongly affected by the oxygen partial pressure. Films grow polycrystalline when there is excess oxygen available, while second phases emerge under oxygen-poor conditions [73]. These high quality epitaxial barium hexaferrite layers on SiC could be used as seed layers for the growth of much thicker hexaferrite films, which often need to be on the order of 10 μm thick for microwave ferrite devices. For further details into the efforts at integrating ferrite films onto semiconductors, see the review article by Chen and Harris [79].

8.3.3   Ge

Germanium has been attracting more attention recently as the fundamental limits of silicon-based technology are being reached. Germanium has higher mobility for both electrons and holes compared to silicon at room temperature. One drawback of germanium is the lack of a stable native oxide for passivation and to serve as a gate dielectric layer. However, this ease of volatilization of the native GeO_2 may also be useful for enabling intimate contact between a functional oxide and the underlying germanium. We will discuss developments in the epitaxial integration of ferroelectric $BaTiO_3$ and buffer material CeO_2 on Ge.

8.3.3.1   BaTiO₃ on Ge

The direct growth of epitaxial $BaTiO_3$ on Ge was originally reported by McKee et al. in 2001 using the same principle as the growth of $SrTiO_3$ on Si [80]. A submonolayer Ba template is used to passivate the Ge surface prior to BTO deposition. Z-contrast STEM of the BTO on Ge shows a defect-free epitaxial interface.

Fig. 8.9 Cross-section
TEM image of BTO film
(*top layer*) grown directly
on Ge substrate (*bottom
layer*) by MBE. (Image
courtesy of David Smith)

They performed C-V and I-V measurements on the BTO/Ge structures and found that
the leakage current is very high even for a 25 nm thick sample. Insertion of a few layers
of BaO reduces the leakage by six orders of magnitude. They also claimed inversion of
the Ge using BTO as the gate dielectric although no ferroelectric hysteresis was
observable. Measurement of the interface trap density shows a very small value of
$<10^{10}\,\mathrm{cm}^{-2}\,\mathrm{eV}^{-1}$ [80].

Using the same ½ monolayer Ba template, Merckling et al. demonstrated the
growth of $BaTiO_3$ on a Ge thick film (1 μm) grown on a Si substrate [81]. The BTO
was grown under an atomic oxygen flux at 650 °C. They monitored the deposition
of the ½ monolayer of Ba in RHEED and found that the half-order diffraction streak
initially decreases in intensity then recovers to its original intensity at ½ monolayer.
Cross-section TEM shows good epitaxy with some crystal disruptions that appear
to be related to steps in the Ge substrate. X-ray diffraction measurements appear to
show two phases of BTO: a c-axis oriented tetragonal phase and a cubic phase. The
width of the XRD rocking curve was measured to be 1.5° indicating significant
mosaicity. No measurement of the electrical properties was reported [81].

We have also demonstrated the epitaxial growth of $BaTiO_3$ on Ge using a ½
monolayer Sr template deposited at 500 °C instead. Growth of BTO was also done
at 600 °C under 5×10^{-6} Torr molecular oxygen. RHEED shows that the film is
crystalline as deposited and proceeds in layer by layer fashion to thicknesses up to
40 nm. STEM imaging of the interface shows no amorphous interlayer and good
epitaxy between the layers (Fig. 8.9). XRD measurements, however, indicate a-axis
oriented growth of BTO on Ge as a result of the thermal expansion mismatch. Band
offset measurements of BTO/Ge structures shows a conduction band offset of
~0.1 eV. The ½ monolayer Sr template on Ge does not behave in the same way
as Ba on Ge or Sr on Si. The Sr on Ge template has been studied by Lukanov
et al. using STM and they find that the ½ monolayer Sr consists of ordered trenches
with a 9–10 unit cell spacing with double atomic layer height [82].

8.3.3.2 CeO$_2$ on Ge

Norton et al. developed a method of growing oxides on semiconductors whose native oxides are not very stable such as Ge and InP [83]. Their method involves decomposing the native oxide using hydrogen during the initial nucleation of the oxide material to be deposited. This technique allows one to deposit oxides on Ge that are stable under the hydrogen chemical potentials used to decompose the native GeO$_2$. The use of hydrogen relaxes the strict base pressure requirements of the deposition chamber in order to achieve a sharp interface, and also allows one to use generally lower substrate temperatures. Epitaxial CeO$_2$ was deposited using this technique on Ge via PLD at 750 °C under a hydrogen partial pressure of 4×10^{-7} Torr. The first 5 nm of the film were deposited under these conditions and additional CeO$_2$ was deposited without hydrogen. Epitaxy was achieved at substrate temperatures as low as 500 °C as confirmed with Z-contrast STEM and XRD ϕ scans. CeO$_2$ is commonly used as a buffer layer for growing other oxide materials and is a way to integrate new functionality into the Ge platform. The Norton group has also used reactive sputtering to deposit the CeO$_2$ layer using a sputtering gas consisting of 4 % H$_2$ in Ar at a pressure of 25 mTorr with 1 mTorr of water vapor [84, 85].

8.3.4 GaAs

GaAs is a compound semiconductor widely used for high frequency electronics as well as in certain diode lasers and light emitting diodes. It has a much higher electron mobility than silicon and has a direct band gap. Its high frequency capability has resulted in its use in mobile phones and radar systems. GaAs, however, does not have a well-defined native oxide because of its compound nature. XPS analysis of oxidized surfaces of GaAs show that the native oxide layer is mainly Ga$_2$O$_3$ with substantial As substitution and that this native oxide is chemically and physically inhomogeneous [86]. Another issue is that As is volatile and some As begins to vaporize from GaAs at temperatures as low as 400 °C [87]. These factors complicate oxide growth on GaAs. In spite of these difficulties, two important developments were made in epitaxial oxide growth on GaAs. One is the growth of MgO and the other is the growth of SrTiO$_3$.

8.3.4.1 MgO on GaAs

MgO readily can be grown in crystalline form at temperatures below 300 °C making it suitable for deposition on GaAs. One of the earliest reports of epitaxial MgO on GaAs was reported by Nashimoto et al. in 1992 [88]. The native oxides of GaAs were thermally desorbed at 600 °C prior to MgO deposition. MgO was deposited by PLD using a Mg metal target and a background oxygen pressure of

5×10^{-6} Torr at a substrate temperature of 350 °C. Epitaxial MgO films as thick as 44 nm were demonstrated. XRD of the optimized film shows a modest rocking curve width of 1.4°. GaAs and MgO have a nominal lattice mismatch of over 25 % but it turns out that MgO still grows reasonably well on GaAs due to a 4:3 coincidence lattice with mismatch of 0.65 %. The good quality of the MgO allowed for the subsequent deposition of ferroelectric $BaTiO_3$ on top even with MgO buffers as thin as 4 nm. The BTO layer appears to be c-axis oriented although no electrical measurements were reported [88].

Depending on the surface treatment of GaAs prior to MgO deposition, other orientations of MgO on (100)-oriented GaAs have been reported. The work of Nashimoto et al. shows (100)-oriented MgO growing when only thermal oxide desorption of the GaAs surface is performed [88]. Hung et al. reported on (110)-oriented MgO forming when the GaAs is treated with ammonium sulfide. Ammonium sulfide allows for MgO deposition without the thermal desorption step [89]. To prevent oxidation of GaAs, MgO was directly evaporated using an e-beam evaporator at a substrate temperature 550 °C. With a 3:4 coincidence lattice along the [110] direction, this process results in a lattice mismatch of 0.2 and 5.8 % along the two perpendicular directions. Tarsa et al. on the other hand, reported the growth of two-domain (111)-oriented MgO on GaAs by PLD when a GaAs surface with residual arsenic oxides was used [90].

8.3.4.2 $SrTiO_3$ on GaAs

In 2004, Motorola reported on the successful direct epitaxy of $SrTiO_3$ on GaAs [91]. The GaAs substrate was prepared by thermally desorbing the native oxide at 600 °C under an As_4 flux, followed by deposition of a 0.5 μm thick homoepitaxial GaAs layer with As termination. Prior to deposition of the $SrTiO_3$, a Ti metal interlayer about half a monolayer thick was deposited on the GaAs at 300 °C. XPS analysis of the interlayer showed evidence of reaction between Ti and surface As. $SrTiO_3$ was deposited at a temperature of 300 °C and molecular oxygen partial pressure in the low 10^{-8} Torr regime with both the temperature and oxygen pressure slowly ramping during the growth of the first few monolayers. Subsequent deposition was performed at a substrate temperature of 550 °C. RHEED analysis during growth showed coherent growth of $SrTiO_3$ to GaAs up to a thickness of 20 Å at which point the $SrTiO_3$ began to relax. Post-deposition characterization of the $SrTiO_3$ on GaAs showed a mean roughness of less than 3 Å and an XRD rocking curve width of 0.42°. Cross-section TEM shows well-crystallized $SrTiO_3$ with a sharp interface to GaAs (Fig. 8.10). It was also reported that growth of $SrTiO_3$ on Ga-terminated surface was not good as secondary phases containing Ga formed. It was also found that the presence of Sr at the GaAs surface prevents epitaxy of $SrTiO_3$ [91]. An ultrathin (0.8 nm) buffer of $SrTiO_3$ on GaAs, was recently used to demonstrate integration of ferroelectric c-axis oriented $BaTiO_3$ on GaAs showing that $SrTiO_3$ functions very well as a buffer layer for growing other oxides on

Fig. 8.10 Cross-section
TEM of epitaxial STO
grown on GaAs. Reprinted
with permission from [91].
Copyright 2004, AIP
Publishing LLC

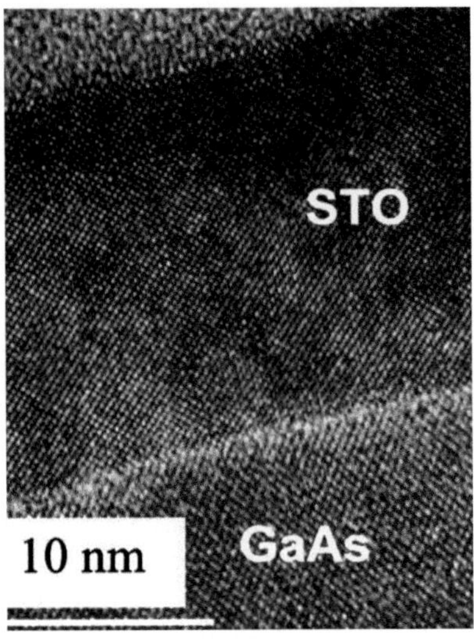

GaAs [92]. The $BaTiO_3$ film was grown to a thickness of 76 Å and showed an x-ray diffraction rocking curve width of 0.58°. Piezoelectric force microscopy confirms the presence of a switchable polarization.

8.3.5   *InP*

InP is a compound semiconductor used in opto-electronic applications. Diode lasers and optical waveguides for the 1.55 μm wavelength are typically constructed out of InGaAsP layers that are lattice-matched with InP substrates. The ability to directly integrate non-linear optical oxide materials in InP can potentially yield more compact and energy efficient circuitry for optical communications.

8.3.5.1 CeO_2 on InP

CeO_2 has been grown on InP using PLD by Ivill et al. using the hydrogen background gas technique to eliminate the native oxide of InP during nucleation of the CeO_2 layer [93]. InP wafers were degreased and loaded into the deposition chamber. The wafers were heated in the presence of hydrogen to a partial pressure of 4×10^{-3} Torr at which In_2O_3 and P_2O_5 are unstable against reduction. One issue with InP is loss of P at temperatures as low as 365 °C. To circumvent this, Ivill et al. used a very rapid heating rate of 120 °C per min to 550 °C and then depositing the first 200 Å at a rate of 1.5 Å/s under the same hydrogen partial pressure used to remove the native oxide layer. Additional CeO_2 was deposited without the hydrogen gas. The combination of rapid heating and hydrogen gas nucleation resulted in epitaxial films with a rocking curve width of 0.9°.

8.3.5.2   Yttria-Stabilized ZrO_2 and $SrTiO_3$ on InP

PLD-grown yttria-stabilized zirconia (YSZ) films have also been demonstrated on InP substrates by Vasco et al. In this case, the InP native oxide was simply desorbed thermally at 550 °C under high vacuum [94]. The YSZ layer was grown at a substrate temperature of 550–600 °C. The same group has also reported on the PLD growth of $SrTiO_3$ on InP. 100-oriented, thick STO can be achieved by PLD at a growth temperature of 630 °C and an oxygen partial pressure of 10^{-3} Torr [95]. This STO layer was subsequently used to deposit a very thick (400 nm) La-doped lead zirconate titanate piezoelectric oxide. Cross-section SEM shows columnar growth of the oxide layers as well as oxidation of the InP/STO interface.

8.4   Epitaxy of Semiconductors on Oxides

8.4.1   GaAs/STO/Si

One of the first successful attempts to use transition metal oxide buffers for integration of different semiconductors has been the demonstration of GaAs growth on a STO/Si pseudosubstrate by Eisenbeiser et al. at Motorola [96]. The transition between the STO layer and the GaAs layer was accomplished by means of a Zintl template layer. This involves forming a thin interlayer with nominal composition of $SrAl_2$ followed by the growth of several monolayers of AlAs before finally transitioning to the GaAs layer. The theoretical mechanism of this approach was analyzed by Demkov et al. [97]. Eisenbeiser et al. showed that the defect density in the doped GaAs active layer after growth of a thick (2 μm) undoped GaAs buffer layer on STO/Si can be as low as 10^5 cm^{-2}. MESFET devices fabricated on this GaAs active layer exhibited transistor performance nearly identical to that of devices fabricated on GaAs substrates [96].

The group of Saint-Girons in Lyon has recently reported a detailed study of nucleation and growth of GaAs islands directly on STO (001) [98]. Using MBE they have found that in the temperature range from 450 to 550 °C, GaAs grows by three-dimensional Volmer-Weber epitaxy. The TEM analysis shows the nucleation of zincblende (001), zincblende (111), and wurtzite (0001) islands. The AFM study of the size distribution revealed that zincblende islands gave the smallest critical nucleation volume and form at the earlier stages of the growth with the (001) orientation, thus having the largest average size. At a higher growth temperature the nucleation of (001)-oriented islands is enhanced (Fig. 8.11). A follow-up study by the same group analyzed the effect of arsenic partial pressures on the growth mode of the GaAs islands on STO. They found that under low arsenic partial pressures, GaAs islands form with (001) orientation, with progressive formation of (111)-oriented GaAs islands as the arsenic partial pressure increases. This results from the competition between the formation of Ga–O and Ga–As bonds at the early stages of the growth [99].

Fig. 8.11 (**a–d**) AFM images of samples grown at different temperature. (**e**) Evolution of the small and large island densities (logarithmic scale) as a function of 1/kT. Reprinted with permission from [98]. Copyright 2009, AIP Publishing LLC

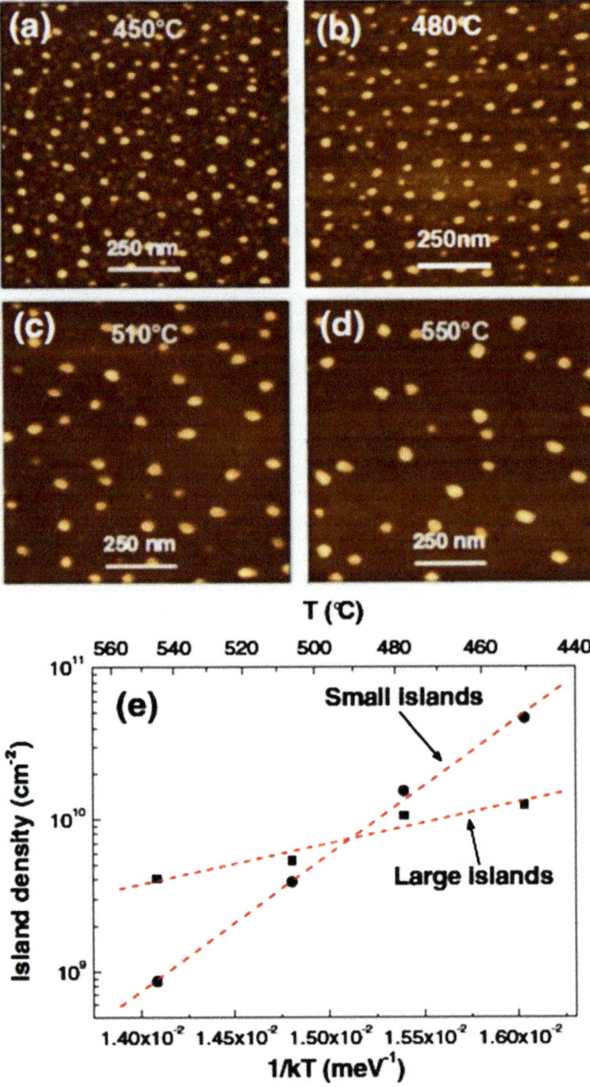

8.4.2   Si(111)/Gd₂O₃/Si(111)

Expanding on their work on integrating rare earth oxides on Si(111), the Osten group has developed a technique for depositing high quality, flat Si(111) layers on top of Gd_2O_3/Si(111) to form sandwiched Si epitaxial layers. Direct growth of Si on Gd_2O_3 and even solid phase epitaxy (grow amorphous then recrystallize) leads to island formation due to the lower surface energy of the oxide compared to Si. To overcome this, they used a technique known as encapsulated solid phase epitaxy

Fig. 8.12 Epitaxial Si(111) grown on Gd$_2$O$_3$/Si using a technique known as encapsulated solid phase epitaxy. Reprinted with permission from [100]. Copyright 2006, AIP Publishing LLC

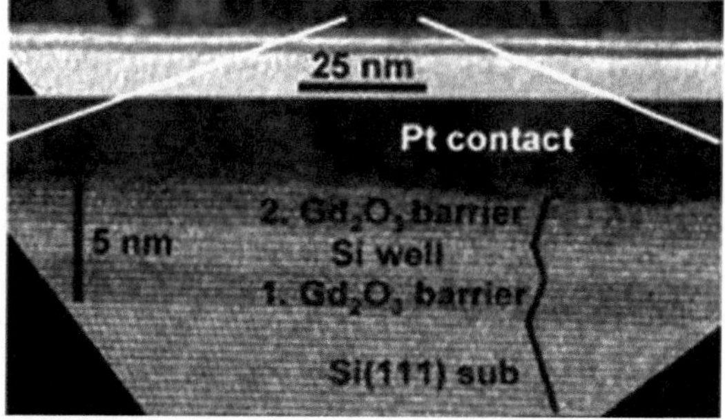

where Si is deposited at low temperature where it covers the underlying Gd$_2$O$_3$ fully but is amorphous. While the substrate is heated to crystallize the Si, additional Gd$_2$O$_3$ is deposited at the same time. This capping layer of Gd$_2$O$_3$ acts to prevent significant surface diffusion of Si resulting in a single crystalline layer of Si sandwiched by Gd$_2$O$_3$ layers (Fig. 8.12) [100]. The topmost Gd$_2$O$_3$ layer can be removed by wet chemical etching and the remaining Si(111) layer shows the expected 7×7 reconstruction of a clean surface after annealing in a low Si flux [101].

This work has been expanded by Translucent by forming solid solutions of various rare earth oxides including La, Er, Nd, and Gd to achieve full control of the resulting lattice constant of the rare earth oxide buffer. This control over the lattice parameter enables the epitaxial growth of Ge and SiGe alloys of any composition on top of Si(111) substrates [102]. The use of rare earth oxide buffer layers has also enabled the fabrication of various epitaxial nanostructures of Si [103, 104] and complicated multilayer structures such as distributed Bragg reflectors [105].

8.4.3   InP/STO/Si

The Lyon group reported a study of the epitaxial growth of InP on SrTiO$_3$ (001) substrates as a first step of monolithic integration of InP with Si [106]. The MBE growth was done using the Riber 32 reactor. The substrates were treated with buffered HF to ensure the TiO$_2$-termination. 100 nm of InP were then deposited at 410 °C, under a phosphorus partial pressure of 4×10^{-6} Torr and at a growth rate of 0.2 ML/s. The STO surface reconstruction was found to have a detrimental influence on the crystalline orientation and morphology of grown InP thin films. On unreconstructed STO, InP grows (111) oriented, the wetting is poor, and the final InP surface is very rough as evidenced by RHEED. On the other hand, on the (2×1) reconstructed STO, InP is (001) oriented and the interface is commensurate, leading to a somewhat better wetting and an improved surface morphology. The improved wetting was attributed to higher surface energy of a (2×1) reconstructed surface as has been suggested by first principles calculations [107]. Overall, however, the grown films still need further improvement in surface morphology and complete elimination of (221)-oriented domains. A detailed crystallographic analysis revealed evidence for the formation of twins during the early stages of the semiconductor growth and its Volmer-Weber nature [108]. The main cause of twin formation during the growth is the existence of a commensurate heterointerface between STO and the twinned InP. The three dimensional growth of InP results in the compliant behavior of the InP/STO interface [109]. The islands are defect-free, oriented with respect to STO, and have their bulk InP lattice parameter. Compliance occurs spontaneously during the growth and does not require any substrate patterning.

Using the crystalline STO/Si pseudosubstrate, Gobaut et al. were able to integrate InAsP/InP quantum well heterostructures grown directly on Si and report their structural and optical properties [110]. Using STO/Si templates improves the structural properties of the III–V heterostructure, sufficiently to allow observing room-temperature photoluminescence from the quantum well.

8.4.4   Ge on SrTiO$_3$

El Kazzi et al. have reported a synchrotron study of the formation of the Ge/STO interface [111]. The substrates were either Nb-doped SrTiO$_3$(001) bulk crystals or 20 nm-thick MBE-grown STO thin films grown on Si(001) wafers. After an ex situ HF etching, both samples were thermally annealed at 550 °C for 3 h in vacuum. This treatment removes most of the surface impurities and provides a TiO$_2$-terminated STO surface. The STO surface is heterogeneously annealed such that the surface temperature increases from 600 °C on one corner to 1,100 °C on the opposite corner. At 600 °C, XPS analysis shows that STO is stoichiometric and fully oxidized, with RHEED showing an unreconstructed surface. At intermediate temperatures, the appearance of Ti^{3+} and Ti^{2+} species is observed in XPS, with RHEED showing a 2 × 1 reconstruction. At 1,100 °C, XPS additionally shows the formation of TiC species and very strong oxygen depletion. On this heterogeneously annealed surface, 0.5 monolayer of Ge was deposited at a substrate temperature of 550 °C at a growth rate of 0.1 monolayer per minute. XPS analysis after Ge deposition shows no Ge on the low temperature corner. Ge is found in the intermediate region and shows Ge with oxidation states of 0, +1, and +2 present. This indicates that Ge bonds to STO through oxygen. Ge is also present in the high temperature corner but with the additional presence of significant intermetallic Ti-Ge bonding. The authors concluded that Ge adsorption can occur only if the STO surface is chemically activated because it is depleted in oxygen. Ge adsorption is strongly dependent on the STO surface's initial composition: the richer the surface is in Ti suboxides, the more Ge wets the STO surface. Time-resolved photoemission measurements of Ge bonding and desorption as a function of substrate temperature show that as temperature increases, Ge adatoms, initially bonded to two (or more) oxygen atoms, gradually aggregate to form Ge clusters bonded via a single oxygen atom, suggesting (111)-oriented clusters.

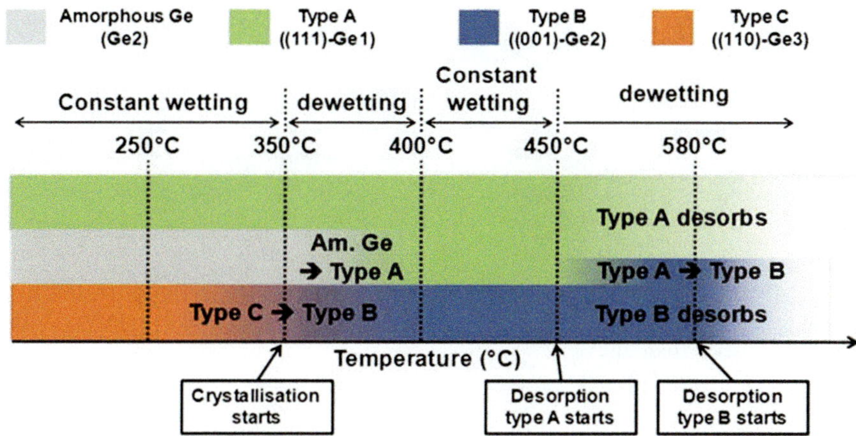

Fig. 8.13 Phase diagram of Ge submonolayer structures deposited directly on TiO_2-terminated STO(001) as a function of temperature. Reprinted with permission from [112]. Copyright 2012, AIP Publishing LLC

In a follow-up work, Gobaut et al. studied the desorption of a submonolayer deposit of Ge on $SrTiO_3$(001) using reflection high energy electron diffraction to analyze the correlation between interface chemistry and crystallographic orientation, particularly the competition between (111) and (001) orientations typical for the semiconductor on perovskite epitaxial systems. Despite poor interface matching, (111)-oriented islands are stabilized at the expense of (001)-oriented islands due to the relatively low energy of their free facets. Such "surface energy driven" crystallographic orientation of the deposit is enhanced by the low adhesion energy characteristic of the $Ge/SrTiO_3$ system (Fig. 8.13) [112].

8.4.5   Ge(100)/SrHfO₃/Si(100)

Seo et al. have demonstrated the growth of epitaxial Ge(100) on Si(100) using perovskite $SrHfO_3$ as the buffer layer [113]. The natural tendency for Ge to form islands on an oxide surface is suppressed by using a two-step growth process with an initial step at low temperature and subsequent growth at higher temperature. To fabricate the heterostructure, Seo et al. first deposited an ~8 nm STO layer on clean Si(100) using the usual ½ monolayer Sr interlayer process. On top of this a similar thickness of $SrHfO_3$ was then deposited using a process similar to [21] (Fig. 8.14a). Ge grows as islands on $SrHfO_3$ with the structure dependent on temperature. The Ge islands are polycrystalline below 500 °C and (100)-oriented above 600 °C. The two-step process involved forming these (100)-oriented islands at high temperature (610 °C) resulting in a spotty RHEED pattern as shown in Fig. 8.14b. When the surface is fully covered by (100)-oriented Ge islands, the

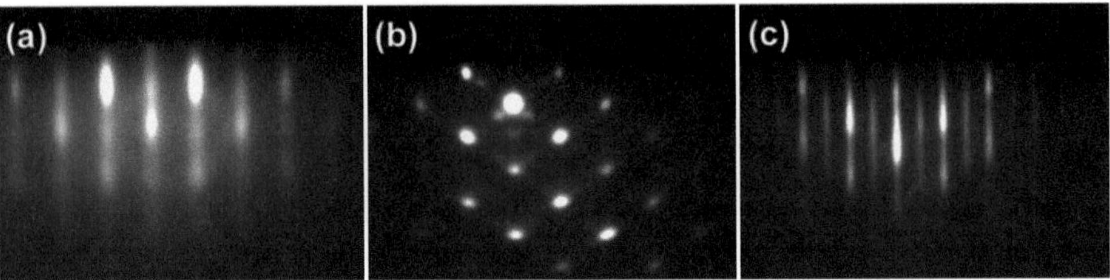

Fig. 8.14 RHEED images of (**a**) epitaxial $SrHfO_3$/Si, (**b**) Ge grown on $SrHfO_3$/Si at 610 °C, and (**c**) additional Ge grown on the surface of (**b**) at 350 °C. The tendency for island formation is suppressed by using the two temperature-step growth process. Reprinted from [113], Copyright 2007, with permission from Elsevier

substrate temperature is lowered to 350 °C and Ge deposition is continued. This results in a streaky RHEED pattern as shown in Fig. 8.14c. High-resolution TEM analysis of the Ge film does show a relatively high density of microtwins and {111} stacking faults.

8.4.6   *InP/Gd₂O₃/Si*

An original approach of monolithic integration of InP based heterostructures on silicon is proposed based on the peculiar properties of the heterointerface between InP and crystalline Gd_2O_3. When grown on a crystalline Gd_2O_3/Si(111) buffer, InP takes its bulk lattice parameter as soon as the growth begins, and the lattice mismatch (7.9 %) is fully accommodated by the formation of a misfit dislocation network at the InP/Gd_2O_3 heterointerface. This plastic compliant effect allows the monolithic growth of good quality InAsP/InP heterostructures on Si, as attested by room-temperature photoluminescence experiments [114].

8.5   Summary

We have provided in this chapter a brief survey of some of the key developments in the epitaxy of complex oxide materials on semiconductor substrates. It is hoped that this short history of the major advances in oxide/semiconductor epitaxy can provide insight into the various substrate preparation and film deposition tricks and techniques that enable such epitaxial systems to be made. We have also described some of the more substantial efforts at making the opposite stack of epitaxial semiconductors on oxides. This area is not as successful mainly as a result of the intrinsic surface energy differences between oxides and semiconductors in general. This is certainly an area that will benefit from more groups studying such systems and developing appropriate fabrication methods.

References

1. K.J. Hubbard, D.G. Schlom, Thermodynamic stability of binary oxides in contact with silicon. J. Mater. Res. **11**, 2757 (1996)
2. D.K. Fork, F.A. Ponce, J.C. Tramontana, T.H. Geballe, Epitaxial MgO on Si(001) for Y-Ba-Cu-O thin-film growth by pulsed laser deposition. Appl. Phys. Lett. **58**, 2294 (1991)
3. G.X. Miao et al., Epitaxial growth of MgO and Fe/MgO/Fe magnetic tunnel junctions on (100)-Si by molecular beam epitaxy. Appl. Phys. Lett. **93**, 142511 (2008)
4. C. Wolverton, K.C. Hass, Phase stability and structure of spinel-based transition aluminas. Phys. Rev. B **63**, 024102 (2000)
5. C. Merckling et al., Pseudomorphic molecular beam epitaxy growth of gamma-Al_2O_3(001) on Si(001) and evidence for spontaneous lattice reorientation during epitaxy. Appl. Phys. Lett. **89**, 232907 (2006)
6. K. Sawada, M. Ishida, T. Nakamura, N. Ohtake, Metalorganic molecular beam epitaxy of γ-Al_2O_3 films on Si at low growth temperatures. Appl. Phys. Lett. **52**, 1672 (1988)
7. C. Merckling et al., Growth of crystalline γ-Al_2O_3 on Si by molecular beam epitaxy: influence of the substrate orientation. J. Appl. Phys. **102**, 024101 (2007)
8. D. Akai, K. Hirabayashi, M. Yokawa, K. Sawada, M. Ishida, Epitaxial growth of Pt(001) thin films on Si substrates using an epitaxial γ-Al_2O_3(001) buffer layer. J. Cryst. Growth **264**, 463 (2004)
9. Y.-C. Jung, H. Miura, K. Ohtani, M. Ishida, High-quality silicon/insulator heteroepitaxial structures formed by molecular beam epitaxy using Al2O3 and Si. J. Cryst. Growth **196**, 88 (1999)
10. L. Wang et al., Wurtzite GaN epitaxial growth on a Si(001) substrate using γ-Al_2O_3 as an intermediate layer. Appl. Phys. Lett. **72**, 109 (1998)
11. D.K. Fork, D.B. Fenner, G.A.N. Connell, J.M. Phillips, T.H. Geballe, Epitaxial yttria-stabilized zirconia on hydrogen-terminated Si by pulsed laser deposition. Appl. Phys. Lett. **57**, 1137 (1990)
12. S.J. Wang, C.K. Ong, L.P. You, S.Y. Xu, Epitaxial growth of yttria-stabilized zirconia oxide thin film on natively oxidized silicon wafer without an amorphous layer. Semicond. Sci. Tech. **15**, 836 (2000)
13. S.J. Wang et al., Crystalline zirconia oxide on silicon as alternative gate dielectrics. Appl. Phys. Lett. **78**, 1604 (2001)
14. M. Ihara et al., Vapor phase epitaxial growth of MgO-Al_2O_3. J. Electrochem. Soc. **129**, 2569 (1982)
15. S. Miura et al., Epitaxial Y-Ba-Cu-O films on Si with intermediate layer by rf magnetron sputtering. Appl. Phys. Lett. **53**, 1967 (1988)
16. D.M. Hwang et al., Epitaxial relations between in situ superconducting $YBa_2Cu_3O_{7-x}$ thin films and $BaTiO_3$/$MgAl_2O_4$/Si substrates. J. Appl. Phys. **68**, 1772 (1990)
17. S.A. Chambers, Y. Liang, Z. Yu, R. Droopad, J. Ramdani, Band offset and structure of SrTiO3/Si(001) heterojunctions. J. Vac. Sci. Tech. A **19**, 934 (2001)
18. M. Sousa et al., Optical properties of epitaxial $SrHfO_3$ thin films grown on Si. J. Appl. Phys. **102**, 104103 (2007)
19. R.A. McKee, F.J. Walker, M.F. Chisholm, Crystalline oxides on silicon: the first five mono-layers. Phys. Rev. Lett. **81**, 3014 (1998)
20. H. Li et al., Two-dimensional growth of high-quality strontium titanate thin films on Si. J. Appl. Phys. **93**, 4521 (2003)
21. C. Rossel et al., $SrHfO_3$ as gate dielectric for future CMOS technology. Microelectron. Eng. **84**, 1869 (2007)
22. C. Rossel et al., Field-effect transistors with $SrHfO_3$ as gate oxide. Appl. Phys. Lett. **89**, 053506 (2006)
23. T.F. Wietler et al., Epitaxial growth of Gd_2O_3 on Ge films grown by surfactant-mediated epitaxy on Si(001) substrates. Solid State Electron. **53**, 833 (2009)

24. J. Wang et al., Crystal structure and strain state of molecular beam epitaxial grown Gd_2O_3 on Si(111) substrates: a diffraction study. Semicond. Sci. Tech. **24**, 045021 (2009)
25. M.E. Hunter, M.J. Reed, N.A. El-Masry, J.C. Roberts, S.M. Bedair, Epitaxial Y_2O_3 films grown on Si(111) by pulsed-laser ablation. Appl. Phys. Lett. **76**, 1935 (2000)
26. S. Guha, N.A. Bojarczuk, V. Narayanan, Lattice-matched, epitaxial, silicon-insulating lanthanum yttrium oxide heterostructures. Appl. Phys. Lett. **80**, 766 (2002)
27. T. Schroeder et al., Structure and thickness-dependent lattice parameters of ultrathin epitaxial Pr_2O_3 films on Si(001). Appl. Phys. Lett. **85**, 1229 (2004)
28. H.J. Osten et al., Molecular beam epitaxy of rare-earth oxides, in *Rare earth Oxide Thin Films: Growth, Characterization, and Application*, ed. by M. Fanciulli, G. Scarel (Springer, Berlin, 2005)
29. A. Fissel et al., Interface formation during molecular beam epitaxial growth of neodymium oxide on silicon. J. Appl. Phys. **99**, 074105 (2006)
30. D.O. Klenov, L.F. Edge, D.G. Schlom, S. Stemmer, Extended defects in epitaxial Sc2O3 films grown on (111) Si. Appl. Phys. Lett. **86**, 051901 (2005)
31. B.E. Park, S. Shouriki, E. Tokumitsu, H. Ishiwara, Fabrication of $PbZr_xTi_{1-x}O_3$ films on Si structures using Y_2O_3 buffer layers. Jpn. J. Appl. Phys. **37**, 5145 (1998)
32. H.N. Lee, Y.T. Kim, S.H. Choh, Characteristics of Pt/SrBi2Ta2O9/Y2O3/Si ferroelectric gate capacitors. J. Kor. Phys. Soc. **34**, 454 (1999)
33. S. Imada, S. Shouriki, E. Tokumitsu, H. Ishiwara, Epitaxial growth of ferroelectric YMnO3 thin films on Si (111) substrates by molecular beam epitaxy. Jpn. J. Appl. Phys. **37**, 6497 (1998)
34. V. Narayanan, S. Guha, N.A. Bojarczuk, F.R. Ross, Growth and characterization of epitaxial $Si/(La_xY_{1-x})_2O_3/Si$ heterostructures. J. Appl. Phys. **93**, 251 (2003)
35. A. Giussani, P. Zaumseil, O. Seifarth, P. Storck, T. Schroeder, A novel engineered oxide buffer approach for fully lattice-matched SOI heterostructures. New J. Phys. **12**, 093005 (2010)
36. A. Wilke et al., Complex interface and growth analysis of single crystalline epi-Si(111)/Y_2O_3/Pr_2O_3/Si(111) heterostructures: Strain engineering by oxide buffer control. Surf. Interface Anal. **43**, 827 (2011)
37. L. Tarnawska et al., Single crystalline Sc_2O_3/Y_2O_3 heterostructures as novel engineered buffer approach for GaN integration on Si (111). J. Appl. Phys. **108**, 063502 (2010)
38. W. Guo et al., Epitaxial ZnO films on (111) Si substrates with Sc_2O_3 buffer layers. Appl. Phys. Lett. **94**, 122107 (2009)
39. Y.Y. Gomeniuk et al., Interface and bulk properties of high-k gadolinium and neodymium oxides on silicon, in *Nanoscaled Semiconductor-on-Insulator Materials, Sensors and Devices*, ed. by A.N. Nazarov, J.P. Raskin (TransTech, Zurich, 2011)
40. J.X. Wang et al., Epitaxial multi-component rare-earth oxide: a high-k material with ultralow mismatch to Si. Mater. Lett. **64**, 866 (2010)
41. A. Laha, E. Bugiel, H.J. Osten, A. Fissel, Crystalline ternary rare earth oxide with capacitance equivalent thickness below 1 nm for high-K application. Appl. Phys. Lett. **88**, 172107 (2006)
42. E.J. Tarsa, J.S. Speck, M. Robinson, Pulsed laser deposition of epitaxial silicon/h-Pr_2O_3/silicon heterostructures. Appl. Phys. Lett. **63**, 539 (1993)
43. H.J. Osten, J.P. Liu, E. Bugiel, H.J. Mussig, P. Zaumseil, Epitaxial growth of praseodymium oxide on silicon. Mater. Sci. Eng. B **87**, 297 (2001)
44. M. Yoshimoto et al., Room-temperature epitaxial growth of CeO_2 thin films on Si(111) substrates for fabrication of sharp oxide/silicon interface. Jpn. J. Appl. Phys. **34**, L668 (1995)
45. H. Koinuma, H. Nagata, T. Tsukahara, S. Gonda, M. Yoshimoto, Ceramic layer epitaxy by pulsed laser deposition in an ultrahigh vacuum system. Appl. Phys. Lett. **58**, 2027 (1991)
46. D.K. Fork, D.B. Fenner, T.H. Geballe, Growth of epitaxial PrO_2 thin films on hydrogen-terminated Si(111) by pulsed laser deposition. J. Appl. Phys. **68**, 4316 (1990)
47. J.W. Seo et al., Interface formation and defect structures in epitaxial $La_2Zr_2O_7$ thin films on (111) Si. Appl. Phys. Lett. **83**, 5211 (2003)

48. D. Dimos, C.H. Mueller, Annu. Rev. Mater. Sci. **28**(1), 397 (1998)
49. C.-R. Cho, J.-Y. Hwang, J.-P. Kim, S.-Y. Jeong, S.-G. Yoon, W.-J. Lee, Jpn. J. Appl. Phys. **43**, L1425 (2004)
50. M.D. Losego, L. Fitting Kourkoutis, S. Mita, H.S. Craft, D.A. Muller, R. Collazo, Z. Sitar, J.-P. Maria, J. Cryst. Growth **311**, 1106 (2009)
51. A. Posadas, J.-B. Yau, C.H. Ahn, J. Han, S. Gariglio, K. Johnston, K.M. Rabe, J.B. Neaton, Appl. Phys. Lett. **87**, 171915 (2005)
52. A. Posadas, J.-B. Yau, C.H. Ahn, Phys. Status Solidi B **243**, 2085 (2006)
53. Y. Chye, T. Liu, D. Li, K. Lee, D. Lederman, T.H. Myers, Appl. Phys. Lett. **88**, 132903 (2006)
54. N. Sai, J. Lee, C.J. Fennie, A.A. Demkov, Appl. Phys. Lett. **91**, 202910 (2007)
55. P.J. Hansen, V. Vaithyanathan, Y. Wu, T. Mates, S. Heikman, U.K. Mishra, R.A. York, D.G. Schlom, J.S. Speck, J. Vac. Sci. Tech. B **23**, 499 (2005)
56. W. Tian, V. Vaithyanathan, D.G. Schlom, Q. Zhan, S.Y. Yang, Y.H. Chu, R. Ramesh, Appl. Phys. Lett. **90**, 172908 (2007)
57. S.-H. Lee, T.W. Noh, Integr. Ferroelectr. **20**, 25 (1998)
58. Y. Tsuchiya, A. Kobayashi, J. Ohta, H. Fujioka, M. Oshima, Phys. Status Solidi A **202**, R145 (2005)
59. P.J. Hansen, Y. Terao, Y. Wu, R.A. York, U.K. Mishra, J.S. Speck, J. Vac. Sci. Tech. B **23**, 162 (2005)
60. L. Hao, J. Zhu, Y. Liu, S. Wang, H. Zeng, X. Liao, Y. Liu, H. Lei, Y. Zhang, W. Zhang, Y. Li, Thin Solid Films **520**, 3035 (2012)
61. H.S. Craft, J.F. Ihlefeld, M.D. Losego, R. Collazo, Z. Sitar, J.-P. Maria, Appl. Phys. Lett. **88**, 212906 (2006)
62. V.E. Henrich, P.A. Cox, *The Surface Science of Metal Oxides* (Cambridge University Press, Cambridge, 1994)
63. E.A. Paisley, T.C. Shelton, S. Mita, R. Collazo, H.M. Christen, Z. Sitar, M.D. Biegalski, J.-P. Maria, Appl. Phys. Lett. **101**, 092904 (2012)
64. E.A. Paisley, M.D. Losego, B.E. Gaddy, J.S. Tweedie, R. Collazo, Z. Sitar, D.L. Irving, J.-P. Maria, Nat. Commun. **2**, 461 (2011)
65. M.D. Losego, S. Mita, R. Collazo, Z. Sitar, J.-P. Maria, J. Cryst. Growth **310**, 51 (2008)
66. A. Schmehl, V. Vaithyanathan, A. Herrnberger, S. Thiel, C. Richter, M. Liberati, T. Heeg, M. Röckerath, L.F. Kourkoutis, S. Mühlbauer, P. Böni, D.A. Muller, Y. Barash, J. Schubert, Y. Idzerda, J. Mannhart, D.G. Schlom, Nat. Mater. **6**, 882 (2007)
67. W.A. Doolittle, A.G. Carver, W. Henderson, J. Vac. Sci. Tech. B **23**, 1272 (2005)
68. J.D. Greenlee, W.L. Calley, W. Henderson, W.A. Doolittle, Phys. Status Solidi C **9**, 155 (2012)
69. T.L. Goodrich, Z. Cai, M.D. Losego, J.-P. Maria, K.S. Ziemer, J. Vac. Sci. Tech. B **25**, 1033 (2007)
70. T.L. Goodrich, J. Parisi, Z. Cai, K.S. Ziemer, Appl. Phys. Lett. **90**, 042910 (2007)
71. A. Posadas, F.J. Walker, C.H. Ahn, T.L. Goodrich, Z. Cai, K.S. Ziemer, Appl. Phys. Lett. **92**, 233511 (2008)
72. T.L. Goodrich, Z. Cai, M.D. Losego, J.-P. Maria, L. Fitting Kourkoutis, D.A. Muller, K.S. Ziemer, J. Vac. Sci. Tech. B **26**, 1110 (2008)
73. Z. Cai, T.L. Goodrich, B. Sun, Z. Chen, V.G. Harris, K.S. Ziemer, J. Phys. D: Appl. Phys. **43**, 095002 (2010)
74. T.L. Goodrich, Z. Cai, K.S. Ziemer, Appl. Surf. Sci. **254**, 3191 (2008)
75. Z. Chen, A. Yang, Z. Cai, K. Ziemer, C. Vittoria, V.G. Harris, IEEE Trans. Magn. **42**, 2855 (2006)
76. Z. Chen, Z. Cai, A. Yang, K.S. Ziemer, C. Vittoria, V.G. Harris, J. Appl. Phys. **103**, 07E513 (2008)
77. Z. Chen, A. Yang, S.D. Yoon, K. Ziemer, C. Vittoria, V.G. Harris, J. Magn. Magn. Mater. **301**, 166 (2006)

78. V.K. Lazarov, P.J. Hasnip, Z. Cai, K. Yoshida, K.S. Ziemer, J. Appl. Phys. **111**, 07A515 (2012)
79. Z. Chen, V.G. Harris, J. Appl. Phys. **112**, 081101 (2012)
80. R.A. McKee, F.J. Walker, M.F. Chisholm, Science **293**, 468 (2001)
81. C. Merckling, G. Saint-Girons, C. Botella, G. Hollinger, M. Heyns, J. Dekoster, M. Caymax, Appl. Phys. Lett. **98**, 092901 (2011)
82. B.R. Lukanov, J.W. Reiner, F.J. Walker, C.H. Ahn, E.I. Altman, Phys. Rev. B **84**, 075330 (2011)
83. D.P. Norton, A. Goyal, J.D. Budai, D.K. Christen, D.M. Kroeger, E.D. Specht, Q. He, B. Saffian, M. Paranthaman, C.E. Klabunde, D.F. Lee, B.C. Sales, F.A. List, Science **274**, 755 (1996)
84. D.P. Norton, J.D. Budai, M.F. Chisholm, Appl. Phys. Lett. **76**, 1677 (2000)
85. M. Patel, K. Kim, M. Ivill, J.D. Budai, D.P. Norton, Thin Solid Films **468**, 1 (2004)
86. G. Hollinger, R. Skheyta-Kabbani, M. Gendry, Phys. Rev. B **49**, 11159 (1994)
87. P.D. Kirchner, J.M. Woodall, J.L. Freeouf, G.D. Pettit, Appl. Phys. Lett. **38**, 427 (1981)
88. K. Nashimoto, D.K. Fork, T.H. Geballe, Appl. Phys. Lett. **60**, 1199 (1992)
89. L.S. Hung, L.R. Zheng, T.N. Blanton, Appl. Phys. Lett. **60**, 3129 (1992)
90. E.J. Tarsa, M. De Graef, D.R. Clarke, A.C. Gossard, J.S. Speck, J. Appl. Phys. **73**, 3276 (1993)
91. Y. Liang, J. Kulik, T.C. Eschrich, R. Droopad, Z. Yu, P. Maniar, Appl. Phys. Lett. **85**, 1217 (2004)
92. R. Contreras-Guerrero, J.P. Veazey, J. Levy, R. Droopad, Appl. Phys. Lett. **102**, 012907 (2013)
93. M. Ivill, M. Patel, K. Kim, H. Bae, S.J. Pearton, D.P. Norton, J.D. Budai, Appl. Phys. Mater. Sci. Process. **75**, 699 (2002)
94. E. Vasco, L. Vázquez, M. Aguiló, C. Zaldo, J. Cryst. Growth **209**, 883 (2000)
95. E. Vasco, C. Polop, C. Coya, A. Kling, C. Zaldo, Appl. Surf. Sci. **208–209**, 512 (2003)
96. K. Eisenbeiser, R. Emrick, R. Droopad, Z. Yu, J. Finder, S. Rockwell, J. Holmes, C. Overgaard, W. Ooms, IEEE Electron. Dev. Lett. **23**, 300 (2002)
97. A.A. Demkov, H. Seo, X. Zhang, J. Ramdani, Appl. Phys. Lett. **100**, 071602 (2012)
98. L. Largeau, J. Cheng, P. Regreny, G. Patriarche, A. Benamrouche, Y. Robach, M. Gendry, G. Hollinger, G. Saint-Girons, Crystal orientation of GaAs islands grown on $SrTiO_3(001)$ by molecular beam epitaxy. Appl. Phys. Lett. **95**, 011907 (2009)
99. J. Cheng, A. Chettaoui, J. Penuelas, B. Gobaut, P. Regreny, A. Benamrouche, Y. Robach, G. Hollinger, G. Saint-Girons, Partial arsenic pressure and crystal orientation during the molecular beam epitaxy of GaAs on $SrTiO_3(001)$. J. Appl. Phys. **107**, 094902 (2010)
100. A. Fissel, D. Kühne, E. Bugiel, H.J. Osten, Appl. Phys. Lett. **88**, 153105 (2006)
101. R. Dargis, A. Fissel, D. Schwendt, E. Bugiel, J. Krügener, T. Wietler, A. Laha, H.J. Osten, Vacuum **85**, 523 (2010)
102. R. Dargis, E. Arkun, A. Clark, R. Roucka, R. Smith, D. Williams, M. Lebby, A.A. Demkov, J. Vac. Sci. Tech. B **30**, 02B110 (2012)
103. A. Laha, E. Bugiel, R. Dargis, D. Schwendt, M. Badylevich, V.V. Afanas'ev, A. Stesmans, A. Fissel, H.J. Osten, Microelectron. J. **40**, 633 (2009)
104. A. Laha, E. Bugiel, A. Fissel, H.J. Osten, Microelectron. Eng. **85**, 2350 (2008)
105. R. Dargis, A. Clark, E. Arkun, R. Roucka, D. Williams, R. Smith, M. Lebby, ECS J. Solid. State. Sci. Tech. **1**, P246 (2012)
106. J. Cheng, P. Regreny, L. Largeau, G. Patriarche, O. Mauguin, K. Naji, G. Hollinger, G. Saint-Girons, Influence of the surface reconstruction on the growth of InP on $SrTiO_3(001)$. J. Cryst. Growth **311**, 1042 (2009)
107. K. Johnston, M.R. Castell, A.T. Paxton, M.W. Finnis, Phys. Rev. B **70**, 085415 (2004)
108. J. Cheng, T. Aviles, A. El-Akra, C. Bru-Chevallier, L. Largeau, G. Patriarche, P. Regreny, A. Benamrouche, Y. Robach, G. Hollinger, G. Saint-Girons, Optically active defects in an

InAsP/InP quantum well monolithically grown on SrTiO3/Si(001). Appl. Phys. Lett. **95**, 232116 (2009)

109. G. Saint-Girons, C. Priester, P. Regreny, G. Patriarche, L. Largeau, V. Favre-Nicolin, G. Xu, Y. Robach, M. Gendry, G. Hollinger, Spontaneous compliance of the InP/SrTiO$_3$ heterointerface. Appl. Phys. Lett. **92**, 241907 (2008)

110. B. Gobaut, J. Penuelas, J. Cheng, A. Chettaoui, L. Largeau, G. Hollinger, G. Saint-Girons, Direct growth of InAsP/InP quantum well heterostructures on Si using crystalline SrTiO$_3$/Si templates. Appl. Phys. Lett. **97**, 201908 (2010)

111. M. El Kazzi, B. Gobaut, J. Penuelas, G. Grenet, M.G. Silly, F. Sirotti, G. Saint-Girons, Ge/SrTiO$_3$(001) interface probed by soft x-ray synchrotron-radiation time-resolved photo-emission. Phys. Rev. B **85**, 075317 (2012)

112. B. Gobaut, J. Penuelas, G. Grenet, D. Ferrah, A. Benamrouche, A. Chettaoui, Y. Robach, C. Botella, M. El Kazzi, M.G. Silly, F. Sirotti, G. Saint-Girons, Ge/SrTiO$_3$(001): correlation between interface chemistry and crystallographic orientation. J. Appl. Phys. **112**, 093508 (2012)

113. J.W. Seo, C. Dieker, A. Tapponnier, C. Marchiori, M. Sousa, J.-P. Locquet, J. Fompeyrine, A. Ispas, C. Rossel, Y. Panayiotatos, A. Sotiropoulos, A. Dimoulas, Microelectron. Eng. **84**, 2328 (2007)

114. G. Saint-Girons, P. Regreny, L. Largeau, G. Patriarche, G. Hollinger, Monolithic integration of InP based heterostructures on silicon using crystalline Gd$_2$O$_3$ buffers. Appl. Phys. Lett. **91**, 241912 (2007)

Chapter 9
Outlook and Parting Thoughts

We think that the current state of oxide electronics is similar to where Si was before Jack Kilby came up with large scale integration. Readily available semiconductor substrates offer an excellent integration platform, as well as an avenue to make hybrid logic/sensor devices. Because many of the properties of functional oxides are extremely sensitive to compositional and structural changes (e.g. atomic levels), the functionality of these hybrid devices can be tuned as never before. Indeed, interactions between functional oxide layers are proving to provide unexpected enhancements in device functionality. Thus an important key in enabling brand new oxide-based technologies is the utilization of semiconductor/oxide epitaxy.

To date, epitaxy offers the best way to control the integration. The problem however, is that this is not your "father's heteroepitaxy"! In addition to the usual thermal and lattice mismatch challenges, functional oxide/semiconductor heteroepitaxy adds the complexity of joining covalent systems to ionic ones, chemical interactions at interfaces, and multi-element materials that are sensitive to atomic-level compositional and structural changes. Because of these challenges this is a very exciting scientific problem of fundamental importance. Undoubtedly, new discoveries will be made in this field. We believe the field will be growing over the coming decade, so many things will change, the preconceived notions will tumble, and new paradigms will be proposed. However, it is important to have a sense of direction and this book is meant to give impetus to large scale oxide integration.

Molecular beam epitaxy (MBE), with its atomic layer control of deposition is an excellent tool for scientific discovery, and we expect many new MBE-grown oxide/semiconductor systems to be synthesized and new physics to be discovered. However, to make a serious impact at the manufacturing level, chemical methods of deposition are of paramount importance. The use of hybrid deposition methods such as the one developed for $SrTiO_3$ [1] is expected to become more important in the near future. As described in more detail in Chap. 4, this hybrid method utilizes a metal-organic precursor for evaporating low vapor pressure titanium allowing one to grow $SrTiO_3$ five times faster than with conventional MBE. We expect the development of new precursors that are tailored for this hybrid metal-organic

A.A. Demkov and A.B. Posadas, *Integration of Functional Oxides with Semiconductors*, DOI 10.1007/978-1-4614-9320-4_9, © The Author(s) 2014

MBE method which will allow high quality growth at higher rates. For certain applications that require thick films on the order of several hundred nanometers or even several micrometers, particularly those for microwave applications involving ferrites and optical waveguide applications using non-linear optical materials, even higher growth rates will be required to be commercially viable. Currently, very high growth rates of several hundred micrometer per hour can be achieved only using metal-organic chemical vapor deposition (MOCVD). However, MOCVD does not have the necessary atomic layer and oxidation control necessary to deposit epitaxial oxides directly on semiconductors. We envision the development of combination techniques where a critical seed layer of an epitaxial complex oxide is grown directly on a semiconductor using MBE, followed by the growth of thicker films of the same or some other oxide using MOCVD. This need for combination methods will likely lead to the development of even more facilities where both MBE and MOCVD are connected in situ. Finally, as the cost of the hardware and maintenance of MBE systems is often prohibitive from the manufacturing stand point, we also expect researchers will develop processes and precursors that will allow for the initial seed layer of the complex oxide itself to be deposited directly using a chemical method. Currently, the only chemical method with atomic layer control of deposition is atomic layer deposition (ALD). Furthermore, ALD is a low temperature process making it conducive for depositing oxides while preventing oxidation of the semiconductor. Currently, ALD is used mainly for deposition of amorphous high-k dielectric materials for the semiconductor industry. We expect that in the near future, ALD will be developed for direct epitaxy of complex oxides on semiconductors. Because ALD and MOCVD can often be performed in the same reactor, such a combination will likely be the future for manufacturable complex oxide on semiconductor epitaxial systems.

As improvements are made in oxide on semiconductor epitaxy the question becomes, what can we do with these oxides on semiconductor systems? The electronic properties of complex oxides by themselves have been the subject of extensive research in a field known as *oxide electronics*. It is probably worth mentioning that the term oxide electronics means different things to different people as oxides find multiple applications. In a seminal 2008 review, Ramesh and Schlom posed an important question "whither oxide electronics?" [2]. In this skillfully written article they followed the trajectory of the fundamental oxide research from Mott, von Hippel and Goodenough to Müller and Bednorz and their discovery of high temperature superconductivity in cuprate oxides. It was the promise of potential technological applications of high temperature superconductivity that resulted in the explosive development of physical deposition methods specifically tailored for oxide growth, namely pulsed laser deposition and molecular beam epitaxy which make the atomic scale engineering of oxide materials and heterostructures possible. Despite the promising applications, it would be a mistake to overlook the fact that more than 20 years after the discovery of high temperature superconductivity, the promising devices such as next-generation Josephson junctions for high-speed, low-power computing, sensitive magnetic field sensors, and high frequency microwave filters are not yet a reality.

There have been many successes, however, utilizing oxides in the area of non-volatile memory. Ferroelectric random access memories (FRAM) are now commercially available offering significantly reduced power consumption and could, in principle, become the "universal memory" of the future. One possible lesson that can be learned from this is that even though oxide materials can be made conductive and even superconductive they are at their best when the intrinsic state of the oxide can be directly exploited as in ferroelectrics.

Ramesh and Schlom also suggested that complex oxides will play a key role in the rapidly emerging area of materials for energy technologies [2]. Some potential areas of application include photovoltaics [3, 4], solar water splitting [5, 6], solar CO_2 capture (artificial photosynthesis) [7], supercapacitors [8], metal-air battery technology [9], solid oxide fuel cells [10], thermoelectric generators [11], and solid state gas sensors [12]. One can envision creating oxide heterostructures and nanostructures integrated with semiconductors that will enable the decoupling of electron and phonon transport, of particular relevance to the design of next-generation thermoelectrics. Another area of research is the conversion of photons to electrons, critical to photovoltaic solar energy conversion, photocatalysis for water splitting and CO_2 conversion to usable fuel. The multitude of exciting opportunities for oxide heterostructures is expected to continue to inspire future generations of researchers in this field for a long time.

In the remainder of the book we discuss particular applications where we feel crystalline epitaxial oxides on semiconductors are likely to have a major impact.

9.1   Oxide Electronics

A field where highly integrated oxides are likely to find applications is in electronic devices exploiting materials with variable internal states based on ion diffusion, filamentary conduction, ferroelectricity, and ferromagnetism, as diagrammed in Fig. 9.1. Such materials have been termed by Ha and Ramanathan as *adaptive oxides* [13]. Adaptive oxide electronic devices are defined as systems that can learn and adapt to various inputs. Usually, if one stays within the realm of Boolean computation, this requires a very complex algorithm and high speed computing. On the other hand, the human brain offers an alternative paradigm where instead of simple but fast switches, fairly slow complex elements are combined in a highly interconnected network. Far reaching goals of adaptive electronics include fabrication of devices that mimic human brain functionality: the strengthening and weakening of synapses emulated by electrically, magnetically, thermally, or optically tunable properties of materials [13].

Oxides may indeed have certain advantages offering novel functionalities, possible compatibility with semiconductor processing, long retention times, fast switching and scalability. Clearly, to achieve such an ambitious program, a large number of oxide devices needs to be integrated and interconnected. Crystalline oxides on semiconductors offer an excellent integration platform. Redox based

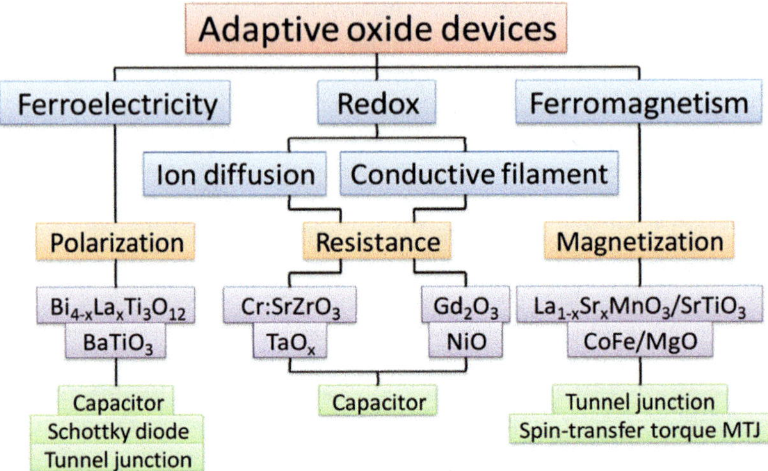

Fig. 9.1 Diagram of possible adaptive oxide devices. The levels of the diagram are switching mechanism (second), internal state (third), representative oxides (fourth), and device structures (fifth). Reprinted with permission from [13]. Copyright 2011, AIP Publishing LLC

functionality has been exploited in so-called memristors or resistive switching devices [14]. In these devices, the oxide can be switched between a high resistance state and a low resistance state by the application of a voltage. Ferromagnetic and multiferroic oxides are also being looked at for the injection and manipulation of spin-polarized currents in a semiconductor channel. This field is more widely known as spintronics and the ability to directly interface such materials with semiconductors is expected to tremendously benefit spintronics devices [15].

Ferroelectric functionality is already being exploited for commercial non-volatile memory devices. Among existing non-volatile memory technologies, ferroelectric memory offers the lowest power consumption, and, at least in principle, may be the "universal" memory of the future, replacing dynamic random access memory (DRAM), electrically erasable programmable read only memory (EEPROM) and flash memory [16]. Thin film ferroelectric devices on semiconductors often suffer from fatigue and imprint caused by the complex interplay between the film's microstructure (in the usual sense), its domain structure and point defects such as oxygen vacancies. The problem of fatigue was initially solved by using conductive oxides as electrodes [17], and later by choosing a different ferroelectric that showed no fatigue even with Pt electrodes [18]. In 1998 Fujitsu was the first company in the industry to introduce embedded ferroelectric random access memory (FRAM) into CMOS logic and to release production quantities a year later. Today it offers a variety of FRAM products from 256K to 4M based on lead-zirconate-titanate (PZT) perovskites that boast of being 30,000 times faster than EEPROM while offering a million times higher endurance and 200 times lower power consumption. In 2012 Texas Instruments brought to market the RF430FRL152H device (also based on PZT) that can be used as a sensor for near

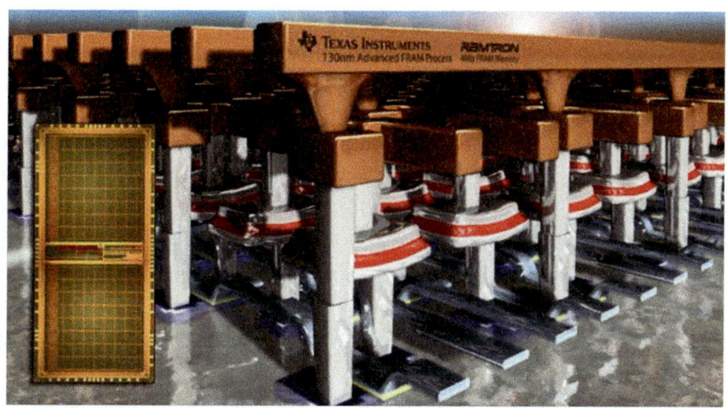

Fig. 9.2 Schematic of the architecture of a Texas Instruments 4-Mb ferroelectric random access memory device (Reproduced with permission from [2])

field communications [19]. The artist's rendition of this architecture is shown in Fig. 9.2. This success did not come cheap, however. In 1999 the Ministry of International Trade and Industry of Japan funded a five-year project titled "Research and Development of Next Generation Ferroelectric Memories" with a total budget of 18 million dollars [16, 20].

Another area of oxide electronics is with the utilization of the phenomenon of metal-insulator transitions that occur in many transition metal oxides. A transformation in crystalline materials from a dielectric to a metal, known as the metal-to-insulator transition (MIT), has been known for many years [21]. One of the best known examples is vanadium oxide. Vanadium can assume several oxidation states and forms several oxides, eight of which exhibit MIT [22]. While in V_2O_3 the transition occurs at 150 K, in VO_2 it is observed at 340 K. In the latter case, more attractive from the application point of view, the conductivity changes by five orders of magnitude, while in the former case it changes by ten!

There are three different types of MITs. In some transitions atomic displacement leads to splitting of the conduction band, while others are purely electronic in nature and the atomic lattice is fixed. MITs caused by electron correlation are known as Mott-Hubbard transitions, and those caused by electron localization induced by disorder are referred to as Anderson transitions. The Peierls transition originates from the change of a single electron wave function, especially around the Fermi surface, caused by the electron-lattice interaction. In some materials, a switch from the metallic or insulating behavior can be controlled by external parameters, such as an electric or magnetic field or homogeneous and inhomogeneous stress, thus making them attractive for possible device applications. An ultrafast switch has been demonstrated utilizing a metal-insulator transition in correlated oxides [23]. The ON and OFF states in this case, are defined as a low-resistance, metallic phase and a high-resistance, insulating phase of the material. In particular, VO_2, where the MIT is believed to be of the Mott type, has been extensively studied [23].

The transition is fast, indeed, occurring on a picosecond time scale. One possible device architecture is a so-called MottFET where the Mott insulator is used as a channel and is switched by the gate voltage. Recent experiments of Ruzmetov et al. have suggested reversible modulation of the VO_2 channel resistance with gate voltage in a three terminal device [24]. In a two-terminal device, due to the current heating of the sample, it is often hard to say whether it is the temperature or the field that drives the transition. A three terminal device offers a convenient platform that allows the separation of these two mechanisms. Overall, the best description of the transistor performance was "the application of the gate voltage caused nontrivial response of the I-V characteristics for a number of studied devices." The best devices were grown on a Si substrate that was used as a back gate and with e-beam evaporated SiO_2 as a gate dielectric. The authors claim that the channel resistance was electrically modulated; however, the exact mechanism remains unclear. This highlights a fundamental difficulty of using a transition metal oxide for a channel material.

A much more promising direction for using MIT in oxides appears to be optical applications. First of all, the MIT can be optically induced. Indeed, optical detectors, sensors, switches and modulators have already been demonstrated [23]. Another exciting avenue for using the MIT in oxides is offered by combining them with metamaterials. For example, using lithography, Driscoll and co-workers fabricated 100 nm thick gold split-ring resonators (SRR) with a 20-μm period on top of a 90-nm-thick VO_2 thin film on a sapphire substrate [25]. The resonance frequency of the SRR metamaterial is highly sensitive to the dielectric properties of the material placed nearby, especially in the vicinity of the SRR gaps, and near the MIT phase transition, VO_2 exhibits a divergent *bulk* permittivity. This modifies the local fields of the SRR within and around the gap region, acting like a tunable dielectric inside a capacitor. This hybrid split-ring resonator VO_2 device demonstrated resonance tuning range of 20 % at microwave frequencies using integrated rf electrical components. Cui et al. have reported VO_2 growth on STO and TiO_2 [26], potentially offering a way to integrate these hybrid devices on Si using an insulating buffer as both STO and TiO_2 can be monolithically integrated on Si (001).

9.2   Integrated Ferrites for rf Applications

Another promising area of applications is high frequency, high power wireless communications. Integration efforts in rf technology go back to the 1980s when microwave monolithic integrated circuits based on GaAs were introduces in radar technology [27]. Ultimately, one would want to integrate active elements such as amplifiers and passive devices, such as circulators, isolators, phase shifters, and delay lines on a single chip. A fully monolithic integrated circuit (MIC) would have both passive and active components and their interconnections fabricated on a common substrate, preferably a semiconductor material. For a long time ferrites (magnetic oxides with spinel, garnet or hexaferrite structure) have been used in bulk

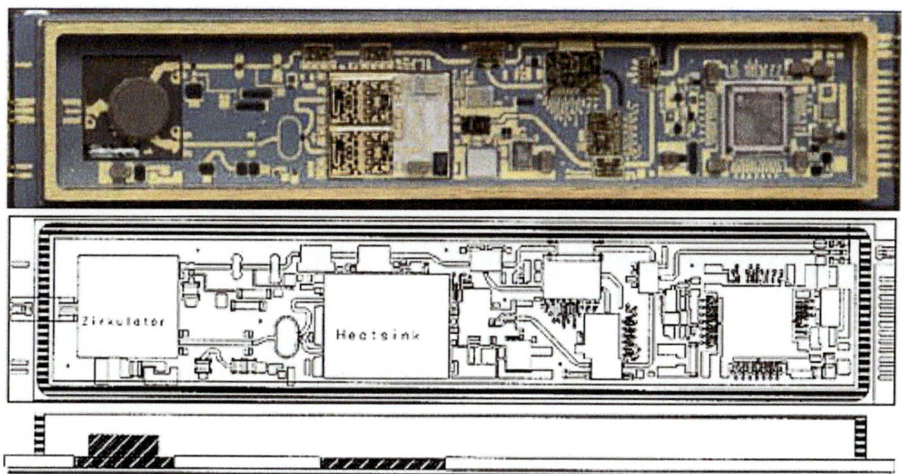

Fig. 9.3 Top view photograph of an Eurofighter Typhoon GaAs-based T/R module for X-band operation. Reprinted with permission from [27]. Copyright 2012, AIP Publishing LLC

passive rf elements. Therefore, one way of making a MIC is to integrate thin film ferrites with semiconductors such as GaAs for the amplifiers and Si for CMOS. Though until now there have been no ferrite film-based integrated microwave devices in either commercial or military markets, much progress has been made. An excellent review of current progress in this field can be found in an article by Chen and Harris [27]. To illustrate the opportunities and challenges in this field we will follow their lead. Consider an X-band (8.0–12.0 GHz) transmit/receive (T/R) module used in the European Union Typhoon-class jet fighters' active phased array radar shown in Fig. 9.3. The T/R module includes passive devices such as circulators, isolators, phase shifters, delay lines, and microwave MICs such as low noise amplifiers, high power amplifiers, CMOS devices and heat sinks. All elements are fabricated on different substrates. This module is 64.5 mm long, 13.5 mm wide, and 4.5 mm in height. Each phased array has potentially thousands of T/R modules. Making the modules smaller, lighter and cheaper to fabricate, while simultaneously enhancing performance and increasing reliability is highly desirable.

Ferrite films have been integrated with Si and GaAs. The main difficulty is the high temperature and oxygen pressure necessary for ferrite processing. Integration of ferrites on SiC and particularly GaN appear to be more promising. The attraction comes from eliminating the need to have GaAs power amplifiers, as GaN and SiC based devices are expected to become commercially available. Thus an entire microwave system can be fully integrated on a single substrate which will be a true breakthrough. A major obstacle in the realization of this program is the large thicknesses of ferrite films used in these devices—often tens of microns thick. Physical vapor deposition methods are ill-suited for these thicknesses so the development of hybrid or purely chemical deposition methods to grow magnetic oxides will be of great value in this field.

The fabrication of film-based ferrite devices on semiconductor substrates is an active area of research and development. At present it appears that wide band gap semiconductors, such as SiC and GaN, have an edge. However, as more insulating magnetic oxides (other than ferrites) are integrated with semiconductors this may very well change.

9.3   Integration of Compound Semiconductors

In the next few years, compound semiconductors may merge with the mainstream CMOS technology since their transport properties are by far superior to those of Si, and, in addition, they offer new functionalities. The ability to tailor compound semiconductors and to integrate them onto foreign substrates can lead to superior or novel functionalities with a potential impact on various areas in electronics, opto-electronics, spintronics, biosensing, and photovoltaics.

The heterogeneous integration of compound semiconductors (CSs) with Si by wafer bonding and thin-layer transfer is a fast developing area of research. The possibility of incorporating CS-based devices into traditional Si technologies has sparked a surge of interest motivated by the novel and improved functionalities potentially achievable by this heterointegration [28]. In this landscape, the much higher charge carrier mobility (as compared with that of Si) and the efficient emission of light by some CSs due to their direct band gap (as compared with the indirect band gap of Si) have been the two major driving forces in the development of heterogeneous devices. Oxide buffer layers offer a unique way to integrate disparate materials. Though the early attempts to integrate GaAs on Si (001) *via* the STO buffer were not successful, they were not a total failure either. In Fig. 9.4 we show GaAs integrated on Si (001) using a buffer layer of STO [29]. As we

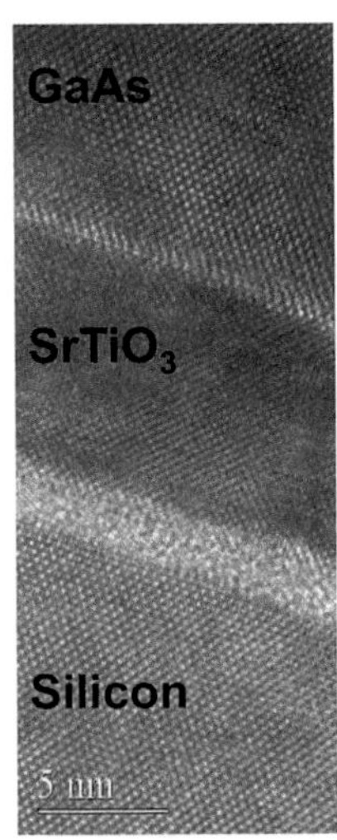

Fig. 9.4 GaAs integrated on Si using epitaxial oxide buffer layer developed by Motorola. Image courtesy of Jamal Ramdani

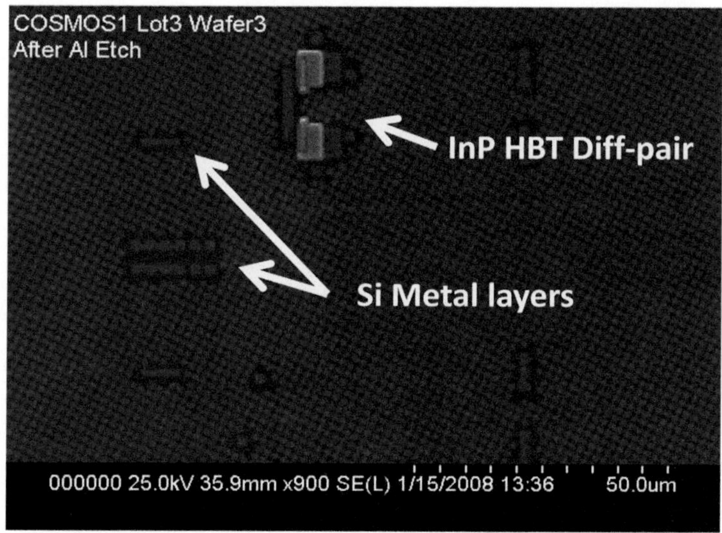

Fig. 9.5 SEM of small area HBT's fabricated on material BCB bonded to silicon CMOS. Reproduced from [31] by permission of ECS - The Electrochemical Society

develop understanding of how to control the interface energy, growth of semiconductors on oxides will be developed. Semiconductor growth on oxides has already been demonstrated for elemental cubic semiconductors with (111) orientation and GaN [30].

The promise of revolutionary advances in science, devices, circuits, and systems is driving the development of monolithic heterogeneous integration processes that combine the performance of compound semiconductor transistors with the complexity of silicon-based integrated circuits. The Defense Advanced Research Projects Agency (DARPA) CoSMOS program advances 3D integration initiatives of mixed semiconductor materials by developing methods to tightly integrate compound semiconductor technologies with state-of-the-art silicon CMOS circuits to achieve unprecedented circuit performance levels. Recently, Patterson et al. of HRL Laboratories have reported the technology for intimate integration of silicon complementary metal-oxide semiconductor (CMOS) devices with 400 GHz InP heterogeneous bipolar transistors (HBTs) to form complex integrated circuits [31]. Using a die to wafer bonding process that preserves the growth orientation of the epitaxial layers bonded to the target substrate, they fabricated large area HBT's on ~1 μm thick InP epitaxial layers transferred to silicon substrates. The scanning electron micrograph shown in Fig 9.5 shows a pair of 0.25 μm emitter HBTs used in a differential amplifier design that incorporates both InP and CMOS transistors.

Herrick et al. used a more direct growth approach to integrating compound semiconductors (CS) and silicon CMOS [32]. They used a unique silicon template

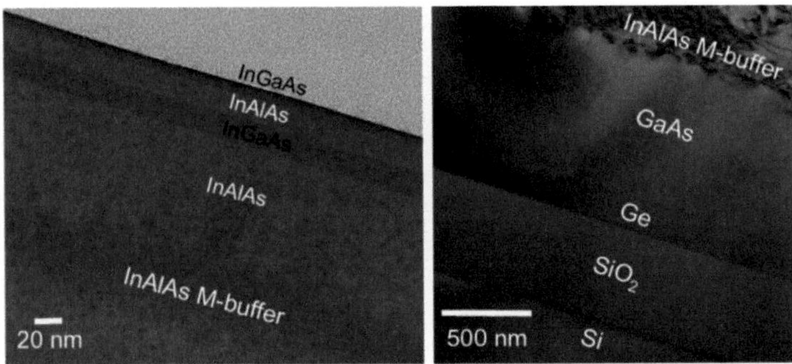

Fig. 9.6 TEM image of InP-HBT grown on GeOI/Si substrate. Reproduced from [32] by permission of ECS - The Electrochemical Society

wafer with an embedded CS template layer of germanium. This enables selective placement of CS devices in arbitrary locations on a silicon CMOS wafer for simple, high yield, monolithic integration and optimal circuit performance. They demonstrated small-area InP HBTs fabricated on a germanium-on-insulator (GeOI) substrate, and an InP-based HBT fabricated on a silicon wafer as shown in Fig. 9.6. HBTs demonstrated a peak current gain cutoff frequency of over 200 GHz. This is an important step in direct growth integration of CS devices with silicon CMOS.

9.4   Photonics

In 2012, IBM announced the achievement of optical components at the 90 nm scale that can be manufactured using standard techniques and incorporated into conventional chips. A colored SEM image of the integrated device is shown in Fig. 9.7 [33]. The following year Intel announced technology to transmit data at speeds of 100 Gb per second along a cable approximately 5 mm in diameter for connecting servers inside data centers. There are no doubts that silicon photonics, a hybrid technology combining semiconductor logic with fast broadband optical communications, has arrived. Because the linear electro-optical effect of ferroelectric oxides is absent in bulk silicon, the development of silicon-based photonics can benefit greatly by using ferroelectric oxides as a novel way to seamlessly connect the electrical and optical domain.

Monolithic integration of optical components on logic devices using crystalline epitaxial oxides on semiconductors offers a promising direction of further development. As we have discussed in Chap. 7, of all perovskite oxides that have been integrated with Si, barium titanate exhibits the largest linear electro-optical coefficients. The next step is to develop optical devices [34] fully integrated on Si(001). We expect that the integration of electro-optical active films on silicon will pave the way towards power-efficient, ultra-compact integrated devices, such as modulators,

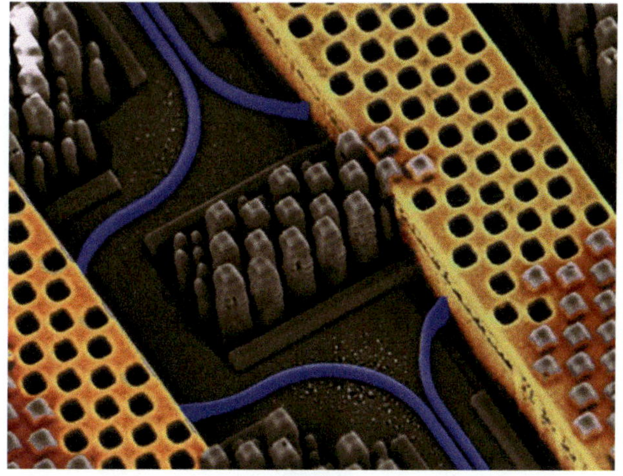

Fig. 9.7 A portion of an IBM chip showing *blue optical waveguides* transmitting high-speed optical signals and *yellow copper wires* carrying high-speed electrical signals. Image taken from [33], with permission

tuning elements and bistable switches. This may bring about the photonics technology revolution—a new approach using light (photons) to move huge amounts of data, at very high speeds, with extremely low power consumption, over a thin optical fiber, rather than using electrical signals over a copper cable.

9.5 Heterogeneous Integration

High-performance and highly integrated systems are not limited to electronic, optical or mechanical systems separately. More and more often, all three of these areas need to be integrated on a single platform. This is known as heterogeneous integration. A modern car offers a convenient example where one would want a hybrid integration technology of CMOS, microelectromechanical systems (MEMS), and photonics circuit devices. In Fig. 9.8 we show a schematic of the safety system of a modern vehicle. It has two types of radars, multiple sensors, local area network (LAN), microprocessors and control units that ensure seamless on-board communications and awareness of the vehicle's surroundings to prevent accidents. All of this needs to be done in real time. In 2010 Lee et al. published a fascinating article describing the 3-D integration technology that combines CMOS large-scale integration (LSI) chips, such as processor, memory, logic, analog and power integrated circuits (ICs), MEMS, and photonics devices into heterogeneous, optoelectromechanical integrated systems [35].

The 3-D integration technology can provide many benefits, namely, increased performance, increased data bandwidth, reduced power, a small form factor, reduced packaging volume, increased yield, and reduced overall costs. The heterogeneous system integration involving CMOS, MEMS, and photonics circuits is attractive because of the promise of high functionality, high-speed communication,

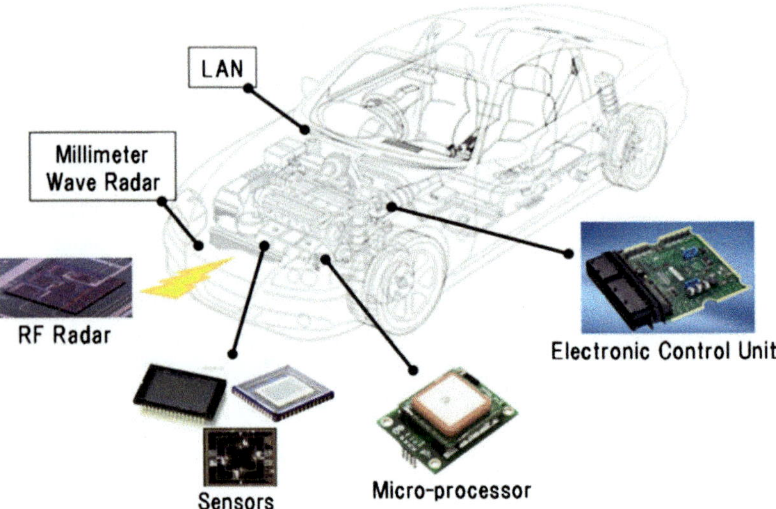

Fig. 9.8 Example of the latest safety-vehicle electronics systems. Reproduced from [35] with permission from IEEE

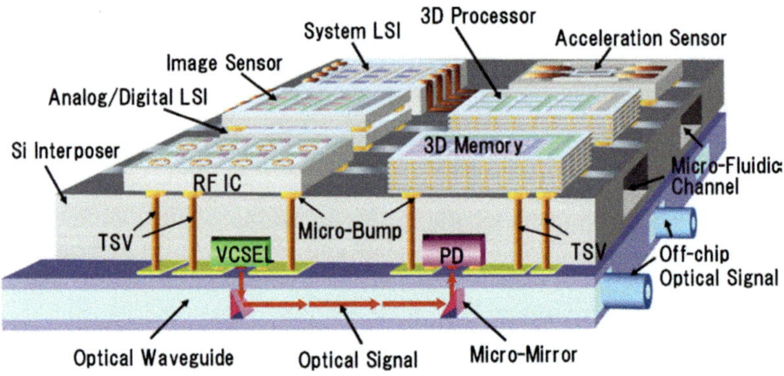

Fig. 9.9 Conceptual structure of the 3-D heterogeneous optoelectronic integrated system-on-silicon for an intelligent vehicle system's variable signal-processing functions depending on the moving speed of the car. Reproduced from [35] with permission from IEEE

and reduced power consumption. Of course, such integrated systems may have many applications and the intelligent vehicle system is only one of them. Among the main challenges of the safety-vehicle electronic systems are a limitation of high-speed signal sensing and data transmission networking, and large power consumption due to the large size and long distance between the components. To achieve a compact, intelligent vehicle system, Lee and co-workers proposed a 3-D heterogeneous optoelectronic integrated system-on-silicon, as shown in Fig. 9.9.

Multiple elements of their design can be achieved with crystalline epitaxial oxides on semiconductors. Optical signals can be handled via the above mentioned silicon photonics aided by integrated optical devices. The integrated oxide MEMS technology is in its infancy, but it can likely be used to develop integrated accelerometers. Chemical sensors can also be integrated using this technology.

In conclusion, we have described the basic physical principles of the oxide/semiconductor epitaxy and a view of the current state of the field. We hope to have shown how this technology can enable large-scale integration of oxide electronic and photonic devices, including possible hybrid semiconductor/oxide systems. There is incredible potential in the realization of multifunctional devices and monolithic heterogeneous integration of materials and devices with a multitude of exciting opportunities for oxides for decades to come.

References

1. B. Jalan, P. Moetakef, S. Stemmer, Appl. Phys. Lett. **95**, 032906 (2009)
2. R. Ramesh, D.G. Schlom, MRS Bull. **33**, 1006 (2008)
3. K.X. Jin, Y.F. Li, Z.L. Wang, H.Y. Peng, W.N. Lin, A.K.K. Kyaw, Y.L. Jin, K.J. Jin, X.W. Sun, C. Soci, T. Wu, AIP Adv. **2**, 042131 (2012)
4. H. Liu, K. Zhao, N. Zhou, H. Lu, M. He, Y. Huang, K.-J. Jin, Y. Zhou, G. Yang, S. Zhao, A. Wang, W. Leng, Appl. Phys. Lett. **93**, 171911 (2008)
5. N.C. Strandwitz, D.J. Comstock, R.L. Grimm, A.C. Nichols-Nielander, J. Elam, N.S. Lewis, J. Phys. Chem. C **117**, 4931 (2013)
6. D.V. Esposito, I. Levin, T.P. Moffat, A.A. Talin, Nat. Mater. **12**, 562 (2013)
7. A.D. Handoko, K. Li, J. Tang, Curr. Opin. Chem. Eng. **2**, 200 (2013)
8. M. Wohlfahrt-Mehrens, J. Schenk, P. Wilde, E. Abdelmula, P. Axmann, J. Garche, J. Power Sources **105**, 182 (2002)
9. F. Cheng, J. Chen, Chem. Soc. Rev. **41**, 2172 (2012)
10. A.J. Jacobson, Chem. Mater. **22**, 660 (2010)
11. A. Weidenkaff, R. Robert, M. Aguirre, L. Bocher, T. Lippert, S. Canulescu, Renew. Energy **33**, 342 (2008)
12. J.W. Fergus, Sensor Actuator B Chem. **123**, 1169 (2007)
13. S.D. Ha, S. Ramanathan, J. Appl. Phys. **110**, 071101 (2011)
14. D.S. Jeong, R. Thomas, R.S. Katiyar, J.F. Scott, H. Kohlstedt, A. Petraru, C.S. Hwang, Rep. Prog. Phys. **75**, 076502 (2012)
15. S.D. Bader, S.S.P. Parkin, Annu. Rev. Cond. Matter Phys. **1**, 71 (2010)
16. H. Ishiwara, J. Semicond. Tech. Sci. **1**, 1 (2001)
17. H.N. Al-Shareef, A.I. Kingon, X. Chen, K.R. Bellur, O. Auciello, J. Mater. Res. **9**, 2968 (1994)
18. C.A. Paz de Araujo, J.D. Cuchiaro, L.D. McMillan, M.C. Scott, J.F. Scott, Nature **374**, 627 (1995)
19. Mena RA, Kozitsky A (2005) Texas instruments white paper <http://www.ti.com/lit/wp/sloy005/sloy005.pdf>
20. H. Ishiwara, Int. J. High. Speed. Electron. Syst. **12**, 315 (2002)
21. N. Mott, *Metal-Insulator Transitions* (Taylor & Francis, London, 1997)
22. F.A. Chudnovsky, Sov. Phys. Tech. Phys. **20**, 999 (1976)
23. Z. Yang, C. Ko, S. Ramanathan, Annu. Rev. Mater. Res. **41**, 337 (2011)
24. D. Ruzmetov, G. Gopalakrishnan, C. Ko, V. Narayanamurti, S. Ramanathan, J. Appl. Phys. **107**, 114516 (2010)

25. T. Driscoll, S. Palit, M.M. Qazilbash, M. Brehm, F. Keilmann et al., Appl. Phys. Lett. **93**, 024101 (2008)
26. Y. Cui, X. Wang, Y. Zhou, R. Gordon, S. Ramanathan, J. Cryst. Growth **338**, 96 (2012)
27. Z. Chen, V.G. Harris, J. Appl. Phys. **112**, 081101 (2012)
28. O. Moutanabbir, U. Gösele, Annu. Rev. Mater. Res. **40**, 469 (2010)
29. K. Eisenbeiser, R. Emrick, R. Droopad, Z. Yu, J. Finder, S. Rockwell, J. Holmes, C. Overgaard, W. Ooms, IEEE Electron. Dev. Lett. **23**, 300 (2002)
30. F.E. Arkun, M. Lebby, R. Dargis, R. Roucka, R.S. Smith, A. Clark, ECS Trans. **50**, 1065 (2013)
31. P.R. Patterson, K. Elliott, J.C. Li, Y. Royter, T. Hussain, ECS Trans. **16**, 221 (2008)
32. K.J. Herrick, T.E. Kazior, J. Laroche, A.W.K. Liu, D. Lubyshev et al., ECS Trans. **16**, 227 (2008)
33. Silicon Integrated Nanophotonics (2012) IBM Research <http://researcher.ibm.com/researcher/view_project.php?id=2757>
34. E.L. Wooten, K.M. Kissa, A. Yi-Yan, E.J. Murphy, D.A. Lafaw, P.F. Hallemeier, D. Maack, D.V. Attanasio, D.J. Fritz, G.J. McBrien, D.E. Bossi, IEEE J. Sel. Top. Quant. Electron. **6**, 69 (2000)
35. K.-W. Lee, A. Noriki, K. Kiyoyama, T. Fukushima, T. Tanaka, M. Koyanagi, IEEE Trans. Electron. Dev. **58**, 748 (2011)

Appendix A
Basic Physical Properties of the Single-Phase Perovskite Oxides at Room Temperature

- Structure type: refers to the particular structural distortion exhibited by the material: Rhombohedral $= R\bar{3}c$, Orthorhombic $=$ Pbmn, Cubic $=$ Pm3m, Layered hexagonal $=$ P6$_3$/mmc, Hexagonal YMnO$_3$ $=$ P6$_3$cm.
- Lattice parameters: a, b, c are given in angstroms (Å).
- Electrical conductivity: The nature of the fully stoichiometric material in terms of electrical conduction. M $=$ Metallic (positive dρ/dT with high conductivity); I $=$ Insulating (negative dρ/dT); SC $=$ Semiconducting (negative dρ/dT but with a small activation energy, which is given in eV); SM $=$ Semi-metallic (positive dρ/dT or temperature-independent ρ with low absolute conductivity).
- Ground state magnetic order: If present, the magnetic order in the ground state is provided. If no magnetic ordering, whether the material is paramagnetic or diamagnetic is indicated. Some very low temperature A-site rare earth ordering is not indicated. P $=$ paramagnetic, D $=$ diamagnetic, AF $=$ antiferromagnetic, F $=$ ferromagnetic.

A.A. Demkov and A.B. Posadas, *Integration of Functional Oxides with Semiconductors*, DOI 10.1007/978-1-4614-9320-4, © The Author(s) 2014

Table A.1 $A^{3+}B^{3+}O_3$ compounds

Material	Structure type	Lattice parameters				Electrical conductivity at room temperature	Ground state magnetic order
		a	b	c	α, β, γ		
$LaAlO_3$	Rhombohedral	5.356			60.11°	I	D
$LaScO_3$	Orthorhombic	5.678	5.787	8.098		I	D
$LaTiO_3$	Orthorhombic	5.546	5.573	7.832		M; SC below 150 K when stoichiometric	AF
$LaVO_3$	Orthorhombic	5.546	5.546	7.827		SC (0.13 eV)	AF (canted)
$LaCrO_3$	Orthorhombic	5.479	5.515	7.757		SC (0.27 eV)	AF
$LaMnO_3$	Orthorhombic	5.537	5.743	7.695		SC (0.2 eV)	AF (canted)
$LaFeO_3$	Orthorhombic	5.553	5.563	7.867		SC (0.6 eV)	AF
$LaCoO_3$	Rhombohedral	5.378			60.80°	SC (0.5 eV)	D
$LaNiO_3$	Rhombohedral	5.395			60.78°	M	P (Pauli)
$LaCuO_3$	Rhombohedral	5.431			60.85°	M	D when stoichiometric
$LaGaO_3$	Orthorhombic	5.494	5.519	7.770		I	D
$LaYO_3$	Orthorhombic	5.877	6.087	8.493		I	D
$LaRuO_3$	Orthorhombic	5.494	5.779	7.855		M	P
$LaRhO_3$	Orthorhombic	5.524	5.679	7.900		SC	D
$LaInO_3$	Orthorhombic	5.723	5.914	8.207		I	D
$LaLuO_3$	Orthorhombic	5.795	6.00	8.35		I	D
$CeAlO_3$	Rhombohedral	5.327			60.27°	I	P
$CeScO_3$	Orthorhombic	5.787	8.047	5.626		I	P
$CeTiO_3$	Orthorhombic	5.513	5.757	7.801		SM	AF
$CeVO_3$	Orthorhombic	5.541	5.541	7.807		SC (0.18 eV)	AF (canted)
$CeCrO_3$	Orthorhombic	5.473	5.473	7.742			AF
$CeMnO_3$	Orthorhombic	5.537	5.557	7.812			AF
$CeFeO_3$	Orthorhombic	5.519	5.536	7.819		SC	AF (canted)
$CeGaO_3$	Cubic	3.87				I	P
$CeLuO_3$	Orthorhombic	5.793	5.997	8.344		I	P

PrAlO$_3$	Rhombohedral	5.308			60.32°	I	P
PrScO$_3$	Orthorhombic	5.615	5.776	8.027		I	P
PrTiO$_3$	Orthorhombic	5.490	5.724	7.798		SC (0.03 eV)	AF
PrVO$_3$	Orthorhombic	5.477	5.545	7.759		SC (0.22 eV)	AF (canted)
PrCrO$_3$	Orthorhombic	5.444	5.484	7.710			AF
PrMnO$_3$	Orthorhombic	5.545	5.787	7.575			AF
PrFeO$_3$	Orthorhombic	5.482	5.578	7.786		SC	AF (canted)
PrCoO$_3$	Orthorhombic	5.331	5.373	7.587		SC	P
PrNiO$_3$	Orthorhombic	5.419	5.380	7.626		M; SC below 130 K	AF
PrGaO$_3$	Orthorhombic	5.458	5.490	7.733		I	P
PrLuO$_3$	Orthorhombic	5.751	5.977	8.320		I	P
PrRhO$_3$	Orthorhombic	5.414	5.747	7.803		SC	P
NdAlO$_3$	Rhombohedral	5.290			60.41°	I	P
NdScO$_3$	Orthorhombic	5.574	5.771	7.998		I	P
NdTiO$_3$	Orthorhombic	5.487	5.707	7.765		SC (0.06 eV)	P
NdVO$_3$	Orthorhombic	5.440	5.589	7.733		SC (0.25 eV)	AF (canted)
NdCrO$_3$	Orthorhombic	5.412	5.494	7.695		SC	AF
NdMnO$_3$	Orthorhombic	5.414	5.829	7.551			AF
NdFeO$_3$	Orthorhombic	5.453	5.584	7.768		I	
NdCoO$_3$	Orthorhombic	5.336	5.336	7.547		SC	P
NdNiO$_3$	Orthorhombic	5.384	5.384	7.615		M; SC below 200 K	AF
NdGaO$_3$	Orthorhombic	5.426	5.502	7.706		I	P
NdLuO$_3$	Orthorhombic	5.737	5.974	8.311		I	P
NdRhO$_3$	Orthorhombic	5.378	5.755	7.774		SC	P
NdInO$_3$	Orthorhombic	5.627	5.891	8.121		I	P
SmAlO$_3$	Orthorhombic	5.291	5.290	7.474		I	P
SmScO$_3$	Orthorhombic	5.53	5.76	7.95		I	P
SmTiO$_3$	Orthorhombic	5.468	5.665	7.737		SC (0.15 eV)	
SmVO$_3$	Orthorhombic	5.371	5.625	7.693		SC (0.28 eV)	AF (canted)
SmCrO$_3$	Orthorhombic	5.372	5.502	7.656		SC	AF

(continued)

Table A.1 (continued)

Material	Structure type	Lattice parameters				Electrical conductivity at room temperature	Ground state magnetic order
		a	b	c	α, β, γ		
$SmMnO_3$	Orthorhombic	5.359	5.843	7.482			AF
$SmFeO_3$	Orthorhombic	5.400	5.597	7.711		SC	AF (canted)
$SmCoO_3$	Orthorhombic	5.289	5.354	7.541		SC	P
$SmNiO_3$	Orthorhombic	5.336	5.431	7.568		SC	AF
$SmGaO_3$	Orthorhombic	5.369	5.520	7.650		I	P
$SmRhO_3$	Orthorhombic	5.321	5.761	7.708		SC	P
$SmInO_3$	Orthorhombic	5.589	5.886	8.082		I	P
$EuAlO_3$	Orthorhombic (metastable)	5.267	5.294	7.459		I	P
$EuScO_3$	Orthorhombic	5.51	5.76	7.94		I	P
$EuVO_3$	Orthorhombic	5.362	5.599	7.651			AF (canted)
$EuCrO_3$	Orthorhombic	5.340	5.515	7.622			AF
$EuMnO_3$	Orthorhombic	5.535	5.853	7.448			AF
$EuFeO_3$	Orthorhombic	5.372	5.606	7.685		SC	AF (canted)
$EuCoO_3$	Orthorhombic	5.246	5.370	7.469		SC	P
$EuNiO_3$	Orthorhombic	5.293	5.466	7.542		SC	AF
$EuGaO_3$	Orthorhombic	5.351	5.528	7.628		I	P
$EuRhO_3$	Orthorhombic	5.298	5.761	7.680		SC	P
$EuInO_3$	Orthorhombic (metastable)	5.567	5.835	8.078		I	P
$GdAlO_3$	Orthorhombic (metastable)	5.250	5.302	7.447		I	AF
$GdScO_3$	Orthorhombic	5.487	5.756	7.925		I	AF
$GdTiO_3$	Orthorhombic	5.407	5.667	7.692		SC (0.19 eV)	F
$GdVO_3$	Orthorhombic	5.345	5.623	7.638		SC (0.34 eV)	AF (canted)
$GdCrO_3$	Orthorhombic	5.312	5.514	7.611			AF
$GdMnO_3$	Orthorhombic	5.317	5.863	7.433			AF

$GdFeO_3$	Orthorhombic	5.349	5.611	7.669	SC	AF (canted)
$GdCoO_3$	Orthorhombic	5.228	5.404	7.436	I	AF
$GdNiO_3$	Orthorhombic	5.258	5.492	7.506	SC	AF
$GdGaO_3$	Orthorhombic	5.322	5.537	7.606	I	P
$GdRhO_3$	Orthorhombic	5.277	5.760	7.658	SC	P
$GdInO_3$	Orthorhombic (metastable)	5.548	5.842	8.071	I	P
$TbAlO_3$	Orthorhombic (metastable)	5.232	5.310	7.420	I	AF
$TbScO_3$	Orthorhombic				I	AF
$TbTiO_3$	Orthorhombic	5.388	5.648	7.676	SC (0.2 eV)	F
$TbVO_3$	Orthorhombic	5.325	5.606	7.614	SC (0.32 eV)	AF (canted)
$TbCrO_3$	Orthorhombic	5.291	5.513	7.557	SC	AF
$TbMnO_3$	Orthorhombic	5.297	5.831	7.403	I (ferroelectric)	AF
$TbFeO_3$	Orthorhombic	5.326	5.602	7.635	SC	AF (canted)
$TbCoO_3$	Orthorhombic	5.200	5.394	7.421	SC	AF
$TbNiO_3$	Orthorhombic				SC	AF
$TbGaO_3$	Orthorhombic	5.307	5.531	7.578	I	AF
$TbRhO_3$	Orthorhombic	5.254	5.749	7.623	SC	AF
$TbInO_3$	Hexagonal $YMnO_3$	6.319		12.295	I	P
$DyAlO_3$	Orthorhombic (metastable)	5.205	5.317	7.395	I	AF
$DyScO_3$	Orthorhombic	5.43	5.71	7.89	I	AF
$DyTiO_3$	Orthorhombic	5.361	5.659	7.647	SC	F
$DyVO_3$	Orthorhombic	5.302	5.602	7.601	SC (0.3 eV)	AF
$DyCrO_3$	Orthorhombic	5.265	5.520	7.559	SC	AF (canted)
$DyMnO_3$	Orthorhombic	5.279	5.843	7.378	I (ferroelectric)	AF
$DyFeO_3$	Orthorhombic	5.302	5.598	7.623	SC	AF (canted)
$DyCoO_3$	Orthorhombic	5.162	5.400	7.398	SC	AF
$DyNiO_3$	Orthorhombic	5.212	5.500	7.445	SC	AF

(continued)

Table A.1 (continued)

Material	Structure type	Lattice parameters				Electrical conductivity at room temperature	Ground state magnetic order
		a	b	c	α, β, γ		
$DyGaO_3$	Orthorhombic	5.282	5.534	7.556		I	AF
$DyRhO_3$	Orthorhombic	5.245	5.731	7.600		SC	AF
$HoAlO_3$	Orthorhombic (metastable)	5.181	5.323	7.374		I	AF
$HoScO_3$	Orthorhombic	5.42	5.71	7.87		I	AF
$HoTiO_3$	Orthorhombic	5.339	5.665	7.626		SC (0.2 eV)	F
$HoVO_3$	Orthorhombic	5.276	5.592	7.576		SC (0.33 eV)	AF
$HoCrO_3$	Orthorhombic	5.243	5.519	7.538		SC	AF (canted)
$HoMnO_3$	Hexagonal $YMnO_3$	6.136		11.42		I (ferroelectric)	AF
$HoFeO_3$	Orthorhombic	5.278	5.591	7.602		SC	AF (canted)
$HoCoO_3$	Orthorhombic	5.157	5.429	7.397		SC	AF
$HoNiO_3$	Orthorhombic	5.181	5.510	7.425		SC	AF
$HoGaO_3$	Orthorhombic	5.251	5.531	7.536		I	AF
$HoRhO_3$	Orthorhombic	5.230	5.726	7.582		SC	AF
$HoInO_3$	Hexagonal $YMnO_3$	6.271		12.263		I	P
$ErAlO_3$	Orthorhombic (metastable)	5.160	5.327	7.354		I	AF
$ErScO_3$	Orthorhombic (metastable)					I	AF
$ErTiO_3$	Orthorhombic	5.318	5.657	7.613		SC (0.24 eV)	F
$ErVO_3$	Orthorhombic	5.262	5.604	7.578		SC (0.33 eV)	AF
$ErCrO_3$	Orthorhombic	5.223	5.516	7.519			AF
$ErMnO_3$	Hexagonal $YMnO_3$	6.117		11.435		I (ferroelectric)	AF
$ErFeO_3$	Orthorhombic	5.267	5.581	7.593		SC	AF (canted)
$ErCoO_3$	Orthorhombic	5.120	5.416	7.340		SC	P
$ErNiO_3$	Orthorhombic	5.160	5.514	7.381		SC	AF
$ErGaO_3$	Orthorhombic	5.239	5.527	7.522		I	AF

$ErRhO_3$	Orthorhombic	5.216	5.712	7.561	SC	AF
$TmAlO_3$	Orthorhombic (metastable)	5.144	5.328	7.334	I	P
$TmScO_3$	Orthorhombic (metastable)				I	P
$TmTiO_3$	Orthorhombic	5.306	5.647	7.607	SC	F
$TmVO_3$	Orthorhombic	5.237	5.573	7.545	SC (0.27 eV)	AF
$TmCrO_3$	Orthorhombic	5.209	5.508	7.500		AF
$TmMnO_3$	Hexagonal $YMnO_3$	6.062		11.40	I (ferroelectric)	AF
$TmFeO_3$	Orthorhombic	5.249	5.572	7.582	SC	AF (canted)
$TmCoO_3$	Orthorhombic	5.104	5.417	7.325	SC	P
$TmNiO_3$	Orthorhombic	5.149	5.495	7.375	SC	AF
$TmGaO_3$	Orthorhombic	5.224	5.515	7.505	I	P
$TmRhO_3$	Orthorhombic	5.203	5.697	7.543	SC	P
$YbAlO_3$	Orthorhombic (metastable)	5.125	5.331	7.315	I	AF
$YbScO_3$	Orthorhombic (metastable)				I	AF
$YbTiO_3$	Orthorhombic	5.293	5.633	7.598	SC (0.24 eV)	F
$YbVO_3$	Orthorhombic	5.223	5.564	7.537	SC (0.35 eV)	AF
$YbCrO_3$	Orthorhombic	5.195	5.510	7.490	SC	AF (canted)
$YbMnO_3$	Hexagonal $YMnO_3$	6.062		11.40	I (ferroelectric)	AF
$YbFeO_3$	Orthorhombic	5.233	5.557	7.570	SC	AF (canted)
$YbCoO_3$	Orthorhombic	5.086	5.419	7.310	SC	AF
$YbNiO_3$	Orthorhombic	5.131	5.496	7.353	SC	AF
$YbGaO_3$	Orthorhombic	5.208	5.510	7.490	I	AF
$LuAlO_3$	Orthorhombic (metastable)	5.101	5.332	7.300	I	D
$LuScO_3$	Orthorhombic (metastable)				I	D

(continued)

Table A.1 (continued)

Material	Structure type	Lattice parameters				Electrical conductivity at room temperature	Ground state magnetic order
		a	b	c	α, β, γ		
$LuTiO_3$	Orthorhombic	5.274	5.633	7.580		SC	F
$LuVO_3$	Orthorhombic	5.214	5.561	7.530		SC (0.26 eV)	AF
$LuCrO_3$	Orthorhombic	5.176	5.497	7.475		SC	AF (canted)
$LuMnO_3$	Hexagonal $YMnO_3$	6.046		11.394		I (ferroelectric)	AF
$LuFeO_3$	Orthorhombic	5.213	5.547	7.565		SC	AF (canted)
$LuCoO_3$	Orthorhombic	5.065	5.418	7.290		SC	D
$LuNiO_3$	Orthorhombic	5.117	5.499	7.356		SC	AF
$LuGaO_3$	Orthorhombic	5.188	5.505	7.484		I	D
$LuRhO_3$	Orthorhombic	5.186	5.670	7.512		SC	D
$InCrO_3$	Orthorhombic	5.170	5.355	7.543			AF
$InMnO_3$	Hexagonal $YMnO_3$	5.876		11.472		I (ferroelectric)	AF
$InGaO_3$	Orthorhombic (metastable)	5.176	5.365	7.548		I	D
$InRhO_3$	Orthorhombic	5.301	5.435	7.586		SC	D
$ScMnO_3$	Hexagonal $YMnO_3$	5.830		11.179		I (ferroelectric)	AF
$YAlO_3$	Orthorhombic (metastable)	5.180	5.330	7.375		I	D
$YScO_3$	Orthorhombic (metastable)	5.431	5.712	7.894		I	D
$YTiO_3$	Orthorhombic	5.340	5.665	7.624		SC (0.23 eV)	F
YVO_3	Orthorhombic	5.284	5.605	7.587		SC (0.18 eV)	AF
$YCrO_3$	Orthorhombic	5.247	5.518	7.540		SC (0.2 eV)	AF (canted)
$YMnO_3$	Hexagonal $YMnO_3$	6.125		11.41		I (ferroelectric)	AF
$YFeO_3$	Orthorhombic	5.283	5.592	7.603		SC	AF (canted)
$YCoO_3$	Orthorhombic	5.143	5.434	7.373		SC	D
$YNiO_3$	Orthorhombic	5.178	5.516	7.419		SC	AF
$YGaO_3$	Orthorhombic	5.257	5.536	7.533		I	D

$YInO_3$	Hexagonal $YMnO_3$	6.260		12.249		I (ferroelectric)	D
$BiAlO_3$	Rhombohedral R3c	5.437			59.25°	I	D
$BiScO_3$	Pseudo-triclinic	4.042	4.127	4.042	90.68° 91.87° 90.68°	I	D
$BiCrO_3$	Pseudo-triclinic	3.906	3.870	3.906	90.55° 89.15° 90.55°	SC	AF
$BiMnO_3$	Pseudo-triclinic	3.935	3.989	3.935	91.47° 90.97° 91.47°	SC	AF
$BiFeO_3$	Rhombohedral R3c	5.634			59.36°	I (ferroelectric)	AF (canted)
$BiCoO_3$	Cubic (metastable)	4.228				SC	AF
$BiNiO_3$	Cubic (metastable)	4.173				SC	AF
$BiYO_3$	Cubic (metastable)	4.2				I	D
$BiRhO_3$	Orthorhombic	5.354	5.813	7.776		SC	D
$BiInO_3$	Orthorhombic	5.723	5.914	8.207		I	D

Table A.2 $A^{2+}B^{4+}O_3$ compounds

Material	Structure type	Lattice parameters				Electrical conductivity	Magnetism
		a	b	c	β		
$SrTiO_3$	Cubic	3.905				I; becomes SC when oxygen-deficient	D
$SrVO_3$	Cubic	3.842				M	P (Pauli)/weak F
$SrVO_{2.5}$	Cubic	3.848					
$SrCrO_3$	Cubic	3.818				M	P (Pauli)
$SrMnO_3$	Layered hexagonal (4H)	5.443		9.070		SC (0.5 eV)	AF
$SrFeO_3$	Cubic	3.85				SM	AF (helical)
$SrFeO_{2.5}$	Brownmillerite (orthorhombic)	5.530	15.540	5.666			
$SrCoO_{3-x}$	Cubic	3.836				M	F
$SrCoO_{2.5}$	Brownmillerite (orthorhombic)	5.456	15.664	5.556		SC	AF
$SrZrO_3$	Orthorhombic	5.792	5.818	8.189		I	D
$SrMoO_3$	Cubic	3.975				M	P (Pauli)
$SrRuO_3$	Orthorhombic	5.53	5.57	7.85		M	F
$SrHfO_3$	Orthorhombic	5.785	5.786	8.182		I	D
$SrSnO_3$	Cubic	4.034				I	D
$BaTiO_3$	Tetragonal (P4mm)	3.994		4.034		I (Ferroelectric)	D
$BaCrO_3$	Layered hexagonal (4H)	5.659		9.359		SC (0.1 eV)	
$BaVO_3$	Layered hexagonal					SC (0.03 eV)	P
$BaMnO_3$	Layered hexagonal (8H)	5.667		18.738			AF
$BaFeO_{2.925}$	Layered hexagonal (12H)	5.691		27.974		SC (0.3 eV)	AF
$BaFeO_{2.75}$	Cubic	3.997				SM	AF
$BaFeO_{2.5}$	Monoclinic	6.969	11.724	23.431	98.74°		AF
$BaCoO_{2.85}$	Layered hexagonal (2H)	5.59		4.83			AF
$BaNiO_3$	Layered hexagonal (2H)	5.629		4.811		I	D
$BaZrO_3$	Cubic	4.20				I	D
$BaMoO_3$	Cubic	4.040				M	P (Pauli)
$BaRuO_3$	Layered hexagonal (9H)	5.75		21.6			

Compound	Structure	a	b	c	angle		
BaSnO$_3$	Cubic	4.116				I	D
BaHfO$_3$	Cubic	4.171				I	D
CaTiO$_3$	Orthorhombic	5.381	5.443	7.645		I	D
CaVO$_3$	Orthorhombic	5.326	5.352	7.547		M	P (Pauli)
CaCrO$_3$	Orthorhombic	5.287	5.316	7.486		M	AF
CaMnO$_3$	Orthorhombic	5.270	5.275	7.464		SC	AF
CaFeO$_3$	Tetragonal	5.325		7.579		M; becomes SC at low temp.	AF (helical)
CaFeO$_{2.5}$	Brownmillerite (Orthorhombic)	5.64	14.68	5.39		SC (0.71 eV)	
CaZrO$_3$	Orthorhombic	5.587	5.758	8.008		I	D
CaMoO$_3$	Orthorhombic	5.45	5.58	7.77		M	P (Pauli)
CaRuO$_3$	Orthorhombic	5.36	5.53	7.67		M	P
CaSnO$_3$	Orthorhombic	5.519	5.668	7.885		I	D
CaHfO$_3$	Orthorhombic	5.568	5.732	7.984		I	D
PbTiO$_3$	Tetragonal	3.904		4.152		I (ferroelectric)	D
PbCrO$_3$	Cubic	4.00				SC (0.27 eV)	AF
PbFeO$_{2.5}$	Brownmillerite-type (tetragonal)	7.79		15.85			
PbZrO$_3$	Orthorhombic	5.872	11.744	8.202		I (antiferroelectric)	D
PbRuO$_3$	Orthorhombic	5.562	5.610	7.862			
PbSnO$_3$	Pseudo-monoclinic	4.076	4.043	4.076	89.75°	I	D
PbHfO$_3$	Pseudo-Orthorhombic	4.136	4.136	4.099		I (antiferroelectric)	D
CdTiO$_3$	Orthorhombic	5.348	5.417	7.615		I	D
CdSnO$_3$	Orthorhombic	5.547	5.577	7.867		I	D
CdHfO$_3$	Pseudo-monoclinic	3.942			91.6°	I	D
EuZrO$_3$	Cubic	4.099				I	
EuTiO$_3$	Cubic	7.810				I	AF

Table A.3 $A^{1+}B^{5+}O_3$ compounds

Material	Structure type	Lattice parameters				Electrical conductivity	Magnetism
		a	b	c	α		
$LiNbO_3$	Rhombohedral R3c (non-perovskite)	5.148		13.863		I (ferroelectric)	D
$LiTaO_3$	Rhombohedral R3c (non-perovskite)	5.154		13.784		I (ferroelectric)	D
$NaNbO_3$	Orthorhombic Pbma	5.566	15.520	5.506		I (antiferroelectric)	D
$NbTaO_3$	Orthorhombic Pc2$_1$n	5.494	7.751	5.513		I (ferroelectric)	D
$KNbO_3$	Orthorhombic Bmm2	5.695	3.971	5.720		I (ferroelectric)	D
$KTaO_3$	Cubic Pm3m	3.9885				I	D

Appendix B
Crystal Structures of the Perovskite Oxides

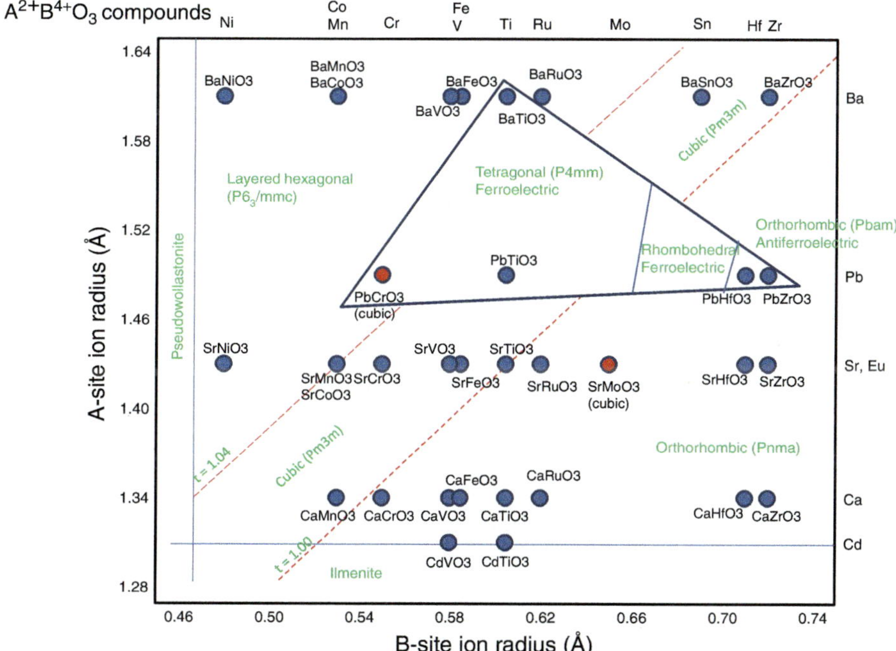

Fig. B.1 Crystal structure adopted by ABO$_3$ compounds at room temperature as a function of the ionic radii of the A and B ions for the case divalent A and tetravalent B cations. Compounds on the *lower right* have an orthorhombically distorted perovskite structure. Compounds on the *upper left* have a layered hexagonal structure consisting partially of face-sharing octahedra. In the *middle* is a band of cubic perovskite materials. For highly porlarizable A ions (Pb and Ba), ferroelectric-type distortions dominate. Materials with a too small A ion result in the ilmenite crystal structure while materials with a too small B ion result in the pseudowollastonite crystal structure

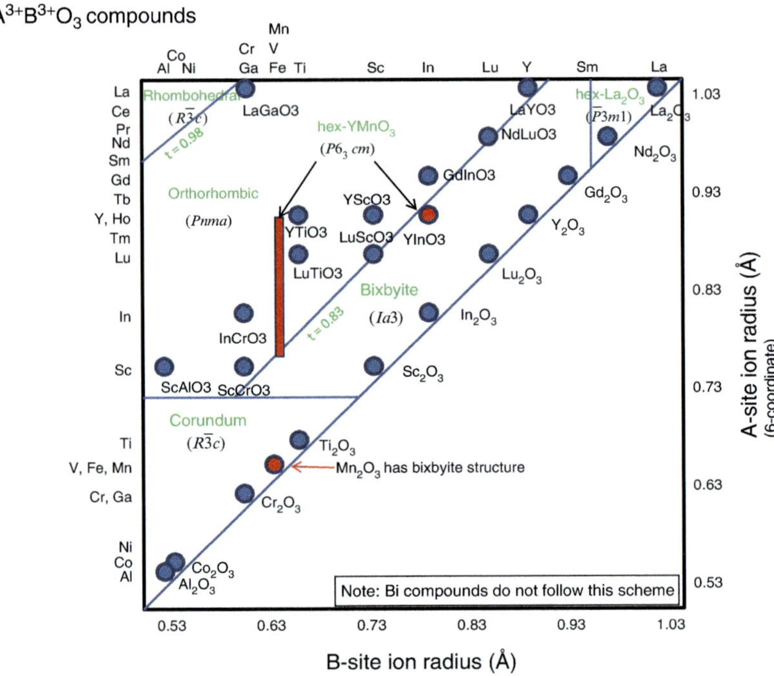

Fig. B.2 Crystal structure adopted by ABO_3 compounds at room temperature as a function of the ionic radii of the A and B ions for the case trivalent A and trivalent B cations. Compounds on the extreme *upper left* (large A and small B ions) adopt a rhombohedrally distorted perovskite structure. Moving towards the *lower right*, ABO_3 compounds adopt the orthorhombically distorted perovskite structure. For materials where the A and B ions are similar in size, non-perovskite crystal structures are formed—corundum for small ions, bixbyite for medium-size ions, and a hexagonal structure (La_2O_3-type) for large ions

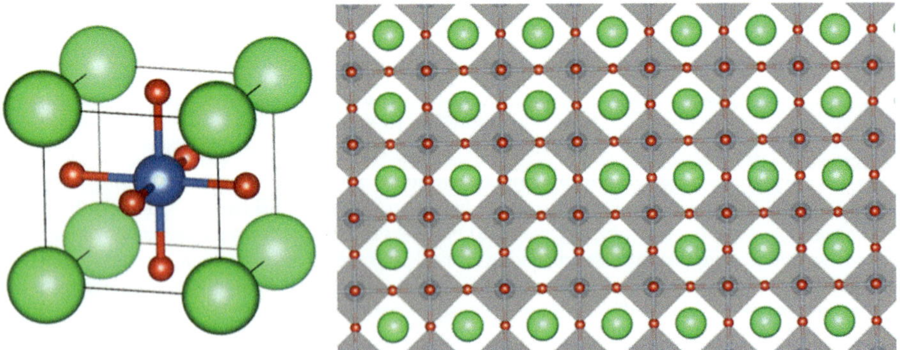

Fig. B.3 Unit cell and octahedral tiling for cubic (undistorted) perovskite with space group $Pm3m$

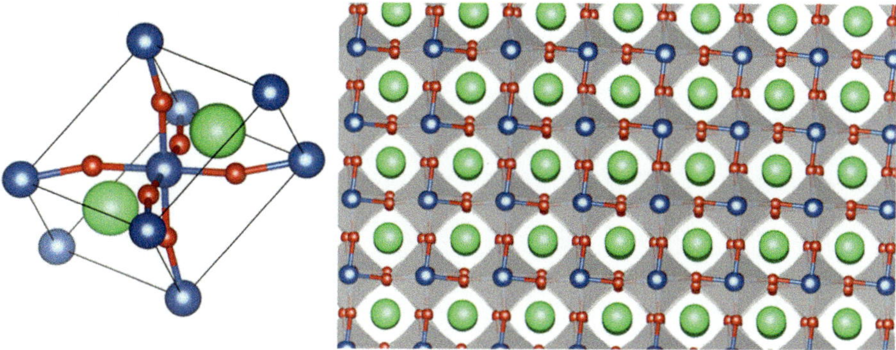

Fig. B.4 Unit cell and octahedral tiling for rhombohedrally-distorted perovskite with space group $R\bar{3}c$

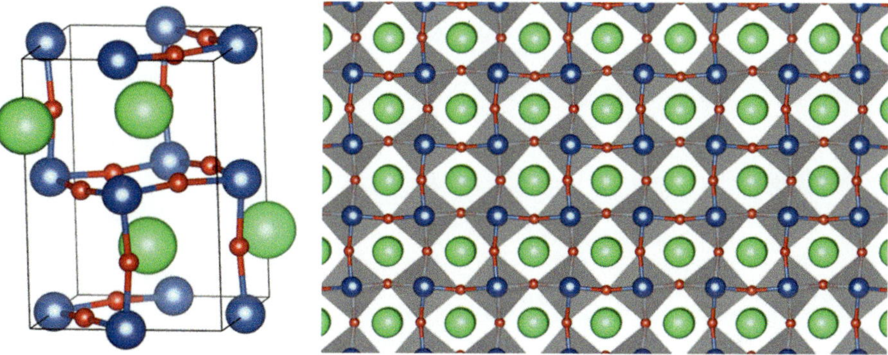

Fig. B.5 Unit cell and octahedral tiling for orthorhombically-distorted perovskite with space group *Pnma*

Appendix C
Basic Properties of Single Crystal Oxide Substrates

A.A. Demkov and A.B. Posadas, *Integration of Functional Oxides with Semiconductors*, DOI 10.1007/978-1-4614-9320-4, © The Author(s) 2014

Substrate material	In-plane (Å)	Out-of-plane (Å)	Thermal expansion coefficient (K^{-1})	Band gap (eV)	Notes
$SrTiO_3$ (100)	3.905	3.905	9×10^{-6}	3.2	Structural transition at 105 K
$LaAlO_3$ (100)	3.79	3.79	10×10^{-6}	5.6	Twinned structure; structural transition at 700 K
$La_{0.2}Sr_{0.8}Al_{0.6}Ta_{0.4}O_3$ (LSAT) (100)	3.868	3.868	8.2×10^{-6}	4.6	Structural transition at 150 K
$DyScO_3$ (110)	3.945 and 3.943	3.939	8.4×10^{-6}	5.7	Paramagnetic
$GdScO_3$ (110)	3.965 and 3.961	3.956	10.9×10^{-6}	5.7	Paramagnetic
$NdGaO_3$ (110)	3.853 and 3.864	3.864	9×10^{-6}	3.8	Paramagnetic; anisotropic
$LaGaO_3$ (110)	3.888 and 3.896	3.896	13.6×10^{-6}	3.8	Anisotropic; structural transition at 420 K
MgO (100)	4.212	2.106 (200)	14×10^{-6}	7.8	Hygroscopic
$MgAl_2O_4$ (100)	8.083 (4.04)	2.021 (200)	7.5×10^{-6}	7.8	
$SrLaAlO_4$ (001)	3.756	2.105 (006)	10×10^{-6}		
$SrLaGaO_4$ (001)	3.84	2.12 (006)	10×10^{-6}		
Yttria-stabilized ZrO_2 (YSZ)	5.12 (3.62)	2.56 (200)	9.2×10^{-6}	5.8	Typically contains 8 mol% Y_2O_3
$YAlO_3$-perovskite (YAP) (110)	3.686 and 3.715	3.713	9×10^{-6}	~7	Anisotropic
$KTaO_3$	3.989	3.989	6.8×10^{-6}	3.8	

Glossary

10Dq	Crystal field splitting
2D	Two-dimensional
AEG	Auger electron spectroscopy
AFM	Antiferromagnetic
AFM	Atomic force microscope/microscopy
ALD	Atomic layer deposition
ALE	Atomic layer epitaxy
APCVD	Atmospheric pressure CVD
APD	Anti-phase domain
ARPES	Angle-resolved photoemission spectroscopy
ARXPS	Angle-resolved X-ray photoelectron spectroscopy
BFO	$BiFeO_3$
BO	Born-Oppenheimer
BST	$(Ba,Sr)TiO_3$
BTO	$BaTiO_3$
BZ	Brillouin zone
CBE	Chemical beam epitaxy
CHA	Concentric hemispherical analyzer
CMA	Cylindrical mirror analyzer
CMOS	Complementary metal oxide semiconductor
COHP	Crystal orbital Hamiltonian population
COS	Crystalline oxides on semiconductors
CS	Compound semiconductors
CT	Charge transfer
C–V	Capacitance–voltage
CVD	Chemical vapor deposition
DARPA	Defense advanced research projects agency
DC	Direct current
DFT	Density functional theory

A.A. Demkov and A.B. Posadas, *Integration of Functional Oxides with Semiconductors*, DOI 10.1007/978-1-4614-9320-4, © The Author(s) 2014

DMFT	Dynamical mean field theory
DOS	Density of states
DRAM	Dynamic random access memory
EDX	Energy dispersive X-ray spectroscopy
EELS	Electron-energy loss spectroscopy
EEPROM	Electrically-erasable programmable read only memory
EO	Electro-optical
EOT	Equivalent oxide thickness
fcc	Face-centered cubic
FeFET	Ferroelectric field effect transistor
FET	Field effect transistor
FM	Ferromagnetic
FRAM	Ferroelectric random access memory
FTIR	Fourier transform infrared spectroscopy
FWHM	Full-width at half-maximum
GeOI	Germanium-on-insulator
GGA	Generalized gradient approximation
GW	DFT approximation including electron Coulomb screening
HAADF	High-angle annular dark field
HBT	Heterojunction bipolar transistor
HEIS	High-energy ion scattering
HRTEM	High-resolution transmission electron microscopy
HS	High-spin
HSE	Heyd-Scuseria-Ernzerhof
HyperSr	Sr bis(triisopropylcyclopentadienyl)
IC	Integrated circuit
IR	Infrared
IS	Intermediate-spin
ISS	Ion scattering spectroscopy
J	Exchange interaction
KS	Kohn-Sham
LAN	Local area network
LAO	$LaAlO_3$
LCO	$LaCoO_3$
LDA	Local density approximation
LED	Light emitting diode
LEED	Low-energy electron diffraction
LEIS	Low-energy ion scattering
LPCVD	Low pressure CVD
LS	Low-spin
LSAT	Lanthanum strontium aluminum tantalum oxide

LSCO	$(La,Sr)CoO_3$
LSDA + U	Local spin density approximation combined with the Hubbard U correction
LSI	Large-scale integration
LZO	$La_2Zr_2O_7$
MBE	Molecular beam epitaxy
MD	Molecular dynamics
MEIS	Medium-energy ion scattering
MEMS	Microelectromechanical systems
MESFET	Metal–semiconductor field effect transistor
MFM	Magnetic force microscopy
MIC	Monolithic integrated circuit
MIT	Metal-to-insulator transition
ML	Monolayer
MOCVD	Metal-organic chemistry vapor deposition
MOMBE	Metal-organic molecular beam epitaxy
MOSFET	Metal-oxide-semiconductor field effect transistor
MottFET	Mott transition field effect transistor
MOVPE	Metal-organic vapor phase epitaxy
MPD	Microchannel plate detector
NM	Nonmagnetic
PAW	Projector augmented wave
PBE	Perdew-Burke-Ernzerhof
pDOS	Partial density of states
PECVD	Plasma enhanced chemical vapor deposition
PFM	Piezo-response force microscopy/piezoelectric force microscopy
PHCVD	Photo-assisted chemical vapor deposition
PL	Photoluminescence
PLD	Pulsed laser deposition
ppb	Parts per billion
ppm	Parts per million
PWSCF	Partial-wave self-consistent field
PZT	Lead zirconate titanate
QP	Quasiparticle
RBS	Rutherford back scattering
RF	Radio frequency
RHEED	Reflection high energy electron diffraction
RP	Ruddlesden-Popper
RTA	Rapid thermal annealing
SAD	Selected area diffraction
SCLS	Surface core-level shift

SEM	Scanning electron microscopy
SIC	Self-interaction corrections
SIMS	Secondary ion mass spectroscopy
SOI	Silicon on insulator
SPA-LEED	Spot profile analysis low-energy electron diffraction
SpinFET	Spin field effect transistor
SPM	Scanning probe microscopy
SQUID	Superconducting quantum interference device
SRR	Split-ring resonators
STEM	Scanning transmission electron microscopy
STM	Scanning tunneling microscopy
STO	$SrTiO_3$
STS	Scanning tunneling spectroscopy
T/R	Transmit/receive
T_c	Curie temperature, critical temperature
TEM	Transmission electron microscope/microscopy
TFT	Tejedor-Flores-Tersoff
TM	Transition metal
TOF-SIMS	Time-of-flight secondary ion mass spectroscopy
TTIP	Titanium isopropoxide
UHV	Ultra-high vacuum
UPS	Ultraviolet photoelectron spectroscopy
UV	Ultraviolet
VBO	Valence band offset
VBT	Valence band top
VPE	Vapor phase epitaxy
XPS	X-ray photoelectron spectroscopy
XRD	X-ray diffraction
XRR	X-ray reflection
YAP	$YAlO_3$-perovskite
YSZ	Yttria-stabilized zirconia
ZB	Zincblende

Index

A

Abinit, 58
Ab-initio, 58–59, 115
Adaptive oxide electronics, 237
Adiabatic approximation, 46
Adsorption, 74, 76, 79, 116, 117, 124, 226
AES. *See* Auger electron spectroscopy (AES)
Al Kα, 94, 129
Alkali metals, 17–19, 34
Alkaline earths, 17–19, 31, 84, 116, 117, 128
Al_2O_3, 177, 206
    Si integration, 177, 206
Anatase, 153, 159–169, 177, 181, 183
Angle-resolved photoemission spectroscopy
    (ARPES), 89, 90, 93, 95, 96
Angle-resolved XPS (ARXPS), 94
Antiferromagnetic, 28, 173, 176, 182, 197, 213
Anti-phase domains (APD), 36, 38
Ar ions, 71, 105, 108
ARPES. *See* Angle-resolved photoemission
    spectroscopy (ARPES)
Atomic absorption spectroscopy, 65
Atomic force microscopy (AFM)
    contact mode, 111
    tapping mode, 111
Atomic layer deposition (ALD)
    precursor, 77–81
    reactor, 78–80, 84
    self-limiting growth, 78–81
Atomic layer epitaxy (ALE), 74
Auger electron spectroscopy (AES), 66, 89, 90,
    93–94, 96

B

Band alignment, 53–55, 144, 160, 161, 167
    calculation, 53–55

Band gap, 5–10, 20, 28, 54, 55, 58, 147, 148,
    159, 160, 165, 175, 195, 207, 209, 215,
    220, 242
Bardeen limit, 147, 148
$(Ba,Sr)TiO_3$/GaN integration, 212
$BaTiO_3$ (BTO)
    domain structure, 37, 187, 193
    experimental growth, 187, 189, 194
    Si integration, 187, 207
    switching, 189, 191
Berry phase, 55
Bethe-Salpeter method, 58
Biaxial strain, 161, 170, 171, 173, 175, 187,
    188, 194
$BiFeO_3$ (BFO), 16, 84, 197–198, 213
    Si integration, 197–198
Bixbyite, 36, 37, 42, 209–210
Bloch functions, 52, 55
Bloch theorem, 56
Bonding and anti-bonding orbitals, 11
BO_6 octahedron, 10, 12
Boolean computation, 237
Born effective charge, 55, 57, 164, 165
Born-Oppenheimer (BO) ansatz, 45–46
Bragg peak, 97, 171

C

CASTEP, 58, 144
Cathodoluminescence, 66
CBE. *See* Chemical beam epitaxy (CBE)
CeO_2, 210, 218, 220, 222
    Si integration, 210
Charge transfer, 2, 8, 31, 117, 119, 121, 127,
    128, 131–133, 150, 162, 165, 168, 169
Charge transfer gap, 10–12
Chemical beam epitaxy (CBE), 74, 76

A.A. Demkov and A.B. Posadas, *Integration of Functional Oxides with
Semiconductors*, DOI 10.1007/978-1-4614-9320-4, © The Author(s) 2014